Advanced Structured Materials

Volume 59

More information about this series at http://www.springer.com/series/8611

Mezhlum A. Sumbatyan
Editor

Wave Dynamics and Composite Mechanics for Microstructured Materials and Metamaterials

Springer

Editor
Mezhlum A. Sumbatyan
Institute of Mathematics, Mechanics
and Computer Sciences
Southern Federal University
Rostov-on-Don
Russia

ISSN 1869-8433 ISSN 1869-8441 (electronic)
Advanced Structured Materials
ISBN 978-981-10-3796-2 ISBN 978-981-10-3797-9 (eBook)
DOI 10.1007/978-981-10-3797-9

Library of Congress Control Number: 2017930275

Printed on acid-free paper

This Springer imprint is published by Springer Nature
The registered company is Springer Nature Singapore Pte Ltd.
The registered company address is: 152 Beach Road, #21-01/04 Gateway East, Singapore 189721, Singapore

Foreword

In the last few years, the study of microstructured materials and metamaterials has become a trend in mechanics. Very often, trends and fashions are fading quickly not leaving deep traces in science.

This is not the case for the effort to design exotic materials, i.e., metamaterials whose mechanical behavior is tailored to meet specific functional requirements. They are indeed potentially of interest in applications where nonstandard mechanical, electromechanical, and acoustical properties, not exhibited by standard materials, are required. Their special behavior is governed by their microstructure, intending with this term their structure at a length scale much smaller (usually, smaller of at least one order of magnitude) than the wavelength of the macroscopic phenomenon they affect, rather than by their constituent materials.

The reader should be warned here: The topic is so important that many groups are working on it, by using different or sometimes very similar approaches. What is somehow puzzling is that the same subject, the same set of scientific and technological problems, and the same methodologies are sometimes labeled with different name (a detailed discussion of this point can be found in[1] or in[2]). Therefore, one finds works on architectured, advanced, multiscale, microstructured, complex, optimized (and so on) materials. Each label characterizes rather a group of researchers and not a really different research field.

In this foreword, and in this book, the preferred nomenclature uses the word "*metamaterial*" as it seems really suggestive: They are materials which go beyond as the Greek prefix "*meta*" exactly suggests this idea.

The interest of presented studies is increased by a circumstance which presented itself only recently: Indeed, the realization of metamaterials has become

---

[1] F. dell'Isola, A. Della Corte, and I. Giorgio "Higher-gradient continua: The legacy of Piola, Mindlin, Sedov and Toupin and some future research perspectives." *Math. Mech. Solids* (2016): doi: 10.1177/1081286515616034.

[2] D. Del Vescovo and I. Giorgio "Dynamic problems for metamaterials: review of existing models and ideas for further research." *Int. J. Engng Sci* **80** (2014): 153–172.

economically viable by means of recent developments in some new manufacturing techniques such as 3D printing, roll-to-roll processing, electrospinning, photolithography and next-generation lithographies (extreme ultraviolet, X-ray, and charged-particle lithographies), dry etching, wet chemical etching, wet bulk micro-machining, thermal and mechanical energy-based removing, and micromolding.

Technology and theoretical disciplines do very often develop closely, one being a stimulus for the others. It is naive to believe that one can still theoretically design new products when disconnected from advanced manufacturing: For a good design, one needs to know the latest manufacturing processes and newest materials. In fact, changes of till-then established theoretical paradigms have been often, if not always, due to the advancement of new technology, which allows for new phenomenological evidence to arise. Given the availability of new technological possibilities as the ones mentioned before, nowadays, the main challenge in the field of metamaterials is the design of purpose-specific solutions to existing problems, for instance, in engineering and biomedical areas, by means of an intelligent exploitation of the properties of "ad hoc" architectured materials.

The desired outcome of the ongoing current research in the field of metamaterials is, therefore, not only the ability to predict the behavior of already existing materials (even with a possibly very complex behavior), but also assigned a certain (preferably exotic) desired behavior at the macroscale, the ability to prescribe constitutive and geometric characteristics at the microscale in order to get the selected macroscopic behavior.

Potential applications may include acoustically active materials, which behave like frequency filter and thus are able either to "cut" or to "pass" some frequency intervals, wind-excited structures, whose reliability can be, for instance, improved by means of targeted anisotropic behaviors and piezo- and flexo-electric induced vibration damping, and bone reconstructive surgery, because implants made of bioresorbable artificial materials guarantee a proper load-carrying capacity and a fast substitution of the device with newly formed bone.

It is well known that the functional adaptation process in bones is strongly related to the external load frequency. Therefore, dynamic properties of bone bioresorbable prostheses play a key role in the bone functional adaptation. For this reason, modal analyses are required to be performed in order to understand how dynamic features evolve with the remodeling process and are influenced by external mechanical factors.

In the design process, the mathematical modeling methodologies play a relevant role: in this book, this vision is clearly shown and exploited.

Mathematical modeling of materials has been developed in the nineteenth century on the basis of some assumptions, which are verified by the majority of natural materials and by standard materials used up to now in engineering. Usually, materials which show sophisticated and often unexpected behaviors are those whose microstructure is very complex, exhibits multiple characteristic length scales, involves coupled multiphysics phenomena, and shows strongly inhomogeneous physical properties at every characteristic length.

Clearly, that hypotheses assumed in classical physics for describing mechanical phenomena are not anymore suitable when one wants to model exotic artificial materials. The construction of the general theoretical framework for the description and prediction of the behavior of advanced architectured materials is the soundest ground for exploring those exotic phenomena.

Wave dynamics and mechanics of composite media with micro- and nanostructure, which might contain arrays of cracks, defects, and, eventually, micro- and nanosize elements coupling physical–mechanical fields of different nature, e.g., piezoelectric elements, are of crucial importance for the investigation of phenomena involved in this kind of materials. New theoretical advancements in these fields are needed to fully exploit the high potential of metamaterials and, in particular, fundamental methods, and models in the theory of wave problems and composite mechanics, which have already been proved to be powerful tools, need to be further developed. They will allow for the investigation of qualitative and quantitative properties exhibited by metamaterials: For instance, they will make possible the synthesis (i.e., the specification of the needed microstructure) when a desired behavior is required.

In conclusion, it is my sincerest hope that this volume enhances networking between some of the brightest scientists working in different fields, ranging from physics and applied mathematics to numerical analysis and materials science. Indeed, the published papers seem to me of the highest quality, as they show novel and relevant improvements of the specialistic literature.

Rome, Italy
October 2016

Francesco dell'Isola

Contents

Contributors

Emilio Barchiesi Department of Structural and Geotechnical Engineering, Universitá di Roma La Sapienza, Rome, Italy

Nikolay V. Boyev Institute of Mathematics, Mechanics and Computer Science, Southern Federal University, Rostov-on-Don, Russia

M. Brigante Department of Structures for Engineering and Architecture, University of Naples - Federico II, Naples, Italy

Victor A. Eremeyev Institute of Mathematics, Mechanics and Computer Science, Southern Federal University, Rostov-on-Don, Russia; The Faculty of Mechanical Engineering, Rzeszów University of Technology, Rzeszów, Poland

Roman A. Gerasimov Southern Federal University, Rostov-on-Don, Russian Federation

G. Iovane Department of Computer Science, University of Salerno, Fisciano (SA), Italy

A.S. Kornievsky Institute of Mathematics, Mechanics and Computer Science, Southern Federal University, Rostov-on-Don, Russia

N.V. Kurbatova Institute of Mathematics, Mechanics and Computer Science, Southern Federal University, Rostov-on-Don, Russia

Andrei V. Maksimov Cherepovets State University, Cherepovets, Russian Federation

Olga G. Maksimova Cherepovets State University, Cherepovets, Russian Federation

D.K. Nadolin Institute of Mathematics, Mechanics and Computer Science, Southern Federal University, Rostov-on-Don, Russia

A.V. Nasedkin Institute of Mathematics, Mechanics and Computer Science, Southern Federal University, Rostov-on-Don, Russia

A.A. Nasedkina Institute of Mathematics, Mechanics and Computer Science, Southern Federal University, Rostov-on-Don, Russia

Konstantin Naumenko Institut für Mechanik, Otto von Guericke University Magdeburg, Magdeburg, Germany

P.A. Oganesyan Institute of Mathematics, Mechanics and Computer Science, Southern Federal University, Rostov-on-Don, Russia

Tatiana O. Petrova Southern Federal University, Rostov-on-Don, Russian Federation

Luca Placidi International Telematic University Uninettuno, Rome, Italy

Vitaly V. Popuzin Institute of Mathematics, Mechanics and Computer Science, Southern Federal University, Rostov-on-Don, Russia

A. Rajagopal Department of Civil Engineering, Indian Institute of Technology Hyderabad, Hyderabad, Telangana, India

M.Yu. Remizov Department of Technical Mechanics, Rostov State University of Civil Engineering, Rostov-on-Don, Russia

A.S. Skaliukh Institute of Mathematics, Mechanics and Computer Science, Southern Federal University, Rostov-on-Don, Russia

A.N. Soloviev Don State Technical University, Department of Theoretical and Applied Mechanics, Rostov-on-Don, Russia

M.A. Sumbatyan Institute of Mathematics, Mechanics and Computer Science, Southern Federal University, Rostov-on-Don, Russia

Alexander E. Tarasov Institute of Mathematics, Mechanics and Computer Science, Southern Federal University, Rostov-on-Don, Russia

Vladimir V. Zotov Institute of Mathematics, Mechanics and Computer Science, Southern Federal University, Rostov-on-Don, Russia

Mathematical Models and Finite Element Approaches for Nanosized Piezoelectric Bodies with Uncoulped and Coupled Surface Effects

Victor A. Eremeyev and A.V. Nasedkin

Abstract In this chapter the dynamic problems for piezoelectric nanosized bodies with account for coupled damping and surface effects are considered. For these problems we propose new mathematical model which generalizes the models of the elastic medium with damping in sense of the Rayleigh approach and with surface effects for the cases of piezoelectric materials. Our model of attenuation and surface effects has coupling properties between mechanical and electric fields, both for the damping terms and constitutive equations for piezoelectric materials on the surface. For solving the problems stated the finite element approximations are discussed. A set of effective finite element schemes is examined for finding numerical solutions of week statements for nonstationary problems, steady-state oscillation problems, modal problems and static problems within the framework of modelling of piezoelectric nanosized materials with damping and surface effects. For transient and harmonic problems, we demonstrate that the proposed models allow the use of the mode superposition method. In addition, we note that for transient and static problems we can use efficient finite element algorithms for solving the systems of linear algebraic equations with symmetric quasi-definite matrices both in the case of uncoupled surface effects and in the case of coupled surface effects.

Keywords Surface elasticity · Piezoelectricity · Finite element method · Rayleigh damping

---

V.A. Eremeyev (✉) · A.V. Nasedkin
Institute of Mathematics, Mechanics and Computer Science, Southern Federal University, Milchakova Street 8a, Rostov-on-Don 344090, Russia
e-mail: eremeyev.victor@gmail.com

A.V. Nasedkin
e-mail: nasedkin@math.sfedu.ru

V.A. Eremeyev
Rzeszów University of Technology, al. Powstańców Warszawy 8, 35959 Rzeszów, Poland

M.A. Sumbatyan (ed.), *Wave Dynamics and Composite Mechanics for Microstructured Materials and Metamaterials*,
Advanced Structured Materials 59, DOI 10.1007/978-981-10-3797-9_1

1 Introduction

The most popular model of the surface elasticity [20, 32, 34] is the Gurtin-Murdoch model [19] which used in nanomechanics, see [8, 9, 38]. The Gurtin–Murdoch model was extended by Steigmann and Ogden [35, 36]. The mentioned models consider elastic media whereas the extension for piezoelectric and magnetoelastic media was introduced in [21, 39]. The existence of surface stresses leads to changes in behaviour of stresses and deformations in the vicinity of stress concentrators, such as crack tips, see [22–24]. In the media with surface stresses new type of waves is possible, that are anti-plane (shear) surface waves [13, 14] which absent in classic elasticity. The mathematical study of boundary-value problems in the elasticity with surface stresses where performed in [1, 2, 10, 11, 33] where existence and uniqueness of weak and strong solutions were proved. Existence and uniqueness of weak solutions for modal problem considering surface effects in solids with coupling of deformations and electric field was proved in [12, 27].

In the one of the first papers in the field of piezoelectricity with surface effects [21] it was proposed new model of a nanosized piezoelectric solid with introduction into the functional of energy surface integrals depending on the surface stresses and strains. Here the surface dielectric permittivities and the surface piezomoduli are introduced as additional material parameters. For the static axially symmetric problem for a piezoelectric ring with thin electrodes at its faces, the surface effects are taken into account in the boundary conditions for stresses. From the mechanical point of view the main idea of [21] consists of introduction of a piezoelectric film at the surface of an nanosized piezoelectric solid. The constitutive equations of the film determine the relation between the surface stresses and the surface electric displacement vector with the strains and the electrical field including its component normal to surface.

Similar approach was applied in analysis of piezoelectric nanowires, beams, and plates in [37, 44–49, 53], of a plate oscillations in [51, 52, 54]. Propagation of the Bleustein–Gulyaev waves in a piezoelectric half-space with thin surface piezoelectric layer are studied in [6] with asymptotic expansions technique. Using atomistic models and multiscale analysis in [31] the mechanical and electrostatic stresses of Piola–Kirchhoff-type are reconstructed. Formulae for the effective shear modulus of a fiber reinforced piezoelectric composite was obtained in [42] using a self-consistent method. Investigations of effective moduli for nanoisized piezoelectric composites were continued in [5, 15, 25, 41, 43] and etc.

Magnetoelectric nanosized composites were analysed in [17, 28, 29, 40] and etc.

Theoretical investigations of piezoelectric and magnetoelectric nanosized materials with surface effects and imperfect interface models were presented in [7, 12, 16–18, 30]. Note that dynamic models for piezoelectric and magnetoelectric nanosized bodies with damping and surface effects were suggested only in [27–29], but for uncoupled dependencies between mechanical and electric or magnetoelectric surface fields.

The main goal of our investigation is to present the mathematical study of the dynamic boundary-value problems for pizoelectric solids with damping and coupling between surface mechanical and electric fields. In this paper we propose the new models for nanosized piezoelectric bodies with surface effects in development of corresponding investigations of nanosized active materials earlier presented in [27–29]. Here we formulate the system of differential equations with damping properties, the special boundary conditions with taking into account the coupled surface effects, damping and the initial conditions for piezoelectric nanosized bodies.

For numerical solution of the dynamic problems with damping and surface effects we propose the finite element approximations and the corresponding generalized matrix problems. We note that the standard finite element software could be used with additional introduction of surface piezoelectric elements with structural membrane option. We demonstrate that the finite element systems for coupled problems for piezoelectric nanosized bodies can be represented in the form of a system of linear algebraic equations with symmetric saddle point quasi-definite matrices. We also describe special efficient approaches to solve the resulting finite element equations for transient, harmonic, modal and static problems.

2 Model of Piezoelectric Materials with Damping and Surface Effects

Let $\Omega \in \mathrm{R}^3$ be a region occupied by a piezoelectric material; $\Gamma = \partial\Omega$ is the boundary of this region; $\mathbf{n}$ is the vector of the external unit normal to Γ; $\mathbf{x} = \{x_1, x_2, x_3\}$ is the vector of the special coordinates; t is the time; $\mathbf{u} = \mathbf{u}(\mathbf{x}, t)$ is the vector of mechanical displacements; $\varphi = \varphi(\mathbf{x}, t)$ is the scalar function of electric potential. The system of differential equations for piezoelectric body with damping effects in the volume Ω can be present in the form

$$\nabla \cdot \boldsymbol{\sigma} + \rho \mathbf{f} = \rho(\ddot{\mathbf{u}} + \alpha_d \dot{\mathbf{u}}), \quad \nabla \cdot \mathbf{D} = q_\Omega, \tag{1}$$

$$\boldsymbol{\sigma} = \mathbf{c} : (\boldsymbol{\varepsilon} + \beta_d \dot{\boldsymbol{\varepsilon}}) - \mathbf{e}^T \cdot \mathbf{E}, \tag{2}$$

$$\mathbf{D} + \zeta_d \dot{\mathbf{D}} = \mathbf{e} : (\boldsymbol{\varepsilon} + \zeta_d \dot{\boldsymbol{\varepsilon}}) + \boldsymbol{\kappa} \cdot \mathbf{E}, \tag{3}$$

$$\boldsymbol{\varepsilon} = (\nabla \mathbf{u} + (\nabla \mathbf{u})^*)/2, \quad \mathbf{E} = -\nabla \varphi, \tag{4}$$

where $\boldsymbol{\sigma}$ and $\boldsymbol{\varepsilon}$ are the second-order stress and strain tensors; $\mathbf{D}$ and $\mathbf{E}$ are the electric flux density vector or the electric displacement vector and the electric field vector; ρ is the mass density of the material; $\mathbf{c} = \mathbf{c}^E$ is the fourth-order tensor of elastic stiffness moduli; $\mathbf{e}$ is the third-order tensor of piezoelectric moduli; $\boldsymbol{\kappa} = \boldsymbol{\kappa}^S = \boldsymbol{\varepsilon}^S$ is the second-order tensor of dielectric permittivity moduli; α_d, β_d, ζ_d are the damping coefficients; $\mathbf{f}$ is the vector of mass forces; q_Ω is the density of free electric charges (usually, $q_\Omega = 0$); $\dot{\mathbf{u}} = \partial\mathbf{u}/\partial t$; $\ddot{\mathbf{u}} = \partial^2\mathbf{u}/\partial^2 t$; $(\ldots)^T$ is the transpose operation; $(\ldots):(\ldots)$ is the double scalar product operation.

We suppose that the material moduli have the usual symmetry properties: $c_{ijkl} = c_{jikl} = c_{klij}$, $e_{ikl} = e_{ilk}$, $\kappa_{kl} = \kappa_{lk}$. In addition to this for the positive definiteness of the intrinsic energy for the piezoelectric medium the following inequalities must be satisfied ($\forall\ \boldsymbol{\varepsilon} = \boldsymbol{\varepsilon}^T, \mathbf{E}$), $\exists\, W_\Omega > 0$:

$$\boldsymbol{\varepsilon}^T : \mathbf{c} : \boldsymbol{\varepsilon} + \mathbf{E}^T \cdot \boldsymbol{\kappa} \cdot \mathbf{E} \geq W_\Omega(\boldsymbol{\varepsilon}^T : \boldsymbol{\varepsilon} + \mathbf{E}^T \cdot \mathbf{E}).$$

In models (1)–(4) for the piezoelectric material, we use a generalized Rayleigh method of damping evaluation [3, 26, 28, 29], which is admissible for many practical applications. When $\zeta_d = 0$ in Eq. (3), we have the model for taking into account of mechanical damping in piezoelectric media which is adopted in the case of elastic and piezoelectric materials in several well-known finite element packages. More complicated model (1)–(4) extends the Kelvin's model to the case of piezoelectric media. It has been shown that the model (2) and (3) with $\beta_d = \zeta_d$ satisfies the conditions of the energy dissipation and has the possibility of splitting the finite element system into independent equations for the separate modes in the case of piezoelectric media, see also Sect. 4.

For nanosized piezoelectric body we assume that on its boundary Γ the surface stress and surface electric flux exist. For these quantities we accept the Gurtin–Murdoch model:

$$\mathbf{n} \cdot \boldsymbol{\sigma} = \nabla^s \cdot \boldsymbol{\sigma}^s + \mathbf{p}, \qquad \mathbf{x} \in \Gamma, \tag{5}$$

$$\mathbf{n} \cdot \mathbf{D} = \nabla^s \cdot \mathbf{D}^s - q, \qquad \mathbf{x} \in \Gamma, \tag{6}$$

where ∇^s is the surface gradient operator, associated with nabla-operator by the formula $\nabla^s = \nabla - \mathbf{n}(\partial/\partial r)$, r is the coordinate, measured by the normal $\mathbf{n}$ to Γ_σ; $\boldsymbol{\sigma}^s$ is the second-order tensor of surface stress; $\mathbf{D}^s$ is the surface electric flux density vector; $\mathbf{p}$ is the vector of mechanical stress; q is the surface density of electric charge.

Here, $\mathbf{p}$ and q are the known (active) or unknown (reactive) surface quantities according to the boundary conditions.

For surface stress $\boldsymbol{\sigma}^s$ and surface electric flux $\mathbf{D}^s$ in general case we take the coupled constitutive relations

$$\boldsymbol{\sigma}^s = \mathbf{c}^s : (\boldsymbol{\varepsilon}^s + \beta_d \dot{\boldsymbol{\varepsilon}}^s) - \mathbf{e}^{sT} \cdot \mathbf{E}^s, \tag{7}$$

$$\mathbf{D}^s + \zeta_d \dot{\mathbf{D}}^s = \mathbf{e}^s : (\boldsymbol{\varepsilon}^s + \zeta_d \dot{\boldsymbol{\varepsilon}}^s) + \boldsymbol{\kappa}^s \cdot \mathbf{E}^s, \tag{8}$$

where

$$\boldsymbol{\varepsilon}^s = (\nabla^s \mathbf{u}^s + (\nabla^s \mathbf{u}^s)^T)/2, \quad \mathbf{u}^s = \mathbf{A} \cdot \mathbf{u}, \quad \mathbf{A} = \mathbf{I} - \mathbf{n} \otimes \mathbf{n}, \quad \mathbf{E}^s = -\nabla^s \varphi, \tag{9}$$

$\mathbf{I}$ is the identity matrix; $\mathbf{c}^s$, $\mathbf{e}^s$, $\boldsymbol{\kappa}^s$ are the surface tensors of elastic stiffness moduli, piezoelectric moduli and dielectric permittivity moduli, respectively.

We suppose that the tensors of surface material moduli have the similar properties that the tensors of volume material moduli, but relatively to surface strains $\boldsymbol{\varepsilon}^s$ and surface electric field vector $\mathbf{E}^s$, i.e. $c^s_{ijkl} = c^s_{jikl} = c^s_{klij}$, $e^s_{ikl} = e^s_{ilk}$, $\kappa^s_{kl} = \kappa^s_{lk}$, and surface energy is positive definiteness ($\forall\ \boldsymbol{\varepsilon}^s = \boldsymbol{\varepsilon}^{sT}, \mathbf{E}^s$), $\exists\, W_\Gamma > 0$:

$$\boldsymbol{\varepsilon}^{sT} : \mathbf{c}^s : \boldsymbol{\varepsilon}^s + \mathbf{E}^{sT} \cdot \boldsymbol{\kappa}^s \cdot \mathbf{E}^s \geq W_\Gamma(\boldsymbol{\varepsilon}^{sT} : \boldsymbol{\varepsilon}^s + \mathbf{E}^{sT} \cdot \mathbf{E}^s).$$

The boundary and the initial conditions should be added to the system of equations (1)–(9). The boundary conditions are mechanical and electric types.

To formulate the mechanical boundary conditions we assume that the boundary Γ is divided in two subsets Γ_σ and Γ_u ($\Gamma = \Gamma_\sigma \cup \Gamma_u$). We will assume that at the part of the boundary Γ_σ there are the surface stresses $\boldsymbol{\sigma}^s$ and the vector of mechanical stress $\mathbf{p}_\Gamma$, i.e.

$$\mathbf{p} = \mathbf{p}_\Gamma, \qquad \mathbf{x} \in \Gamma_\sigma, \tag{10}$$

and so, in accordance with (5)

$$\mathbf{n} \cdot \boldsymbol{\sigma} = \nabla^s \cdot \boldsymbol{\sigma}^s + \mathbf{p}_\Gamma, \qquad \mathbf{x} \in \Gamma_\sigma. \tag{11}$$

On the part Γ_u we pose known the mechanical displacements vector $\mathbf{u}_\Gamma$

$$\mathbf{u} = \mathbf{u}_\Gamma, \qquad \mathbf{x} \in \Gamma_u, \tag{12}$$

and so, in (6) $\mathbf{p}$ is unknown reactive surface load vector on Γ_u.

To set the electric boundary conditions we assume that the surface Γ is also subdivided in two subsets: Γ_D and Γ_φ ($\Gamma = \Gamma_D \cup \Gamma_\varphi$).

The regions Γ_D does not contain electrodes, and we pose known the surface density of electric charge q_Γ

$$q = q_\Gamma, \qquad \mathbf{x} \in \Gamma_D, \tag{13}$$

i.e. in accordance with (6) on Γ_D hold the following conditions

$$\mathbf{n} \cdot \mathbf{D} = \nabla^s \cdot \mathbf{D}^s - q_\Gamma, \qquad \mathbf{x} \in \Gamma_D, \tag{14}$$

and usually, $q_\Gamma = 0$.

The subset Γ_φ is the union of $M + 1$ regions $\Gamma_{\varphi j}$ ($\Gamma_\varphi = \Gamma_V \cup \Gamma_Q$, $\Gamma_V = \cup_j \Gamma_{\varphi j}, j \in J_V, J_V = \{0, m, m+1, \dots, M\}$, $\Gamma_Q = \cup_j \Gamma_{\varphi j}, j \in J_Q, J_Q = \{1, 2, \dots, m\}$), that does not border on each other and are covered with infinitely thin electrodes. At these regions we set the following boundary conditions

$$\varphi = \Phi_j, \quad \mathbf{x} \in \Gamma_{\varphi j}, \quad j \in J_Q, \tag{15}$$

$$\int_{\Gamma_{\varphi j}} q\, d\Gamma = Q_j, \quad I_j = \pm\dot{Q}_j, \qquad \mathbf{x} \in \Gamma_{\varphi j}, \quad j \in J_Q, \tag{16}$$

$$\varphi = V_j, \qquad \mathbf{x} \in \Gamma_{\varphi j}, \quad j \in J_V, \quad \Gamma_{j0} \neq \emptyset, \tag{17}$$

where the variables Φ_j, V_j do not depend on $\mathbf{x}$; Q_j is the overall electric charge on $\Gamma_{\varphi j}$, the sign "±" in (16) is chosen in accordance with the accepted direction of the current I_j in the electric circuit, and by using (6) we can rewrite the relation (16) in the other form

$$\int_{\Gamma_{\varphi j}} \mathbf{n} \cdot \mathbf{D}\, d\Gamma - \int_{\Gamma_{\varphi j}} \nabla^s \cdot \mathbf{D}^s\, d\Gamma = -Q_j, \quad I_j = \pm \dot{Q}_j, \qquad \mathbf{x} \in \Gamma_{\varphi j}, \quad j \in J_Q. \tag{18}$$

For transient problems it is also necessary to pose initial conditions, which can be written as

$$\mathbf{u} = \mathbf{u}_*(\mathbf{x}), \quad \dot{\mathbf{u}} = \mathbf{r}_*(\mathbf{x}), \quad t = 0, \quad \mathbf{x} \in \Omega, \tag{19}$$

where $\mathbf{u}_*(\mathbf{x})$ and $\mathbf{r}_*(\mathbf{x})$ are the known initial values of the corresponding fields.

Formulas (1)–(19) represent the statement of the transient problem for piezoelectric body with the generalized Rayleigh damping and with account for coupled surface effects for mechanical and electric fields.

If in (7) and (8) we assume $\mathbf{e}^s = 0$, i.e. surface piezomoduli are equal to zero, then we obtain the uncoupled constitutive equations, when the mechanical surface stress $\boldsymbol{\sigma}^s$ depend only on the surface strain $\boldsymbol{\varepsilon}^s$, and surface electric flux $\mathbf{D}^s$ depend only on the surface electric field vector $\mathbf{E}^s$. This more simple model we are investigated for different dynamic problems earlier in [12, 27–29].

3 Week Formulations of Dynamic Problem

In order to formulate the weak or generalized statement of dynamic transient problem for nanosized piezoelectric solid we scalar multiply Eq. (1) by some sufficiently differentiable vector-function $\mathbf{v}$ and functions χ, which satisfies following principal boundary conditions, i.e.

$$\mathbf{v} = 0, \qquad \mathbf{x} \in \Gamma_u, \tag{20}$$

$$\chi = X_j, \quad \mathbf{x} \in \Gamma_{\varphi j}, \quad j \in J_Q, \tag{21}$$

$$\chi = 0, \qquad \mathbf{x} \in \Gamma_{\varphi j}, \quad j \in J_V, \tag{22}$$

where X_j are the arbitrary constant values on $\Gamma_{\varphi j} \subset \Gamma_Q$.

By integrating the obtained equations over Ω and by using the standard technique of the integration by parts with Eqs. (2)–(18) and (20)–(22), we obtain

$$\rho(\mathbf{v}, \ddot{\mathbf{u}}) + d(\mathbf{v}, \dot{\mathbf{u}}) + c(\mathbf{v}, \mathbf{u}) + e_u(\varphi, \mathbf{v}) = \tilde{L}_u(\mathbf{v}), \tag{23}$$

$$-e_{\varphi}(\chi, \mathbf{u}+\zeta_d\dot{\mathbf{u}}) + \kappa(\chi, \varphi) = \tilde{L}_{\varphi}(\chi) + \zeta_d \frac{\partial}{\partial t}\tilde{L}_{\varphi}(\chi), \tag{24}$$

where

$$\rho(\mathbf{v}, \mathbf{u}) = \int_{\Omega} \rho \mathbf{v}^T \cdot \mathbf{u}\, d\Omega, \quad d(\mathbf{v}, \mathbf{u}) = \alpha_d \rho(\mathbf{v}, \mathbf{u}) + \beta_d c(\mathbf{v}, \mathbf{u}), \tag{25}$$

$$c(\mathbf{v}, \mathbf{u}) = c_{\Omega}(\mathbf{v}, \mathbf{u}) + c_{\Gamma}(\mathbf{v}, \mathbf{u}), \quad \kappa(\chi, \varphi) = \kappa_{\Omega}(\chi, \varphi) + \kappa_{\Gamma}(\chi, \varphi), \tag{26}$$

$$e_u(\varphi, \mathbf{v}) = e_{\Omega}(\varphi, \mathbf{v}) + e_{\Gamma\sigma}(\varphi, \mathbf{v}), \quad e_{\varphi}(\chi, \mathbf{u}) = e_{\Omega}(\chi, \mathbf{u}) + e_{\Gamma D}(\chi, \mathbf{u}), \tag{27}$$

$$c_{\Omega}(\mathbf{v}, \mathbf{u}) = \int_{\Omega} \boldsymbol{\varepsilon}(\mathbf{v}) : \mathbf{c} : \boldsymbol{\varepsilon}(\mathbf{u})\, d\Omega, \quad c_{\Gamma}(\mathbf{v}, \mathbf{u}) = \int_{\Gamma_{\sigma}} \boldsymbol{\varepsilon}^s(\mathbf{v}) : \mathbf{c}^s : \boldsymbol{\varepsilon}^s(\mathbf{u})\, d\Gamma, \tag{28}$$

$$\kappa_{\Omega}(\chi, \varphi) = \int_{\Omega} \mathbf{E}(\chi) \cdot \boldsymbol{\kappa} \cdot \mathbf{E}(\varphi)\, d\Omega, \quad \kappa_{\Gamma}(\chi, \varphi) = \int_{\Gamma_D} \mathbf{E}^s(\chi) \cdot \boldsymbol{\kappa}^s \cdot \mathbf{E}^s(\varphi)\, d\Gamma, \tag{29}$$

$$e_{\Omega}(\varphi, \mathbf{v}) = -\int_{\Omega} \mathbf{E}(\varphi) \cdot \mathbf{e} : \boldsymbol{\varepsilon}(\mathbf{v})\, d\Omega, \; e_{\Gamma\sigma}(\varphi, \mathbf{v}) = -\int_{\Gamma_{\sigma}} \mathbf{E}^s(\varphi) \cdot \mathbf{e}^s : \boldsymbol{\varepsilon}^s(\mathbf{v})\, d\Gamma, \tag{30}$$

$$e_{\Gamma D}(\chi, \mathbf{u}) = -\int_{\Gamma_D} \mathbf{E}^s(\chi) \cdot \mathbf{e}^s : \boldsymbol{\varepsilon}^s(\mathbf{u})\, d\Gamma, \tag{31}$$

$$\tilde{L}_u(\mathbf{v}) = \int_{\Omega} \mathbf{v} \cdot \rho \mathbf{f}\, d\Omega + \int_{\Gamma_{\sigma}} \mathbf{v} \cdot \mathbf{p}_{\Gamma}\, d\Gamma, \tag{32}$$

$$\tilde{L}_{\varphi}(\chi) = \int_{\Omega} \chi q_{\Omega}\, d\Omega + \int_{\Gamma_D} \chi q_{\Gamma} + \sum_{k \in J_Q} X_k Q_k\, d\Gamma. \tag{33}$$

The weak form of the initial conditions (19) can be represent by the relations

$$\rho(\mathbf{v}, \mathbf{u}) = \rho(\mathbf{v}, \mathbf{u}_*), \quad \rho(\mathbf{v}, \dot{\mathbf{u}}) = \rho(\mathbf{v}, \mathbf{r}_*), \quad t = 0. \tag{34}$$

Further we present the functions $\mathbf{u}$ and φ as

$$\mathbf{u} = \mathbf{u}_0 + \mathbf{u}_b, \quad \varphi = \varphi_0 + \varphi_b, \tag{35}$$

where $\mathbf{u}_0$, φ_0 satisfy "homogeneous" boundary mechanical and electric conditions and $\mathbf{u}_b$, φ_b are the given functions satisfying the inhomogeneous boundary conditions, i.e.

$$\mathbf{u}_0 = 0, \quad \mathbf{u}_b = \mathbf{u}_{\Gamma}, \qquad \mathbf{x} \in \Gamma_u, \tag{36}$$

$$\varphi_0 = \Phi_{0j}, \quad \varphi_b = \Phi_{bj}, \quad \Phi_j = \Phi_{0j} + \Phi_{bj}, \quad \mathbf{x} \in \Gamma_{\varphi j}, \quad j \in J_Q, \tag{37}$$

$$\varphi_0 = 0, \quad \varphi_b = V_j, \qquad \mathbf{x} \in \Gamma_{\varphi j}, \quad j \in J_V. \tag{38}$$

Using (35), we can modify the system (23) and (24) into the form

$$\rho(\mathbf{v}, \ddot{\mathbf{u}}_0) + d(\mathbf{v}, \dot{\mathbf{u}}_0) + c(\mathbf{v}, \mathbf{u}_0) + e_u(\varphi_0, \mathbf{v}) = L_u(\mathbf{v}), \tag{39}$$

$$- e_\varphi(\chi, \mathbf{u}_0 + \zeta_d \dot{\mathbf{u}}_0) + \kappa(\chi, \varphi_0) = L_\varphi(\chi), \tag{40}$$

where

$$L_u(\mathbf{v}) = \tilde{L}_u(\mathbf{v}) - \rho(\mathbf{v}, \ddot{\mathbf{u}}_b) - d(\mathbf{v}, \dot{\mathbf{u}}_b) - c(\mathbf{v}, \mathbf{u}_b) - e_u(\varphi_b, \mathbf{v}), \tag{41}$$

$$L_\varphi(\chi) = \tilde{L}_\varphi(\chi) + \zeta_d \frac{\partial}{\partial t} \tilde{L}_\varphi(\chi) + e_\varphi(\chi, \mathbf{u}_b + \zeta_d \dot{\mathbf{u}}_b) - \kappa(\chi, \varphi_b). \tag{42}$$

We denote with the Hilbert vector space H_u the closure of the set of vector functions $\mathbf{v} \in \mathrm{C}^1$, satisfying homogeneous principal boundary condition (20), with the norm generated by bilinear form defined in the first relation (26) and in (28).

We also denote with the Hilbert space H_φ the closure of the set of function $\varphi \in C^1$, satisfying boundary condition (21) and (22), in the norm generated by scalar production from the second relation (26) and (29).

Finally, we introduce the functional spaces $Q_u = L^2(0, T; H_u)$ and $Q_\varphi = L^2(0, T; H_\varphi)$, where for Banach space X with norm $||.||_X$ the space $L^2(0, T; X)$ is the space of class functions $t \to f(t)$ from $[0, T]$ into X which satisfy the condition

$$||f||^2_{L^2(0,T;X)} = \int_0^T ||f||^2_X \, dt < \infty.$$

Now we can define generalized or weak solution of dynamic problem (1)–(19) using these functional spaces.

Definition. The functions $\mathbf{u} = \mathbf{u}_0 + \mathbf{u}_b$, $\mathbf{u}_0 \in Q_u$; $\varphi = \varphi_0 + \varphi_b$, $\varphi_0 \in Q_\varphi$ are the weak solution of dynamic problem for the piezoelectric body with damping and coupled surface effects, if Eqs. (39) and (40) with (25)–(33), (41) and (42) are satisfied for $\forall t \in [0, T]$; $\mathbf{v} \in H_u$, $\chi \in H_\varphi$, and the initial conditions (34) are also hold.

It is important to mark that after transfer Eqs. (23) and (24) to (39) and (40), we obtain

$$e_{\Gamma\sigma}(\varphi_0, \mathbf{v}) = e_{\Gamma\sigma D}(\varphi_0, \mathbf{v}) = - \int_{\Gamma_{\sigma D}} \mathbf{E}^s(\varphi_0) \cdot \mathbf{e}^s : \boldsymbol{\varepsilon}^s(\mathbf{v}) \, d\Gamma, \tag{43}$$

$$e_{\Gamma D}(\chi, \mathbf{u}_0) = e_{\Gamma\sigma D}(\chi, \mathbf{u}_0) = - \int_{\Gamma_{\sigma D}} \mathbf{E}^s(\chi) \cdot \mathbf{e}^s : \boldsymbol{\varepsilon}^s(\mathbf{u}_0) \, d\Gamma, \tag{44}$$

where $\Gamma_{\sigma D} = \Gamma_\sigma \cap \Gamma_D$, as far as $\mathbf{E}^s(\varphi_0) = 0$ for $\mathbf{x} \in \Gamma_\sigma \setminus \Gamma_{\sigma D}$, and $\boldsymbol{\varepsilon}^s(\mathbf{u}_0) = 0$ for $\mathbf{x} \in \Gamma_D \setminus \Gamma_{\sigma D}$.

These equations allow us to obtain later the symmetric systems, because from (43) and (44) with (27), (30) and (31) it is follow that

$$e_u(\varphi_0, \mathbf{v}) = e_\varphi(\varphi_0, \mathbf{v}), \quad e_\varphi(\chi, \mathbf{u}_0) = e_u(\chi, \mathbf{u}_0).$$

4 Finite Element Approaches

4.1 Nonstationary Problems

For numerical solving the problems (34), (39) and (40) we will use the finite element method. Let Ω_h be the region of the corresponding finite element mesh: $\Omega_h \subset \Omega$, $\Omega_h = \cup_k \Omega_{ek}$, where Ω_{ek} is a separate volume or surface finite element with number k. On this mesh we shall find the approximation to the weak solution $\{\mathbf{u}_{0h} \approx \mathbf{u}_0, \varphi_{0h} \approx \varphi_0\}$ in the form

$$\mathbf{u}_h(\mathbf{x}, t) = \mathbf{N}_u^T(\mathbf{x}) \cdot \mathbf{U}(t), \quad \varphi_h(\mathbf{x}, t) = \mathbf{N}_\varphi^T(\mathbf{x}) \cdot \boldsymbol{\Phi}(t), \tag{45}$$

where $\mathbf{N}^T$ is the matrix of the shape functions for displacements, $\mathbf{N}_\varphi^T$ is the row vector of the shape functions for electric potential, $\mathbf{U}(t)$, $\boldsymbol{\Phi}(t)$ are the global vectors of nodal displacements, electric potential and magnetic potential, respectively.

We represent the projecting functions $\mathbf{v}$ and χ in finite-dimensional spaces by the formulae

$$\mathbf{v}^T = \delta\mathbf{U}^T \cdot \mathbf{N}_u(\mathbf{x}), \quad \chi = \mathbf{N}_\varphi^T \cdot \delta\boldsymbol{\Phi} = \delta\boldsymbol{\Phi}^T \cdot \mathbf{N}_\varphi, \tag{46}$$

Note that we can use the same nodal degrees of freedom and the shape functions for volume and surface elements, or rather, in accordance with (9) we will consider the surface shape functions for displacements as a reduction of volume shape function on the surface elements by formula $\mathbf{N}_u^{sT} = \mathbf{A} \cdot \mathbf{N}_u^T$, and $\mathbf{N}_\varphi^s = \mathbf{N}_\varphi$.

Substituting (45) and (46) into the problem (39) and (40) with (25)–(33), (41) and (42) for Ω_h, $\Gamma_h = \partial\Omega_h$, $\Gamma_{\sigma h}$, Γ_{Dh}, $\Gamma_{\sigma Dh}$, we obtain the finite element system of ordinary differential equations with respect to time

$$\mathbf{M}_{uu} \cdot \ddot{\mathbf{U}} + \mathbf{C}_{uu} \cdot \dot{\mathbf{U}} + \mathbf{K}_{uu} \cdot \mathbf{U} + \mathbf{K}_{u\varphi} \cdot \boldsymbol{\Phi} = \mathbf{F}_u, \tag{47}$$

$$-\mathbf{K}_{u\varphi}^* \cdot (\mathbf{U} + \zeta_d \dot{\mathbf{U}}) + \mathbf{K}_{\varphi\varphi} \cdot \boldsymbol{\Phi} = \mathbf{F}_\varphi, \tag{48}$$

with the initial conditions

$$\mathbf{U}(0) = \mathbf{U}_*, \quad \dot{\mathbf{U}}(0) = \mathbf{R}_*, \tag{49}$$

which are derived from the corresponding initial conditions (19) or (34).

Here,

$$\mathbf{M}_{uu} = \int_{\Omega_h} \rho \mathbf{N}_u \cdot \mathbf{N}_u^T \, d\Omega, \quad \mathbf{C}_{uu} = \alpha_d \mathbf{M}_{uu} + \beta_d \mathbf{K}_{uu}, \tag{50}$$

$$\mathbf{K}_{uu} = \mathbf{K}_{\Omega uu} + \mathbf{K}_{\Gamma uu}, \quad \mathbf{K}_{u\varphi} = \mathbf{K}_{\Omega u\varphi} + \mathbf{K}_{\Gamma u\varphi}, \quad \mathbf{K}_{\varphi\varphi} = \mathbf{K}_{\Omega\varphi\varphi} + \mathbf{K}_{\Gamma\varphi\varphi}, \tag{51}$$

$$\mathbf{K}_{\Omega uu} = \int_{\Omega_h} \mathbf{B}_u^T \cdot \mathbf{c} \cdot \mathbf{B}_u \, d\Omega, \quad \mathbf{K}_{\Gamma uu} = \int_{\Gamma_{\sigma h}} \mathbf{B}_u^{sT} \cdot \mathbf{c}^s \cdot \mathbf{B}_u^s \, d\Gamma, \tag{52}$$

$$\mathbf{K}_{\Omega u\varphi} = \int_{\Omega_h} \mathbf{B}_u^T \cdot \mathbf{e}^T \cdot \mathbf{B}_\varphi \, d\Omega, \quad \mathbf{K}_{\Gamma u\varphi} = \int_{\Gamma_{\sigma Dh}} \mathbf{B}_u^{sT} \cdot \mathbf{e}^{sT} \cdot \mathbf{B}_\varphi^s \, d\Gamma, \tag{53}$$

$$\mathbf{K}_{\Omega\varphi\varphi} = \int_{\Omega_h} \mathbf{B}_\varphi^T \cdot \boldsymbol{\kappa} \cdot \mathbf{B}_\varphi \, d\Omega, \quad \mathbf{K}_{\Gamma\varphi\varphi} = \int_{\Gamma_{Dh}} \mathbf{B}_\varphi^{sT} \cdot \boldsymbol{\kappa}^s \cdot \mathbf{B}_\varphi^s \, d\Gamma, \tag{54}$$

$$\mathbf{B}_u^{(s)} = \mathbf{L}(\nabla^{(s)}) \cdot \mathbf{N}_u^{(s)T}, \quad \mathbf{B}_\varphi^{(s)} = \nabla \mathbf{N}_\varphi^{(s)T}, \quad \mathbf{L}^T(\nabla^{(s)}) = \begin{bmatrix} \partial_1^{(s)} & 0 & 0 & 0 & \partial_3^{(s)} & \partial_2^{(s)} \\ 0 & \partial_2^{(s)} & 0 & \partial_3^{(s)} & 0 & \partial_1^{(s)} \\ 0 & 0 & \partial_3^{(s)} & \partial_2^{(s)} & \partial_1^{(s)} & 0 \end{bmatrix}. \tag{55}$$

The vectors $\mathbf{F}_u$, $\mathbf{F}_\varphi$ in (47) and (48) are obtained from the corresponding right parts of the weak statements (39) and (40) with (25)–(33), (41) and (42) and the finite element representations (46).

In (52)–(54) we use vector-matrix forms for the moduli [4]: $\mathbf{c}^{(s)}$ is the 6×6 matrix of elastic moduli, $c_{\alpha\beta}^{(s)} = c_{ijkl}^{(s)}$; $\alpha, \beta = 1, \ldots, 6$; $i, j, k, l = 1, 2, 3$ with the correspondence law $\alpha \leftrightarrow (ij)$, $\beta \leftrightarrow (kl)$, $1 \leftrightarrow (11)$, $2 \leftrightarrow (22)$, $3 \leftrightarrow (33)$, $4 \leftrightarrow (23) = (32)$, $5 \leftrightarrow (13) = (31)$, $6 \leftrightarrow (12) = (21)$; $\mathbf{e}^{(s)}$ is the 3×6 matrix of piezoelectric moduli ($e_{i\beta}^{(s)} = e_{ikl}^{(s)}$).

We note that in (50)–(54) the matrices of mass and stiffness $\mathbf{M}_{uu}$, $\mathbf{K}_{\Omega uu}$, and nodal mechanical force vector $\mathbf{F}_u$ are formed in the same way as for purely elastic body, and the matrices $\mathbf{K}_{\Omega u\varphi}$, $\mathbf{K}_{\Omega\varphi\varphi}$ and nodal electric force vector $\mathbf{F}_\varphi$ are identical to the corresponding matrices and vector for piezoelectric bodies. The matrices $\mathbf{K}_{\Gamma uu}$, $\mathbf{K}_{\Gamma u\varphi}$ and $\mathbf{K}_{\Gamma\varphi\varphi}$ are defined by the surface mechanical, piezoelectric and dielectric effects. The matrix $\mathbf{K}_{\Gamma uu}$ is analogous to the stiffness matrix for surface elastic membranes and the matrix $\mathbf{K}_{\Gamma\varphi\varphi}$ is the matrix of dielectric permittivities for surface dielectric films. Hence, for implementing the finite element piezoelectric analysis for the bodies with surface effects it is necessary to have surface piezoelectric elements with structural membrane option along with ordinary solid piezoelectric finite elements.

4.2 Static Problems

In the case of static problems all dependencies on time t are absent, and the finite element system (47) and (48) reduces to the form

$$\mathbf{K} \cdot \mathbf{a} = \mathbf{F}, \tag{56}$$

where

$$\mathbf{K} = \begin{bmatrix} \mathbf{K}_{uu} & \mathbf{K}_{u\varphi} \\ \mathbf{K}_{u\varphi}^T & -\mathbf{K}_{\varphi\varphi} \end{bmatrix}, \quad \mathbf{a} = \begin{bmatrix} \mathbf{U} \\ \boldsymbol{\Phi} \end{bmatrix}, \quad \mathbf{F} = \begin{bmatrix} \mathbf{F}_u \\ -\mathbf{F}_\varphi \end{bmatrix}. \tag{57}$$

The matrix $\mathbf{K}$ in (56) and (57) is symmetric and quasi-definite. Thus, problem (56) possesses the main calculating properties of finite element matrices for the theory of piezoelectricity, and therefore it can be solved by the same effective algorithms as the analogous problems for ordinary piezoelectric media. For example, we can use the set of algorithms for finite element analysis with symmetric and quasi-definite matrices represented in ACELAN package [26, 29]: the degree of freedom rotations, boundary condition settings, LDL^T-factorization or Cholesky method for solving the system of linear algebraic equations, and others.

4.3 *Steady-State Oscillation Problems*

When all external loads and degree of freedom constraints vary with the same harmonic law $\exp[j\omega t]$, i.e. $\mathbf{F}_u = \tilde{\mathbf{F}}_u \exp[j\omega t]$, $\mathbf{F}_\varphi = \tilde{\mathbf{F}}_\varphi \exp[j\omega t]$, we have the behavior of steady-state or harmonic oscillations ($\mathbf{a} = \tilde{\mathbf{a}} \exp[j\omega t]$, $\mathbf{U} = \tilde{\mathbf{U}} \exp[j\omega t]$, $\boldsymbol{\Phi} = \tilde{\boldsymbol{\Phi}} \exp[j\omega t]$). In this case, as it is obvious from (47) and (48), we have a system of linear algebraic equations (56) for the amplitude values $\tilde{\mathbf{a}}$ with

$$\mathbf{K} = \begin{bmatrix} -\omega^2\mathbf{M}_{uu} + j\omega\mathbf{C}_{uu} + \mathbf{K}_{uu} & \mathbf{K}_{u\varphi} \\ \mathbf{K}_{u\varphi}^T & -\mu\mathbf{K}_{\varphi\varphi} \end{bmatrix}, \quad \tilde{\mathbf{a}} = \begin{bmatrix} \tilde{\mathbf{U}} \\ \tilde{\boldsymbol{\Phi}} \end{bmatrix}, \quad \tilde{\mathbf{F}} = \begin{bmatrix} \tilde{\mathbf{F}}_u \\ -\mu\tilde{\mathbf{F}}_\varphi \end{bmatrix}, \tag{58}$$

where $\mu = (1 + j\omega\zeta_d)^{-1}$.

Then, the well-known algorithm for large symmetric complex matrices can be applied for solution of Eq. (56) with (58).

4.4 *Modal Problems*

We can find the resonance frequencies $f_k = \omega_k/(2\pi)$ for nanosized piezoelectric body using the finite element approaches from the solution of the generalized eigenvalue problem, obtained from (56) and (58) with $\mathbf{C}_{uu} = 0$, $\zeta_d = 0$, $\tilde{\mathbf{F}} = 0$, given by

$$\mathbf{K}_{uu} \cdot \tilde{\mathbf{U}} + \mathbf{K}_{u\varphi} \cdot \tilde{\boldsymbol{\Phi}} = \omega^2\mathbf{M}_{uu} \cdot \tilde{\mathbf{U}}, \tag{59}$$

$$-\mathbf{K}_{u\varphi}^T \cdot \tilde{\mathbf{U}} + \mathbf{K}_{\varphi\varphi} \cdot \tilde{\boldsymbol{\Phi}} = 0. \tag{60}$$

Eigenvalue problem (59) and (60) can be represented in the more compact form

$$\overline{\mathbf{K}}_{uu} \cdot \tilde{\mathbf{U}} = \omega^2 \mathbf{M}_{uu} \cdot \tilde{\mathbf{U}}, \tag{61}$$

where

$$\overline{\mathbf{K}}_{uu} = \mathbf{K}_{uu} + \mathbf{K}_{u\varphi} \cdot \mathbf{K}_{\varphi\varphi}^{-1} \cdot \mathbf{K}_{u\varphi}^{T}, \quad \tilde{\boldsymbol{\Phi}} = \mathbf{K}_{\varphi\varphi}^{-1} \cdot \mathbf{K}_{u\varphi}^{T} \cdot \tilde{\mathbf{U}}. \tag{62}$$

By virtue of positive definiteness of the intrinsic bulk and surface energies, the generalized stiffness matrix $\overline{\mathbf{K}}_{uu}$ is nonnegative definite ($\overline{\mathbf{K}}_{uu} \geq 0$), and mass matrix $\mathbf{M}_{uu}$ is positive definite ($\mathbf{M}_{uu} > 0$), because $\rho(\mathbf{x}) \geq \rho_0 > 0$. Then, by analogy with the classical eigenvalue problems for elastic body of usual size, the eigenvalues $\lambda_k = \omega_k^2$ ($k = 1, 2, \ldots, n$; n is the order of matrices $\overline{\mathbf{K}}_{uu}$ and $\mathbf{M}_{uu}$) are real and non-negative. The eigenvectors, corresponding to them, which we will denote by $\mathbf{W}_k = \tilde{\mathbf{U}}_k$, form basis in R^n. The system of these eigenvectors can be chosen orthonormal with respect to the mass matrix $\mathbf{M}_{uu}$ and orthogonal with respect to the generalized stiffness matrix $\overline{\mathbf{K}}_{uu}$

$$\langle \mathbf{W}_k, \mathbf{W}_m \rangle = \mathbf{W}_k^T \cdot \mathbf{M}_{uu} \cdot \mathbf{W}_m = \delta_{km}, \quad \mathbf{W}_k^T \cdot \overline{\mathbf{K}}_{uu} \cdot \mathbf{W}_m = \omega_m^2 \delta_{km}. \tag{63}$$

Thus, the coupled eigenvalue problems (59) and (60) with respect to the triple of unknowns $\{\omega, \tilde{\mathbf{U}}, \tilde{\boldsymbol{\Phi}}\}$ are the generalized eigenvalue problems (61) and (62) with respect to the pairs $\{\omega, \tilde{\mathbf{U}}\}$.

4.5 Mode Superposition Method for Steady-State Oscillation Problems

In the case of harmonic problem with $\beta_d = \zeta_d$ on solution of Eq. (56) with (58), we obtain

$$(-\omega^2 \mathbf{M}_{uu} + j\omega \overline{\mathbf{C}}_{uu} + \overline{\mathbf{K}}_{uu}) \cdot \tilde{\mathbf{U}} = \overline{\mathbf{F}}_u, \tag{64}$$

$$\overline{\mathbf{C}}_{uu} = \alpha_d \mathbf{M}_{uu} + \beta_d \overline{\mathbf{K}}_{uu}, \quad \overline{\mathbf{F}}_u = \tilde{\mathbf{F}}_u - \mathbf{K}_{u\varphi} \cdot \tilde{\boldsymbol{\Phi}}_{st}, \tag{65}$$

$$\tilde{\boldsymbol{\Phi}} = \tilde{\boldsymbol{\Phi}}_{st} + (1 + j\omega\beta_d) \mathbf{K}_{\varphi\varphi}^{-1} \cdot \mathbf{K}_{u\varphi}^{T} \cdot \tilde{\mathbf{U}}, \quad \tilde{\boldsymbol{\Phi}}_{st} = \mathbf{K}_{\varphi\varphi}^{-1} \cdot \tilde{\mathbf{F}}_{\varphi}. \tag{66}$$

If $\mathbf{u}_\Gamma = 0$ in (12), we will find the solution of problem (64) in the form of an expansion in eigenvectors (modes) $\mathbf{W}_k$ of eigenvalue problem (61) with the same homogeneous principal mechanical boundary conditions

$$\tilde{\mathbf{U}} = \sum_{k=1}^{n} z_k \mathbf{W}_k. \tag{67}$$

Substituting (67) into (64) and multiplying the obtained equation scalarly by $\mathbf{W}_m^*$ and taking into account the orthogonality relations (63) and (65), we obtain

$$z_k = \frac{1}{\omega_k^2 - \omega^2 + 2j\xi_k\omega_k\omega} P_k, \quad P_k = \mathbf{W}_k^T \cdot \overline{\mathbf{F}}_u, \quad \xi_k = \alpha_d \frac{1}{2\omega_k} + \beta_d \frac{\omega_k}{2}. \tag{68}$$

Thus, using the method of mode superposition, the solutions of the harmonic problems are determined by (66)–(68).

The advantages and disadvantages of the mode-expansion method are well known from experience of solving problems of structural analysis. Consequently, an important advantage of the method is the possibility of a direct determination of the damping coefficient ξ_k of the individual modes without using the last formula from (68). These factors can be specified from the experimentally measured value of the coupled mechanical and electric quality factor Q_k of the mode with number k: $\xi_k = 1/(2Q_k)$.

4.6 Mode Superposition Method for Nonstationary Problems

For transient problems with homogeneous principal boundary conditions and $\beta_d = \zeta_d$ we can also apply the method of mode superposition. Having solved Eq. (48) for $\boldsymbol{\Phi}$ and converted Eq. (47), we obtain

$$\mathbf{M}_{uu} \cdot \ddot{\mathbf{U}} + \overline{\mathbf{C}}_{uu} \cdot \dot{\mathbf{U}} + \overline{\mathbf{K}}_{uu} \cdot \mathbf{U} = \mathbf{F}_u - \mathbf{K}_{u\varphi} \cdot \boldsymbol{\Phi}_{qst}, \tag{69}$$

$$\boldsymbol{\Phi} = \boldsymbol{\Phi}_{qst} + \mathbf{K}_{\varphi\varphi}^{-1} \cdot \mathbf{K}_{u\varphi}^T \cdot (\mathbf{U} + \beta_d \dot{\mathbf{U}}), \tag{70}$$

where $\boldsymbol{\Phi}_{qst}$ is determined from the separate quasistatic problem

$$\boldsymbol{\Phi}_{qst} = \mathbf{K}_{\varphi\varphi}^{-1} \cdot \mathbf{F}_{\varphi}. \tag{71}$$

We will find the solution $\mathbf{U}$ of problem (69) in the form of an expansion in modes (67), where $z_k = z_k(t)$. Substituting this expansion into Eq. (69), multiplying the resulting equality scalarly by $\mathbf{W}_m^T$, and using the orthogonality relation (63), we derive scalar differential equations for the individual functions $z_k(t)$. Solving these equations with corresponding initial conditions, we obtain

$$z_k = \frac{1}{\overline{\omega}_k} \int_0^t P_k(\tau) e^{-\xi_k\omega_k(t-\tau)} \sin[\overline{\omega}_k(t-\tau)]\, d\tau + A_k(0) e^{-\xi_k\omega_k t} \sin(\overline{\omega}_k t + \delta_k), \tag{72}$$

$$P_k = \mathbf{W}_k^T \cdot (\mathbf{F}_u - \mathbf{K}_{u\varphi} \cdot \boldsymbol{\Phi}_{qst}), \quad \overline{\omega}_k = \omega_k \sqrt{1 - \xi_k^2}, \tag{73}$$

$$A_k(0) = \sqrt{z_k^2(0) + \frac{(\dot{z}_k(0) + \xi_k \omega_k z_k(0))^2}{\overline{\omega}_k^2}}, \quad \delta_k = \operatorname{arctg} \frac{z_k(0)\overline{\omega}_k}{\dot{z}_k(0) + \xi_k \omega_k z_k(0)}, \tag{74}$$

$$z_k(0) = \mathbf{W}_k^T \cdot \mathbf{M}_{uu} \cdot \mathbf{U}_*, \quad \dot{z}_k(0) = \mathbf{W}_k^T \cdot \mathbf{M}_{uu} \cdot \mathbf{R}_*. \tag{75}$$

Hence, using mode superposition method, the solution of problem (47) and (48) with homogeneous principal mechanical boundary conditions and $\beta_d = \zeta_d$ is given by (67) and (72)–(75) for $\mathbf{U}$ and by (70) and (71) for $\boldsymbol{\Phi}$.

4.7 The Newmark Scheme for Solving Non-stationary Problems

The mode superposition method requires the equality of the damping parameters for different media and the homogeneity of the principal boundary conditions. Methods of direct integration with respect to time are more general. We will use the Newmark method for integrating Cauchy problem (47)–(49) in a formulation in which the velocities and accelerations in the time layers are not explicitly given [50].

This variant of the Newmark scheme base on the average expressions on time layer t_i for the vector functions $\mathbf{a}_i = \mathbf{a}(t_i)$, $\mathbf{a} = \{\mathbf{U}, \boldsymbol{\Phi}\}$ and its derivatives ($t_i = i\tau$, $\tau = \Delta t$ is the constant time step size)

$$\mathbf{a}_i \approx \beta \mathbf{a}_{j+1} + \beta_1 \mathbf{a}_j + \beta_2 \mathbf{a}_{j-1}, \tag{76}$$

$$\dot{\mathbf{a}}_i \approx (\gamma \mathbf{a}_{i+1} + \gamma_1 \mathbf{a}_i + \gamma_2 \mathbf{a}_{i-1})/\tau, \tag{77}$$

$$\ddot{\mathbf{a}}_i \approx (\mathbf{a}_{i+1} - 2\mathbf{a}_i + \mathbf{a}_{i-1})/\tau^2, \tag{78}$$

where $\beta_1 = 1/2 + \gamma - 2\beta$, $\beta_2 = 1/2 - \gamma + \beta$, $\gamma_1 = 1 - 2\gamma$, $\gamma_2 = \gamma - 1$, β and γ are the parameters of the Newmark method.

Writing Eqs. (47) and (48) for the time layer t_i, and using representations (76)–(78), we obtain the system of linear equations for $\mathbf{a}_{i+1}$, if we count the known values at the previous time layers t_i and t_{i-1}

$$\mathbf{K}^{\text{eff}} \cdot \mathbf{a}_{i+1} = \mathbf{F}_{i+1}^{\text{eff}}(\mathbf{a}_i, \mathbf{a}_{i-1}), \tag{79}$$

where

$$\mathbf{K}^{\text{eff}} = \begin{bmatrix} \mathbf{K}_{uu}^{\text{eff}} & \mathbf{K}_{u\varphi} \\ \mathbf{K}_{u\varphi}^T & -\lambda \mathbf{K}_{\varphi\varphi} \end{bmatrix}, \quad \mathbf{a}_{i+1} = \begin{bmatrix} \mathbf{U}_{i+1} \\ \boldsymbol{\Phi}_{i+1} \end{bmatrix}, \quad \mathbf{F}^{\text{eff}} = \begin{bmatrix} \mathbf{F}_{u,i+1}^{\text{eff}} \\ -\mathbf{F}_{\varphi,i+1}^{\text{eff}} \end{bmatrix}, \tag{80}$$

$$\mathbf{K}_{uu}^{\text{eff}} = \mathbf{K}_{uu} + \frac{\gamma}{\beta\tau}\mathbf{C}_{uu} + \frac{1}{\beta\tau^2}\mathbf{M}_{uu}, \quad \lambda = (1 + \frac{\zeta_d\gamma}{\beta\tau})^{-1}, \tag{81}$$

$$\mathbf{F}_{u,i+1}^{\text{eff}} = \mathbf{F}_{u,i+1} + \frac{\beta_1}{\beta}\mathbf{F}_{u,i} + \frac{\beta_2}{\beta}\mathbf{F}_{u,i-1} - (\frac{\beta_1}{\beta}\mathbf{K}_{uu} + \frac{\gamma_1}{\beta\tau}\mathbf{C}_{uu} - \frac{2}{\beta\tau^2}\mathbf{M}_{uu}) \cdot \mathbf{U}_i - \tag{82}$$

$$-(\frac{\beta_2}{\beta}\mathbf{K}_{uu} + \frac{\gamma_2}{\beta\tau}\mathbf{C}_{uu} + \frac{1}{\beta\tau^2}\mathbf{M}_{uu}) \cdot \mathbf{U}_{i-1} - \frac{\beta_1}{\beta}\mathbf{K}_{uu} \cdot \boldsymbol{\Phi}_i - \frac{\beta_2}{\beta}\mathbf{K}_{uu} \cdot \boldsymbol{\Phi}_{i-1},$$

$$\mathbf{F}_{\varphi,i+1}^{\text{eff}} = \lambda\mathbf{F}_{\varphi,i+1} + \frac{\lambda\beta_1}{\beta}\mathbf{F}_{\varphi,i} + \frac{\lambda\beta_2}{\beta}\mathbf{F}_{\varphi,i-1} + \frac{\lambda(\beta_1 + \zeta_d\gamma_1)}{\beta\tau}\mathbf{K}_{u\varphi}^T \cdot \mathbf{U}_i + \tag{83}$$

$$+\frac{\lambda(\beta_2 + \zeta_d\gamma_2)}{\beta\tau}\mathbf{K}_{u\varphi}^T \cdot \mathbf{U}_{i-1} - \frac{\lambda\beta_1}{\beta}\mathbf{K}_{\varphi\varphi}^T \cdot \boldsymbol{\Phi}_i - \frac{\lambda\beta_2}{\beta}\mathbf{K}_{\varphi\varphi}^T \cdot \boldsymbol{\Phi}_{i-1}.$$

The matrix $\mathbf{K}^{\text{eff}}$ can be factorized using the LDL^T–factorization method, and only the systems of linear algebraic equations with lower and upper triangular matrices can be solved in each time layer.

Note that the Newmark scheme is absolutely stable when $\beta \geq (1/2 + \gamma)^2/4$; $\gamma \geq 1/2$, and, when $\beta \geq 1/4$; $\gamma = 1/2$, it does not have an approximation viscosity [50]. The Newmark scheme in form (79)–(83) does not explicitly use velocities and accelerations, and this makes it preferable in the case of the transient problems for piezoelectric nanosized solids with account for coupled damping and surface effects.

5 Concluding Remarks

Thus, we have proposed a new model that describes the behavior of the piezoelectric materials, taking into account the damping properties and surface effects at the nanoscale.

The novelty of the model consists in taking into account the volumetric and surface damping properties, as well as coupled surface material phenomena which are important at the nanoscale. To describe the size effects, we use recently popular theory of surface stresses and its generalization to piezoelectric media. Under this generalization, we also consider the coupled surface mechanical and electric fields. Other new feature is the account for the damping properties in the sense of a generalization of the conventional for the structural analysis Rayleigh damping method for the coupled mechanical and electric fields. We also added the terms, describing the attenuation, in the constitutive equations for the surface mechanic and electric fields. When taking the coupled damping into account, the basic idea was that the method of mode superposition can be applied for transient and harmonic problems for piezoelectric bodies at the nanoscale.

After the initial-boundary value problem setting for the piezoelectric nanosized bodies, we have obtained a weak or generalized formulation of this problem in terms of energy functional spaces. We note that the basic mathematical properties of this problem are quite close to the particular case of the relevant problems for piezoelectric media with uncoupled surface effects. Therefore, most of the previously established results, valid for piezoelectric media with uncoupled surface effects, can be extended to the more general case of piezoelectric media with coupled surface piezoelectric effects.

In order to solve the formulated problems numerically, we use the finite element approaches. Based on the weak formulations of the problems, we derive the finite element equations for transient, harmonic, modal and static problems for piezoelectric media with coupled surface effects. As we can see from the finite element systems, the account for the surface mechanical, piezoelectric and electrical effects gives the additional components in the stiffness, piezoelectric and permittivity matrices. Therefore, for the computer analysis of nanoscale piezoelectric bodies one can use the well-known finite element software with added surface piezoelectric elements with membrane option.

We also demonstrated that for the dynamic problems our models with the same damping coefficients allow to use the method of mode superposition which is efficient tool for the analysis of the influence of individual modes and for the analysis of the same model under various external loads. In the cases of transient and static problems, we showed that the resulting finite element systems have symmetric quasi-definite matrices typical for problems with a saddle point.

Acknowledgements This work was supported by the Russian Science Foundation (grant number 15-19-10008).

References

1. Altenbach, H., Eremeyev, V.A., Lebedev, L.P.: On the existence of solution in the linear elasticity with surface stresses. ZAMM **90**(3), 231–240 (2010)
2. Altenbach, H., Eremeyev, V.A., Lebedev, L.P.: On the spectrum and stiffness of an elastic body with surface stresses. ZAMM **91**(9), 699–710 (2011)
3. Belokon, A.V., Nasedkin, A.V., Soloviev, A.N.: New schemes for the finite-element dynamic analysis of piezoelectric devices. J. Applied Math. Mech. (PMM) **66**(3), 481–490 (2002)
4. Berlincourt, D.A., Curran, D.R., Jaffe, H.: Piezoelectric and piezomagnetic materials. Physical Acoustics. Part A, vol. 1, pp. 233–256. Academic Press, NY (1964)
5. Chen, T.: Exact size-dependent connections between effective moduli of fibrous piezoelectric nanocomposites with interface effects. Acta Mech. **196**, 205–217 (2008)
6. Chen, W.Q.: Surface effect on Bleustein-Gulyaev wave in a piezoelectric half-space. Theoret. Appl. Mech. Lett. **1**, 041001 (2011)
7. Dai, Sh., Gharbi, M., Sharma, P., Park, H.S.: Surface piezoelectricity: size effects in nanostructures and the emergence of piezoelectricity in non-piezoelectric materials. J. Appl. Phys. **110**, 104305-1–104305-7 (2011)
8. Duan, H.L., Wang, J., Karihaloo, B.L.: Theory of elasticity at the nanoscale. In: Advances in Applied Mechanics, vol. 42, pp. 1–68. Elsevier (2008)
9. Eremeyev, V.A.: On effective properties of materials at the nano- and microscales considering surface effects. Acta Mech. **227**, 29–42 (2016)

10. Eremeyev, V.A., Lebedev, L.P.: Existence of weak solutions in elasticity. Math. Mech. Solids **18**(2), 204–217 (2013)
11. Eremeyev, V.A., Lebedev, L.P.: Mathematical study of boundary-value problems within the framework of Steigmann-Ogden model of surface elasticity. Continuum Mech. Therm. **28**(1–2), 407–422 (2016)
12. Eremeev, V.A., Nasedkin, A.V.: Natural vibrations of nanodimensional piezoelectric bodies with contact-type boundary conditions. Mech. Solids **50**(5), 495–507 (2015)
13. Eremeyev, V.A., Rosi, G., Naili, S.: Surface/interfacial anti-plane waves in solids with surface energy. Mech. Res. Commun. **74**, 8–13 (2016)
14. Fan, H., Yang, J., Xu, L.: Piezoelectric waves near an imperfectly bonded interface between two half-spaces. Appl. Phys. Lett. **88**(20), 203509 (2006)
15. Gu, S.-T., He, Q.C., Pensée, V.: Homogenization of fibrous piezoelectric composites with general imperfect interfaces under anti-plane mechanical and in-plane electrical loadings. Mech. Mater. **88**, 12–29 (2015)
16. Gu, S.-T., Liu, J.-T.: He. Q.-C.: Piezoelectric composites: imperfect interface models, weak formulations and benchmark problems. Comp. Mater. Sci. **94**, 182–190 (2014)
17. Gu, S.-T., Liu, J.-T.: He. Q.-C.: The strong and weak forms of a general imperfect interface model for linear coupled multifield phenomena. Int. J. Eng. Sci. **85**, 31–46 (2014)
18. Gu, S.-T., Qin, L.: Variational principles and size-dependent bounds for piezoelectric inhomogeneous materials with piezoelectric coherent imperfect interfaces. Int. J. Eng. Sci. **78**, 89–102 (2014)
19. Gurtin, M.E., Murdoch, A.I.: A continuum theory of elastic material surfaces. Arch. Ration. Mech. Anal. **57**(4), 291–323 (1975)
20. Hamilton, J.C., Wolfer, W.G.: Theories of surface elasticity for nanoscale objects. Surf. Sci. **603**, 1284–1291 (2009)
21. Huang, G.Y., Yu, S.W.: Effect of surface piezoelectricity on the electromechanical behaviour of a piezoelectric ring. Phys. Status Solidi B **243**(4), R22–R24 (2006)
22. Kim, C., Ru, C., Schiavone, P.: A clarification of the role of crack-tip conditions in linear elasticity with surface effects. Math. Mech. Solids **18**(1), 59–66 (2013)
23. Kim, C.I., Schiavone, P., Ru, C.Q.: Effect of surface elasticity on an interface crack in plane deformations. Proc. R. Soc. A. **467**(2136), 3530–3549 (2011)
24. Kushch, V.I., Chernobai, V.S., Mishuris, G.S.: Longitudinal shear of a composite with elliptic nanofibers: local stresses and effective stiffness. Int. J. Eng. Sci. **84**, 79–94 (2014)
25. Malakooti, M.H., Sodano, H.A.: Multi-Inclusion modeling of multiphase piezoelectric composites. Compos. Part B **47**, 181–189 (2013)
26. Nasedkin, A.V.: Some finite element methods and algorithms for solving acousto-piezoelectric problems. In: Parinov, I.A. (ed.) Piezoceramic Materials and Devices, pp. 177–218. Nova Science Publications, NY (2010)
27. Nasedkin, A.V., Eremeyev, V.A.: Harmonic vibrations of nanosized piezoelectric bodies with surface effects. ZAMM **94**(10), 878–892 (2014)
28. Nasedkin, A.V., Eremeyev, V.A.: Modeling of nanosized piezoelectric and magnetoelectric bodies with surface effects. AIP Conf. Proc. **1627**, 70–75 (2014)
29. Nasedkin, A.V., Eremeyev, V.A.: Some models for nanosized magnetoelectric bodies with surface effects. In: Parinov, I.A., Chang, S.-H., Topolov, V.Y. (eds.) Advanced Materials-Manufacturing, Physics, Mechanics and Applications, Springer Proceedings in Physics, vol. 175, pp. 373–391. Springer, Heidelberg (2016)
30. Pan, X.H., Yu, S.W., Feng, X.Q.: A continuum theory of surface piezoelectricity for nanodielectrics. Sci. China Phys. Mech. Astron. **54**(4), 564–573 (2011)
31. Park, H.S., Devel, M., Wang, Z.: A new multiscale formulation for the electromechanical behavior of nanomaterials. Comput. Methods Appl. Mech. Eng. **200**, 2447–2457 (2011)
32. Povstenko, Y.Z.: Theoretical investigation of phenomena caused by heterogeneous surface-tension in solids. J. Mech. Phys. Solids **41**, 1499–1514 (1993)
33. Schiavone, P., Ru, C.Q.: Solvability of boundary value problems in a theory of plane-strain elasticity with boundary reinforcement. Int. J. Eng. Sci. **47**(11), 1331–1338 (2009)

34. Shuttleworth, R.: The surface tension of solid. Proc. Phys. Soc. A **63**, 444–457 (1950)
35. Steigmann, D.J., Ogden, R.W.: Plane deformations of elastic solids with intrinsic boundary elasticity. Proc. R. Soc. A **453**(1959), 853–877 (1997)
36. Steigmann, D.J., Ogden, R.W.: Elastic surface-substrate interactions. Proc. R. Soc. A **455**(1982), 437–474 (1999)
37. Wang, G.F., Feng, X.Q.: Effect of surface stresses on the vibration and buckling of piezoelectric nanowires. EPL (Lett. J. Explor. Front. Phys.) **91**(5), 56007 (2010)
38. Wang, J., Huang, Z., Duan, H., Yu, S., Feng, X., Wang, G., Zhang, W., Wang, T.: Surface stress effect in mechanics of nanostructured materials. Acta Mech. Solida Sinica **24**(1), 52–82 (2011)
39. Wang, K.F., Wang, B.L., Kitamura, T.: A review on the application of modified continuum models in modeling and simulation of nanostructures. Acta Mech. Sinica **32**(1), 83–100 (2016)
40. Wang, W., Li, P., Jin, F.: Two-dimensional linear elasticity theory of magneto-electro-elastic plates considering surface and nonlocal effects for nanoscale device applications. Smart Mater. Struct. **25**(9), 095026 (2016)
41. Wang, Z., Zhu, J., Jin, X.Y., Chen, W.Q., Zhang, Ch.: Effective moduli of ellipsoidal particle reinforced piezoelectric composites with imperfect interfaces. J. Mech. Phys. Solids **65**, 138–156 (2014)
42. Xiao, J.H., Xu, Y.L., Zhang, F.C.: Size-dependent effective electroelastic moduli of piezoelectric nanocomposites with interface effect. Acta Mech. **222**(1–2), 59–67 (2011)
43. Xiao, J.H., Xu, Y.L., Zhang, F.C.: Evaluation of effective electroelastic properties of piezoelectric coated nano-inclusion composites with interface effect under antiplane shear. Int. J. Eng. Sci. **69**, 61–68 (2013)
44. Yan, Z., Jiang, L.Y.: Surface effects on the electromechanical coupling and bending behaviours of piezoelectric nanowires. J. Phys. D Appl. Phys. **44**, 075404 (2011)
45. Yan, Z., Jiang, L.Y.: The vibrational and buckling behaviors of piezoelectric nanobeams with surface effects. Nanotechnology **22**, 245703 (2011)
46. Yan, Z., Jiang, L.Y.: Electromechanical response of a curved piezoelectric nanobeam with the consideration of surface effects. J. Phys. D Appl. Phys. **44**, 365301 (2011)
47. Yan, Z., Jiang, L.Y.: Surface effects on the electroelastic responses of a thin piezoelectric plate with nanoscale thickness. J. Phys. D Appl. Phys. **45**, 255401 (2012)
48. Yan, Z., Jiang, L.Y.: Surface effects on the vibration and buckling of piezoelectric nanoplates. EPL (Europhys. Lett.) **99**(2), 27007 (2012)
49. Yan, Z., Jiang, L.Y.: Vibration and buckling analysis of a piezoelectric nanoplate considering surface effects and in-plane constraints. Proc. R. Soc. A **468**, 3458–3475 (2012)
50. Zienkewicz, O.C., Morgan, K.: Finite Elements and Approximation. Wiley, NY (1983)
51. Zhang, C., Chen, W., Zhang, C.: On propagation of anti-plane shear waves in piezoelectric plates with surface effect. Phys. Lett. A **376**, 3281–3286 (2012)
52. Zhang, C., Chen, W., Zhang, C.: Two-dimensional theory of piezoelectric plates considering surface effect. Eur. J. Mech. A. Solids **41**, 50–57 (2013)
53. Zhang, J., Wang, Ch., Adhikari, S.: Surface effect on the buckling of piezoelectric nanofilms. J. Phys. D Appl. Phys. **45**, 285301 (2012)
54. Zhang, L.L., Liu, J.X., Fang, X.Q., Nie, G.Q.: Size-dependent dispersion characteristics in piezoelectric nanoplates with surface effects. Phys. E **57**, 169–174 (2014)

On the Theory of Acoustic Metamaterials with a Triple-Periodic System of Interior Obstacles

M.A. Sumbatyan and M.Yu. Remizov

Abstract The paper is devoted to the calculation of the reflection and the transmission coefficients, when a plane longitudinal wave is incident on a three-dimensional grating with a periodic array of rectangular cracks in the elastic material. In the one-mode frequency range the problem is reduced to a system of integral equations, which can be solved for various sizes of the cracks to give an explicit representation for the wave field inside the cracked structure, as well as the values of the reflection and the transmission coefficients.

Keywords Acoustic metamaterials · Periodic system of obstacles · Integral equation · Hypersingular kernel · Reflection coefficient · Transmission coefficient

1 Introduction

The study of elastic waves penetration through periodic gratings is an important subject in many practical applications in the field of mechanical, acoustical and electromagnetic sciences. In practice, analytical results can be obtained under assumption of low frequency with a weak interaction regime, where some approximated results can be established in an analytical form [1, 3, 5, 15].

The papers of Scarpetta, Sumbatyan and Tibullo [11–14] provide explicit analytical formulas for reflection and transmission coefficients in the one-mode case for acoustic waves penetrating through a doubly and triple-periodic arrays of arbitrary-shaped apertures and volumetric obstacles in wave propagation through a periodic array of screens in elastic solids.

M.A. Sumbatyan (✉)
Institute of Mathematics, Mechanics and Computer Science, Southern Federal University, Milchakova Street 8a, Rostov-on-Don 344090, Russia
e-mail: masumbatyan@sfedu.ru

M.Yu. Remizov
Department of Technical Mechanics, Rostov State University of Civil Engineering, Socialisticheskaya Street 164, Rostov-on-Don 344006, Russia
e-mail: Remizov72@mail.ru

M.A. Sumbatyan (ed.), *Wave Dynamics and Composite Mechanics for Microstructured Materials and Metamaterials*,
Advanced Structured Materials 59, DOI 10.1007/978-981-10-3797-9_2

In the previous paper we studied the 3-D normal penetration of an elastic wave into a plane screen with an infinite doubly periodic system of cracks [7] with a one-mode frequency assumption. Earlier, the problems of high-frequency diffraction processes by cracks in an elastic material were analyzed in [8–10].

The present work continues to study the 3-D problem for the couple of such plane screens what forms a triple-periodic system. As shown in [7], the qualitative properties of such a geometrical structure are analogous to that in the in-plane problem. Thus, as a particular case, we study here the two arrays with a periodic system of cracks in each of them. The wave process is harmonic in time and all physical quantities contain the factor $e^{-i\omega t}$, which is further omitted, for the sake of brevity. In the same way as in [7] the following assumption is accepted: (a) only one-mode propagation (with normal incidence) is considered, $ak_2 < \pi$, where $2a$ is a period of the grating, k_2 is the transverse wave number; (b) the vertical cracked planes are sufficiently distant from each other so that the ratio D/a is comparatively large, where D is the distance between the two arrays.

The aim of the present work is to generalize the results obtained before extending explicit analytical expressions for the reflection and transmission coefficients for the system of parallel plane screens (using the properties of the kernel of a hyper-singular integral equation) based on the context of the in-plane problem for wave propagation through elastic solids with a periodic array of cracks.

The obtained results are meaningful in the aspect of the so-called "acoustic materials" which become nowadays an important component of the modern technology. The acoustic properties of such materials are prescribed not by the physical substance they are made from, but by their internal structure. Among other helpful properties of these materials, we note the cutoff of the transmitted acoustic energy on certain frequency intervals, i.e. they work as acoustic filters. Typically, this is attained for a certain periodic internal structure, like a triple-periodic system of relatively rigid spheres embedded in the epoxy matrix [4]. The results of the present study show that such a cutoff is an intrinsic feature caused even by simpler kinds of the periodicity. For example, in some sense this can be attained even in the 2D problem with a pair of parallel arrays, each of them containing a periodic linear system of coplanar cracks.

2 Mathematical Formulation of the Problem

Let us consider an unbounded (two-dimensional) elastic medium, which consists of 2 identical periodic systems of cracks, located at $x = 0, D$ with the period $2a$ along axes y, while the size of each crack is $2b$. The distance between the systems of cracks, forming the second period is D. If we study the incidence of a plane longitudinal wave upon the grating along the positive direction of axis x, then the problem is obviously equivalent (due to a symmetry) to a single waveguide of width $2a$ along axis y, see Fig. 1. Hence, if the incident wave of a unit amplitude is assumed to propagate normally to the planes along axis x, then the Lamè potentials in the various regions, satisfying the Helmholtz equation, are:

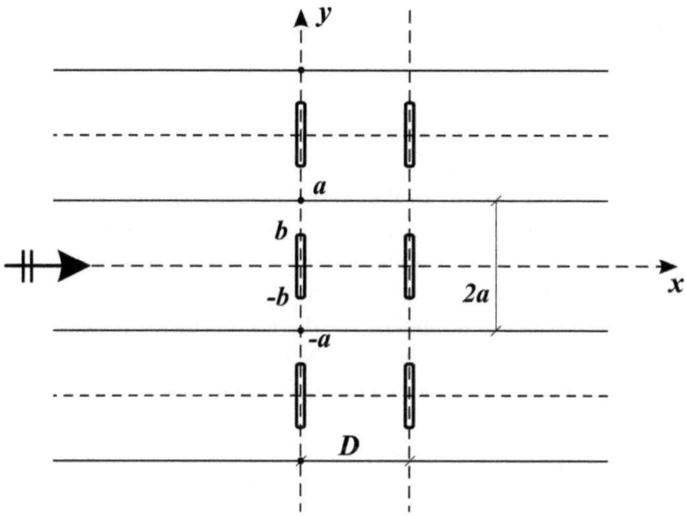

Fig. 1 Propagation of the longitudinal incident wave through a pair of the periodic arrays of cracks

$$\varphi^l = e^{ik_1x} + Re^{-ik_1x} + \sum_{n=1}^{\infty} A_n e^{q_n x} \cos\left(\frac{\pi n y}{a}\right),$$

$$\psi^l = \sum_{n=1}^{\infty} B_n^1 e^{r_n x} \sin\left(\frac{\pi n y}{a}\right), \quad x < 0 \tag{1a}$$

$$\varphi^1 = e^{ik_1x} + F_0^1 \cos[k_1 x] + H_0^1 \cos[k_1(x-D)] +$$

$$+ \sum_{n=1}^{\infty} \left\{F_n^1 ch[q_n x] + H_n^1 ch[q_n(x-D)]\right\} \cos\left(\frac{\pi n y}{a}\right),$$

$$\psi^1 = \sum_{n=1}^{\infty} \left\{G_n^1 ch[r_n x] + P_n^1 ch[q_n(x-D)]\right\} \sin\left(\frac{\pi n y}{a}\right), \quad 0 < x < D, \tag{1b}$$

$$\varphi^r = Te^{ik_1(x-D)} + \sum_{n=1}^{\infty} C_n e^{-q_n(x-D)} \cos\left(\frac{\pi n y}{a}\right),$$

$$\psi^r = \sum_{n=1}^{\infty} D_n e^{-r_n(x-D)} \sin\left(\frac{\pi n y}{a}\right), \quad x > D. \tag{1c}$$

All capital letters are some unknown constants and

$$q_n = \left(a_n^2 - k_1^2\right)^{1/2}, \quad r_n = \left(a_n^2 - k_2^2\right)^{1/2}, \qquad a_n = \frac{\pi n}{a} \tag{2}$$

where k_1, k_2 are the longitudinal and the transverse wave numbers, c_1, c_2—corresponding wave speeds in the elastic material, R and T are the reflection and the transmission coefficients, respectively. Let the consideration be restricted to the one-mode case: $0 < k_2 a < \pi$, then $q_n > 0,\ r_n > 0$ for all $n = 1, 2, \ldots$. Besides, we assume that the cracks arrays are sufficiently distant from each other, this involves $D/a \gg 1$. For $n = 0 \quad q_0 = -ik_1$ and $r_0 = -ik_2$, according to the radiation condition.

The components of the stress tensor can be expressed in terms of the Lamè wave potentials, two of them are represented in the following form:

$$\sigma_{xx} = -c_1^2 k_1^2 \varphi - 2c_2^2 \left(\frac{\partial^2 \varphi}{\partial y^2} - \frac{\partial^2 \psi}{\partial y \partial x}\right), \quad \sigma_{xy} = c_2^2 \left(2\frac{\partial^2 \varphi}{\partial x \partial y} - \frac{\partial^2 \psi}{\partial x^2} + \frac{\partial^2 \psi}{\partial y^2}\right). \tag{3}$$

The displacement field $\mathbf{u} \equiv (u_x, u_y)$ is given by a representation of the Green-Lamè type, as follows:

$$u_x = \frac{\partial \varphi}{\partial x} + \frac{\partial \psi}{\partial y}; \qquad u_y = \frac{\partial \varphi}{\partial y} - \frac{\partial \psi}{\partial x}. \tag{4}$$

In the considered structure, a longitudinal plane wave of the form

$$\varphi_0 = e^{ik_1 x}, \qquad \psi_0 = 0 \tag{5}$$

is entering from $x = -\infty$, generating the scattered fields before the first array ($x < 0$) between the two of them ($0 < x < D$), and after the second one ($x > D$).

Due to the natural periodicity in the vertical direction, let us restrict the consideration by the one cell $|y| < a$ only. Accepting the continuity of the displacement field outside the cracks at each vertical periodic system, let us introduce the following unknown functions $g_x^{(s)}(y)$, $g_y^{(s)}(y)$, $s = 1, 2$ by

$$x = 0: \ u_x^l - u_x^1 = \begin{cases} g_x^1(y); & |y| < b, \\ 0; & b < |y| < a, \end{cases} \tag{6a}$$

$$x = 0: \ u_y^l - u_y^1 = \begin{cases} g_y^1(y); & |y| < b, \\ 0; & b < |y| < a, \end{cases} \tag{6b}$$

$$x = D: \ u_x^1 - u_x^r = \begin{cases} g_x^2(y); & |y| < b, \\ 0; & b < |y| < a, \end{cases} \tag{6c}$$

$$x = D: \ u_y^1 - u_y^r = \begin{cases} g_y^2(y); & |y| < b, \\ 0; & b < |y| < a. \end{cases} \tag{6d}$$

Now Eqs. (4), (6) can be used to represent expressions for all constants appearing in potentials (1) in terms of $g_x(y), g_y(y)$. By integration of (6) over the domains $|y| < a$, one obtains

$$-ik_1R - H_0^1k_1\sin(k_1D) = \frac{1}{2a}\int\limits_{-b}^{b} g_x^1(\eta)d\eta, \tag{7a}$$

$$-F_0^1k_1\sin(k_1D) + ik_1e^{ik_1D} - ik_1T = \frac{1}{2a}\int\limits_{-b}^{b} g_x^2(\eta)d\eta. \tag{7b}$$

The orthogonality of the trigonometric functions reduces Eq. (6) to the following relations:

$$(A_n + H_n^1 sh(q_nD))q_n + (B_n - P_n^1 ch(r_nD))a_n = \frac{1}{a}\int\limits_{-b}^{b} g_x^1(\eta)\cos(a_n\eta)d\eta, \tag{8a}$$

$$(-A_n + H_n^1 ch(q_nD))a_n - (B_n + P_n^1 sh(r_nD))r_n = \frac{1}{a}\int\limits_{-b}^{b} g_y^1(\eta)\sin(a_n\eta)d\eta, \tag{8b}$$

$$(C_n + F_n^1 sh(q_nD))q_n - (D_n - G_n^1 ch(r_nD))a_n = \frac{1}{a}\int\limits_{-b}^{b} g_x^2(\eta)\cos(a_n\eta)d\eta, \tag{8c}$$

$$(C_n - F_n^1 ch(q_nD))a_n - (D_n + G_n^1 sh(r_nD))r_n = \frac{1}{a}\int\limits_{-b}^{b} g_y^2(\eta)\sin(a_n\eta)d\eta. \tag{8d}$$

When crossing each of the two arrays, $x = 0$ and $x = D$, one can see that the normal stresses on crack's left and right faces are equal due to the boundary conditions, and outside crack's domain—due to continuity of the stress. This implies the continuity of the stress field for arbitrary y:

$$\sigma_{xx}^l = \sigma_{xx}^1, \quad \sigma_{xy}^l = \sigma_{xy}^1, \qquad x = 0, \ |y| < a, \tag{9a}$$

$$\sigma_{xx}^1 = \sigma_{xx}^r, \quad \sigma_{xy}^1 = \sigma_{xy}^r, \qquad x = D, \ |y| < a. \tag{9b}$$

This leads to the following relations:

$$\begin{gathered}-k_2^2[F_0^1 + H_0^1\cos(k_1D) - R] - \sum_{n=1}^{\infty}\{k_2^2[H_n^1 ch(q_nD) - A_n] - \\ -2a_n^2[H_n^1 ch(q_nD) - A_n] + 2a_nr_n[P_n^1 sh(r_nD) + B_n]\}\cos(a_ny) = 0,\end{gathered} \tag{10a}$$

$$\begin{gathered}\sum_{n=1}^{\infty}\{2a_nq_n[H_n^1 sh(q_nD) + A_n] - r_n^2[P_n^1 ch(r_nD) - B_n] - \\ -a_n^2[P_n^1 ch(r_nD) - B_n]\}\sin(a_ny) = 0,\end{gathered} \tag{10b}$$

$$-k_2^2[F_0^1\cos(k_1D)+H_0^1+e^{ik_1D}-T]-\sum_{n=1}^{\infty}\{k_2^2[F_n^1 ch(q_nD)-C_n]-$$
$$-2a_n^2[F_n^1 ch(q_nD)-C_n]-2a_nr_n[G_n^1 sh(r_nD)+D_n]\}\cos(a_ny)=0, \tag{10c}$$

$$-\sum_{n=1}^{\infty}\{2a_nq_n[F_n^1 sh(q_nD)+C_n]+r_n^2[G_n^1 ch(r_nD)-D_n]+$$
$$+a_n^2[G_n^1 ch(r_nD)-D_n]\}\sin(a_ny)=0. \tag{10d}$$

Now Eq. (8) together with the relations obtained from (10), with the use of the orthogonality of the trigonometric functions, form the following systems of linear algebraic equations for $x=0,D$:

$$\begin{aligned}\gamma_0E_2-2c_2^2a_nr_nE_3&=0,\\ -a_nE_2-r_nE_3&=G_y^1,\end{aligned} \tag{11a}$$

$$\begin{aligned}q_nE_1+a_nE_4&=G_x^1,\\ -2a_nq_nE_1-(r_n^2+a_n^2)E_4&=0,\end{aligned} \tag{11b}$$

$$\begin{aligned}q_nE_5+a_nE_8&=G_x^2,\\ -2a_nq_nE_5-(r_n^2+a_n^2)E_8&=0,\end{aligned} \tag{11c}$$

$$\begin{aligned}-a_nE_6-r_nE_7&=G_y^2,\\ \gamma_0E_6-2c_2^2a_nr_nE_7&=0,\end{aligned} \tag{11d}$$

where

$$G_x^{1,2}=\frac{1}{a}\int_{-b}^{b}g_x^{1,2}(\eta)\cos(a_n\eta)d\eta,\quad G_y^{1,2}=\frac{1}{a}\int_{-b}^{b}g_y^{1,2}(\eta)\sin(a_n\eta)d\eta,$$

$$\gamma_0=c_1^2k_1^2-2c_2^2a_n^2$$

and the new unknown quantities E_m, $m=1,\ldots,8$ are defined as follows:

$$\begin{aligned}E_1&=H_n^1 sh(q_nD)+A_n, & E_2&=-H_n^1 sh(q_nD)+A_n,\\ E_3&=P_n^1 sh(r_nD)+B_n, & E_4&=-P_n^1 ch(r_nD)+B_n,\end{aligned}$$

$$\begin{aligned}E_5&=F_n^1 sh(q_nD)+C_n, & E_6&=F_n^1 ch(q_nD)-C_n,\\ E_7&=G_n^1 sh(r_nD)+D_n, & E_8&=G_n^1 ch(r_nD)-D_n.\end{aligned} \tag{12}$$

Once the solutions for both the systems ($x=0$ and $x=D$) are constructed, one can easily find the eight unknown constants in the following form:

$$A_n = -\frac{r_n^2 + a_n^2}{2k_2^2 q_n} G_x^1 - \frac{a_n}{k_2^2} G_y^1, \quad B_n = \frac{a_n}{k_2^2} G_x^1 - \frac{\gamma_0}{2k_1^2 c_1^2 r_n} G_y^1,$$

$$H_n^1 = \frac{1}{2sh(q_n D)} \left(-\frac{r_n^2 + a_n^2}{k_2^2 q_n} G_x^1 + \frac{2a_n}{k_2^2} G_y^1 \right),$$

$$P_n^1 = \frac{1}{2sh(r_n D)} \left(-\frac{2a_n}{k_2^2} G_x^1 - \frac{\gamma_0}{k_1^2 c_1^2 r_n} G_y^1 \right),$$

$$F_n^1 = \frac{1}{2sh(q_n D)} \left(-\frac{r_n^2 + a_n^2}{k_2^2 q_n} G_x^2 - \frac{2a_n}{k_2^2} G_y^2 \right),$$

$$G_n^1 = \frac{1}{2sh(r_n D)} \left(\frac{2a_n}{k_2^2} G_x^2 - \frac{\gamma_0}{k_1^2 c_1^2 r_n} G_y^2 \right),$$

$$C_n = -\frac{r_n^2 + a_n^2}{2k_2^2 q_n} G_x^2 + \frac{a_n}{k_2^2} G_y^2, \quad D_n = -\frac{a_n}{k_2^2} G_x^2 - \frac{\gamma_0}{2k_1^2 c_1^2 r_n} G_y^2, \tag{13}$$

By integration of Eq. (10) over domain $|y| < a$, together with relations (7), one obtains the remaining unknown constants F_0^1, H_0^1, R, T from the following algebraic system:

$$\left(\begin{array}{cccc|c} 0 & -k_1 \sin(k_1 D) & -ik_1 & 0 & J_1 \\ -k_1 \sin(k_1 D) & 0 & 0 & -ik_1 & J_2 - ik_1 e^{ik_1 D} \\ 1 & \cos(k_1 D) & -1 & 0 & 0 \\ \cos(k_1 D) & 1 & 0 & -1 & -e^{ik_1 D} \end{array} \right), \tag{14}$$

where we put

$$J_{1,2} = \frac{1}{2a} \int_{-a}^{a} g_x^{1,2}(\eta) d\eta. \tag{15}$$

The solution to system (14) is given as follows:

$$F_0^1 = \frac{J_1 e^{ik_1 D}}{2k_1 \sin(k_1 D)} - \frac{J_2}{2k_1 \sin(k_1 D)}, \quad H_0^1 = -\frac{J_1}{2k_1 \sin(k_1 D)} + \frac{J_2 e^{ik_1 D}}{2k_1 \sin(k_1 D)},$$

$$R = -\frac{J_1}{2ik_1} - \frac{J_2 e^{ik_1 D}}{2ik_1}, \qquad T = -\frac{J_1 e^{ik_1 D}}{2ik_1} - \frac{J_2}{2ik_1} + e^{ik_1 D}. \tag{16}$$

It can easily be shown from the previous equations that the two unknown functions $g_y^{1,2}$ are trivial: $g_y^{1,2}(y) \equiv 0$. Omitting some routine mathematical transformations with the assumptions (a), (b) of Sect. 1, from the equations

$$\sigma_{xx}^{l} + \sigma_{xx}^{1} = 0, \ (x = 0, |y| < b), \qquad \sigma_{xx}^{1} + \sigma_{xx}^{r} = 0, \ (x = D, |y| < b), \tag{17}$$

one finally obtains the following system of two integral equations for the unknown functions $g_x^{1,2}(y)$, holding over the crack $|y| < b$:

$$\frac{1}{2a}\int_{-b}^{b} g_x^1(\eta)\left\{\frac{1}{2ik_1} - \frac{1}{k_2^4}\sum_{n=1}^{\infty}\frac{R_n}{q_n}\cos\left[a_n(y-\eta)\right]\right\}d\eta +$$

$$+\frac{e^{ik_1D}}{4aik_1}\int_{-b}^{b} g_x^2(\eta)d\eta = 1, \tag{18a}$$

$$\frac{e^{ik_1D}}{4aik_1}\int_{-b}^{b} g_x^1(\eta)d\eta +$$

$$+\frac{1}{2a}\int_{-b}^{b} g_x^2(\eta)\left\{\frac{1}{2ik_1} - \frac{1}{k_2^4}\sum_{n=1}^{\infty}\frac{R_n}{q_n}\cos\left[a_n(y-\eta)\right]\right\}d\eta = e^{ik_1D}, \tag{18b}$$

where the numerator in the kernels takes the form of the Rayleigh function

$$R_n = (2a_n^2 - k_2^2)^2 - 4r_nq_na_n^2. \tag{19}$$

3 The Properties of the Basic Integral Equation

Let us start from the study of the auxiliary integral equation ($|y| < b$):

$$\frac{1}{2ak_2^2}\int_{-b}^{b} h(\eta)K(y-\eta)d\eta = 1, \quad K(y)=\sum_{n=1}^{\infty}L_n\cos(a_ny), \quad L_n=\frac{R_n}{q_n}. \tag{20}$$

Notice that $L_n \approx -2(1 - c_2^2/c_1^2)a_n, \ n \to \infty$. Hence, the sum defining the kernel can be transformed as follows:

$$K(y-\eta) = -2\left(1-\frac{c_2^2}{c_1^2}\right)\sum_{n=1}^{\infty} a_n \cos[a_n(y-\eta)] +$$

$$+\sum_{n=1}^{\infty}\left[L_n + 2\left(1-\frac{c_2^2}{c_1^2}\right)a_n\right]\cos[a_n(y-\eta)], \tag{21}$$

$$\sim K(y-\eta) = -2\left(1-\frac{c_2^2}{c_1^2}\right)I(y-\eta) + K_r(y-\eta). \tag{22}$$

Now the second term in the kernel, K_r, is a certain regular function. The first one consists of a regular and a singular part: $I(y) = I_r(y) + I_s(y)$.

Let us introduce the dimensionless variable $\tilde{y} = (y-\eta)/a$, then one rewrites:

$$\frac{a}{\pi} I(\tilde{y}) = \sum_{n=1}^{\infty} n \cos(\pi n \tilde{y}). \tag{23}$$

By using the generalized value of the series in (23), see [6]:

$$\sum_{n=1}^{\infty} n\cos(\pi n\tilde{y}) = \lim_{\varepsilon\to+0}\sum_{n=1}^{\infty} e^{-\varepsilon n} n\cos(\pi n\tilde{y}) =$$

$$= -\frac{1}{4\sin^2(\pi\tilde{y}/2)},\quad \left(\sim -\frac{1}{\pi^2\tilde{y}^2},\quad y\to 0\right) \tag{24}$$

one obtains the kernel of the basic integral equation (21) in the following form:

$$K(y-\eta) = K_r(y-\eta) - 2\left(1-\frac{c_2^2}{c_1^2}\right)[I_r(y-\eta) + I_s(y-\eta)], \tag{25}$$

where the singular and the regular parts of $I(\tilde{y})$ are, respectively:

$$I_s = -\frac{a}{\pi(y-\eta)^2},\quad I_r = \frac{a}{\pi(y-\eta)^2} - \frac{\pi}{4a\sin^2[\pi(y-\eta)/2a]}. \tag{26}$$

This results in the following form of the basic integral equation:

$$\frac{1}{2ak_2^2}\int_{-b}^{b} h(\eta)\left[\Phi_r(y-\eta) + \frac{2a(1-c_2^2/c_1^2)}{\pi(y-\eta)^2}\right]d\eta = 1,\qquad |y| < b,$$

$$\Phi_r(y-\eta) = -2\left(1-\frac{c_2^2}{c_1^2}\right) I_r(y-\eta) + K_r(y-\eta). \tag{27}$$

The obtained singular behavior of the kernel for small arguments contains a 1D hyper-singular kernel arising in the theory of cracks, well known in the linear elasticity theory in unbounded media [16].

In order to provide the stability of the numerical treatment in the performed numerical experiments, we apply a discrete quadrature formulas for the 1D hyper-singular kernel, known as a "method of discrete vortices" [2]. Transforming the left part of (27) to a discrete form, one obtains

$$\frac{1}{2ak_2^2}\sum_{k=1}^{N}\int_{\eta_{k-1}}^{\eta_k} h(\eta)\left[\Phi_r(y_l-\eta) + \frac{2a(1-c_2^2/c_1^2)}{\pi(y_l-\eta)^2}\right]d\eta =$$
$$= \frac{1}{2ak_2^2}\sum_{k=1}^{N} h(\eta_k)\left[\varepsilon_1\Phi_r(y_l-\eta_k) + \int_{\eta_{k-1}}^{\eta_k}\frac{2a(1-c_2^2/c_1^2)}{\pi(y_l-\eta)^2}d(\eta-y_l)\right] =$$

$$\frac{1}{2ak_2^2}\sum_{k=1}^{N} h(\eta_k)\left[\varepsilon_1\Phi_r(y_l-\eta_k) - \frac{2a(1-c_2^2/c_1^2)}{\pi(\eta_k-y_l)} + \frac{2a(1-c_2^2/c_1^2)}{\pi(\eta_{k-1}-y_l)}\right], \tag{28}$$

where

$$\eta_k = -b + k\varepsilon_1, \qquad y_l = -b + (l-0.5)\varepsilon_1,$$
$$l, k = 1, \ldots, N, \qquad \varepsilon_1 = 2b/N.$$

Finally we have the system of algebraic equations with respect to the quantities $h(\eta_k)$:

$$\frac{1}{2ak_2^2}\sum_{k=1}^{N} h(\eta_k)\left[\varepsilon_1\Phi_r(y_l-\eta_k) - \frac{2a(1-c_2^2/c_1^2)}{\pi}\left(\frac{1}{\eta_k-y_l} - \frac{1}{\eta_{k-1}-y_l}\right)\right] = 1. \tag{29}$$

Let us give a short description how an efficient treatment of the regular part of the kernel, function K_r in Eq. (21), can be arranged. This is based on the asymptotic estimate that the qualitative behavior of the regular kernel is as follows:

$$K_r(y,z) \sim \sum_{n=1} \frac{\cos(a_n y)}{a_n}, \tag{30}$$

Of course, this series can be calculated explicitly by using the tables [6]. However, some problems arise in this case when integrating over the small sub-intervals (η_{k-1}, η_k) in an explicit form. The simpler alternative way is to apply the explicit integration just to the series (30) itself:

$$\int_{\eta_{k-1}}^{\eta_k} \frac{\cos[a_n(\eta - y_l)]}{a_n} d\eta = \frac{\sin[a_n(\eta_k - y_l)] - \sin[a_n(\eta_{k-1} - y_l)]}{a_n^2}. \tag{31}$$

Note that with such a treatment of the considered particular term of the regular kernel, the factor ε_1 in (29) should be omitted in front of this particular term. It can easily be seen from Eq. (31) that the convergence is rapid enough along the index n. In practice, few hundred terms are sufficient, to guarantee 1% relative error in all diagrams demonstrated below.

4 Calculation of the Wave Characteristics

Let us set

$$H = \int_{-b}^{b} h(t)dt. \tag{32}$$

In terms of the even function $h(y)$, we deduce from system (18):

$$g_x^1(y) = \left\{ [1/(4aik_1)]J_1 + [e^{ik_1D}/[4aik_1]]J_2 - 1 \right\} k_2^2 h(y,z),$$

$$g_x^2(y) = \left\{ [1/(4aik_1)]J_2 + [e^{ik_1D}/[4aik_1]]J_1 - e^{ik_1D} \right\} k_2^2 h(y,z), \tag{33}$$

where the quantities J_1, J_2 are defined in Eq. (15).

The integration of Eq. (33) over the segment $[-b, b]$ leads to the following system of linear algebraic equations for the unknown constants J_1, J_2:

$$\left\{ k_2^{-2} - H/(4aik_1) \right\} J_1 - \left\{ He^{ik_1D}/[4aik_1] \right\} J_2 = -H,$$

$$-\left\{ He^{ik_1D}/[4aik_1] \right\} J_1 + \left\{ k_2^{-2} - H/(4aik_1) \right\} J_2 = -He^{ik_1D}. \tag{34}$$

Therefore, as soon as system (34) is solved, all necessary constants and the wave field can be found. In particular, for the reflection and transmission coefficients we obtain

$$R = -\frac{1}{4aik_1} J_1 - \frac{e^{ik_1D}}{4aik_1} J_2,$$

$$T = -\frac{e^{ik_1D}}{4aik_1} J_1 - \frac{1}{4aik_1} J_2 + e^{ik_1D}. \tag{35}$$

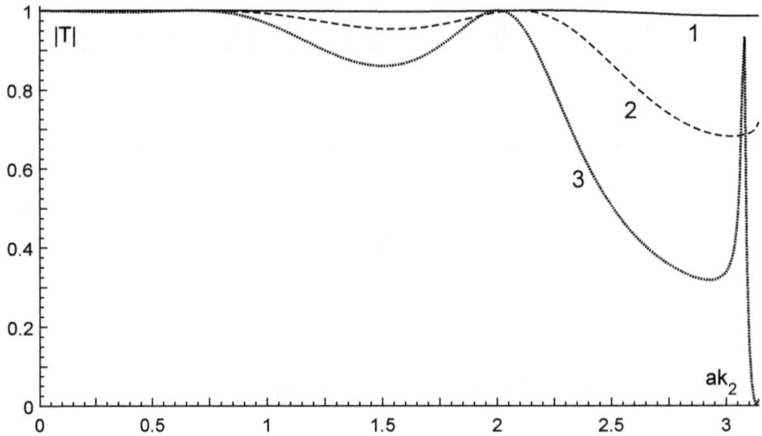

Fig. 2 Transmission coefficient versus frequency parameter: $D = 4.0$, line 1—$b/a = 0.15$, line 2—$b/a = 0.3$, line 3—$b/a = 0.4$

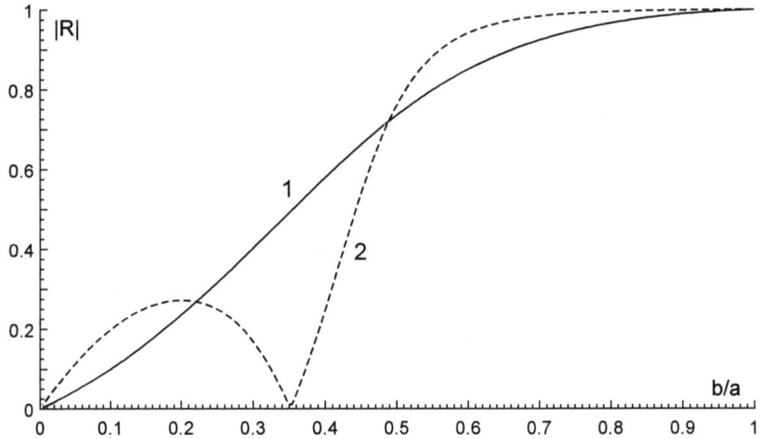

Fig. 3 Reflection coefficient versus crack's size: $D = 4.0$, line 1—$ak_2 = 0.935$, line 2—$ak_2 = 1.870$

The explicit expressions for the scattering parameters developed above complete the semi-analytical approach in the one-mode range. Some examples on the reflection and transmission coefficients versus frequency parameter, crack's size, and the distance between the two vertical arrays, for the elastic media with the wave speeds ratio $c_1/c_2 = 1.870$, are plotted in Figs. 2, 3, 4 and 5.

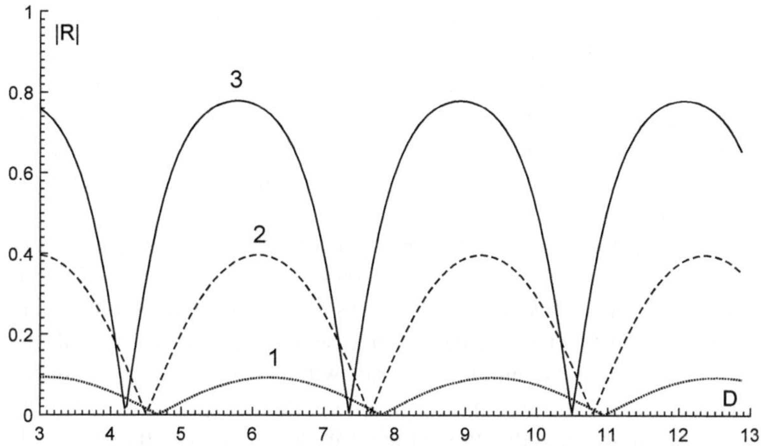

Fig. 4 Reflection coefficient versus distance between the two arrays of periodic systems of cracks: $ak_2 = 1.870$, line 1—$b/a = 0.15$, line 2—$b/a = 0.30$, line 3—$b/a = 0.45$

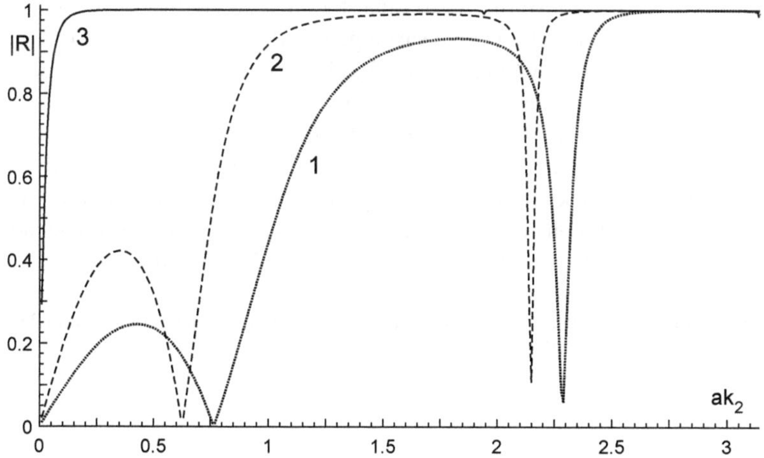

Fig. 5 Reflection coefficient versus frequency parameter: $D = 3.0$, line 1—$b/a = 0.6$, line 2—$b/a = 0.8$, line 3—$b/a = 0.999$

5 Conclusions

The obtained results are analyzed on the subject of the cutoff properties of the acoustic metamaterials possessing an internal periodic geometric structure, as described in the Introduction.

It is obvious from Fig. 2 that the number of frequency intervals with suppressed transmission, which is really a certain cutoff, grows with the increasing of the relative crack's size. If ak_2 and D are fixed, then the reflection coefficient $|R|$ versus relative crack's size b/a may show various behavior. It is a monotonically increasing function for $ak_2 = 0.935$, but when $ak_2 = 1.870$ the monotonic character of the diagram takes place only after the frequency passes a certain critical value, see Fig. 3. If b/a and ak_2 are fixed, then the behavior of function $|R(D)|$ is always wavy. For all that, the higher maxima correspond to the higher crack's length. It means that the strongest reflection takes place for large cracks; this is quite natural from the physical point of view, see Fig. 4.

Figure 5 demonstrates that the property of acoustic filters is attained for not only long cracks, where it is natural, but also for cracks of moderate length. Of course, the extremely long crack with $b/a = 0.999$ shows almost absolute cutoff for almost all frequencies in the considered one-mode range, which is physically natural. However, it is also interesting that the middle-size crack with $b/a = 0.6$ has a pair of relatively long frequency intervals where the reflection coefficient approaches the unit value. And it should also be noted that in the higher frequency part of the one-mode range with $ak_2 \approx \pi$ the cracks of all demonstrated lengths provide the cutoff. With this noting, let us outline that longer cracks make this upper cutoff frequency interval longer too.

It follows from the above discussion that the desired control of the acoustic filtering in the considered grating can be arranged by the appropriate choice of crack's length, respective frequency interval, and finally—but the distance between the two vertical arrays containing periodic systems of cracks.

The method developed in the present work permits efficient treatment of a more complex wave problem, when the number of vertical arrays containing the studied periodic system of cracks may be arbitrary but finite. This only requires to solve an alternative finite-dimensional system of linear algebraic equations, instead of the 2×2 system (34). This case will be studied in the authors' next work.

Acknowledgements The first author is thankful to the Russian Science Foundation (RSCF), for its support by Project 15-19-10008.

References

1. Achenbach, J.D., Li, Z.L.: Reflection and transmission of scalar waves by a periodic array of screens. Wave Motion **8**, 225–234 (1986)
2. Belotserkovsky, S.M., Lifanov, I.K.: Method of Discrete Vortices. CRC Press, Boca Raton, Florida (1992)
3. Kok, Y-L., Galagher, Jr. N.C., Ziolkovski, R.V.: Dual series solution to the scattering of plane wave from a binary conducting grating. IEEE Trans. Anten. Prop. **AP-37**, 901–917 (1989)
4. Liu, Z., Zhang, X., Mao, Y., Zhu, Y.Y., Yang, Z., Chan, C.T., Sheng, P.: Locally resonant sonic materials. Science **289**(5485), 1734–1736 (2000)
5. Miles, J.W.: On Rayleigh scattering by a grating. Wave Motion **4**, 285–292 (1982)
6. Prudnikov, A.P., Brychkov, Y.A., Marichev, O.I.: Integrals and Series, vols. 1, 2. Gordon & Breach, Amsterdam (1986)
7. Remizov, M.Yu, Sumbatyan, M.A.: 3-D one-mode penetration of elastic waves through a doubly periodic array of cracks. Math. Mech. Solids (in press)

8. Remizov, M.Yu, Sumbatyan, M.A.: Asymptotic analysis in the anti-plane high-frequency diffraction by interface cracks. Appl. Math. Lett. **34**(1), 72–75 (2014)
9. Remizov, M.Yu., Sumbatyan, M.A.: A semi-analytical method of solving problems of the high-frequency diffraction of elastic waves by cracks. J. Appl. Math. Mech. **77**(4), 452–456 (2013)
10. Remizov, M.Yu., Sumbatyan, M.A.: A semi-analytical approach in the high-frequency diffraction by cracks. Mech. Res. Comm. **38**(8), 607–609 (2011)
11. Sumbatyan, M.A.: Low-frequency penetration of acoustic waves through a periodic arbitrary-shaped grating: the three-dimensional problem. Wave Motion **22**, 133–144 (1995)
12. Scarpetta, E., Sumbatyan, M.A.: Explicit analytical results for one-mode oblique penetration into a periodic array of screens. IMA J. Appl. Math. **56**, 109–120 (1996)
13. Scarpetta, E., Sumbatyan, M.A.: On wave propagation in elastic solids with a doubly periodic array of cracks. Wave Motion **25**, 61–72 (1997)
14. Scarpetta, E., Tibullo, V.: On the three-dimensionl wave propagation through cascading screens having a periodic system of arbitrary openings. Int. J. Eng. Sci. **46**, 105–111 (2008)
15. Shenderov, E.L.: Sound penetration through a rigid screen of finite thickness with apertures. Soviet Phys. Acoust. **16**(2) (1970)
16. Sneddon, I.N., Lowengrub, M.: Crack Problems in the Classical Theory of Elasticity. Wiley, London (1969)

Analytical and Computer Methods to Evaluate Mechanical Properties of the Metamaterials Based on Various Models of Polymeric Chains

Roman A. Gerasimov, Olga G. Maksimova, Tatiana O. Petrova, Victor A. Eremeyev and Andrei V. Maksimov

Abstract The formation of polymer coating on a solid substrate is investigated by means of computer simulation (Monte-Carlo method). The sticking coefficient depending on different factors affecting the adhesion of monomer units is calculated. Mechanical properties are stimulated on the base of the hybrid discrete-continuous model, which describes the system consisting of flexible substrate and polymer coating. At different temperatures and intermolecular interactions constants, the dependencies of Young modulus on the deformation degree are calculated. Ferroelectric properties of the polymer coating depending on frequency and amplitude of external electric field, temperature and interchain interactions are investigated.

Keywords Metamaterials · Ferroelectric polymer · Monte-Carlo method · Stockmayer potential · Hybrid discrete-continual model · Tension and bending deformations · Hysteresis

---

R.A. Gerasimov (✉) · T.O. Petrova · V.A. Eremeyev
Southern Federal University, 105/42 Bolshaya Sadovaya Str., Rostov-on-Don 344006, Russian Federation
e-mail: roman-gerasimoff@yandex.ru

T.O. Petrova
e-mail: to_87@bk.ru

V.A. Eremeyev
e-mail: eremeyev.victor@gmail.com

O.G. Maksimova · A.V. Maksimov
Cherepovets State University, 5 Lunacharskii Av., Cherepovets 162600, Russian Federation
e-mail: og62@mail.ru

A.V. Maksimov
e-mail: a_v_maximov@mail.ru

M.A. Sumbatyan (ed.), *Wave Dynamics and Composite Mechanics for Microstructured Materials and Metamaterials*,
Advanced Structured Materials 59, DOI 10.1007/978-981-10-3797-9_3

1 Introduction

Physics of surface and interfacial phenomena is an active research field, which continuously allows obtaining new important results. Currently, the field of knowledge related to the control of surface properties of new materials is intensively developing. Special attention is paid to the problems of regulation of the substrate surface wetting aimed at imparting them superhydrophobic and superhydrophilic properties [1–6]. Among physical and chemical methods of surface modification, it is essentially to note such as plasma treatment, electrodeposition, lithography, etching, polymer grafting [7–9]. The methods of polymer chains grafting are particularly attractive, as they allow to selectively changing some properties of substrate surface: adhesion, wettability, tribological properties; without changes of physical properties of the substrate to give its surface the specific properties of the grafted polymer [3, 8]. A wide range of grafted polymer chains allows selecting necessary characteristics of surface of phase interface and obtain in new hybrid materials. Moreover, this method allows obtaining rough and porous surfaces, what gives the opportunity to imitate the lotus leaf structure. In this case, wetting of the surface is defined both by chemical nature of the grafted compounds and by microtexture of the coating [3].

The grafted polymer layers have a broad prospect of application for preventing capillary condensation and icing, protection against corrosion and biofouling, hydroprotection of reinforced concrete structures, creating waterproof textiles. The surface-modified materials may have the ability to self-cleaning and non-wetting, to responding the changes at environmental conditions (temperature, light radiation, pH or solvent composition), may have antibacterial and antistatic properties.

The surfaces modified by grafted polymers are the most promising fields of application in nanotechnologies and micromechanics. Materials of this kind are widely used in different industries: they are used by creation of sensitive membranes and filters, biochips and biosensors; for enhancing foreign bodies' biocompatibility, regulation of adhesion, friction, surface wettability, as protective coatings for glass, metal, wood, ceramics, and textiles. Modification of textile fibers by polymer coatings allows obtaining new materials with preset properties. Formed on the surface, nano-rough polymer film gives textile materials with both hydro- and oleophobic properties allowing their application as protective clothing [10]. Such artificial materials the properties of which are mainly determined by their structure (in this case, by the surface structure) are called "metamaterials".

Recently, on the way to implementation of "smart" surfaces, the synthesis of polymer brushes has gained particular attention [11]. Polymer brush is a monolayer of polymer chains linked with some impermeable surface by terminal groups. There are two fundamentally different methods to create such polymer brushes [12]. By chemical method, terminal groups are "sewn" to some surface by chemical bonding. The example of such systems is not only flat brushes in which the chains are grafted to flat surface (Fig. 1a), but also some regularly branched polymers (stars with a large number of branches can be considered as spherically convex brushes,

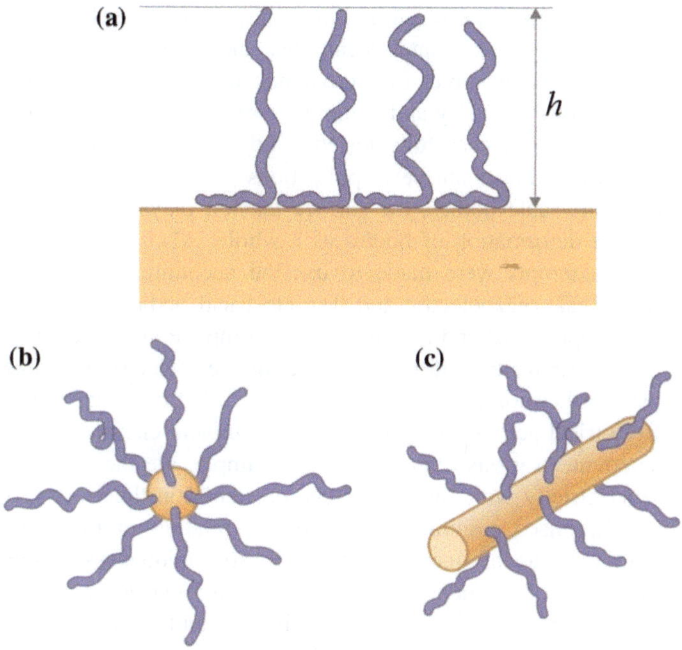

Fig. 1 The flat polymer brush with the chain height h (**a**) and regularly branched polymers as examples of brushes: spherical brush (**b**), cylindrical brush (**c**)

and comb-shape polymers, as brushes around the convex cylindrical surface (Fig. 1b and c).

By the other method of brush formation, a very important property of polymer molecules is used, which is the basis for the organization of complex biological structures, namely, the ability of macromolecules to self-organization. A necessary condition for the self-organization is the existence of different groups in macromolecules. Usually, there are monomer units with different groups. In the simplest cases it is sufficiently to have only two kinds of monomer units in each polymer molecule (two-component copolymer). The driving force of self-organization is intermolecular interaction. There are both the intra- and inter-component interactions and interaction with the environment, for example with a solvent, a solid phase surface, i.e., with particles which may be present in the system, etc. The first effects determine self-organization, when it occurs without a solvent (or at a low solvent content). Intermolecular interactions are much weaker than the covalent ones. Therefore, chemical structure is not changed during the self-organization, and everything is defined by intermolecular interactions in the system with the given chemical structure.

Also, in recent years, delicate experimental methods have made possible to reveal many interesting properties of solids, determined both by the influence of the solid body surface, and primarily due to its modification during the adsorption of active elements and sputtering of ultrathin multilayer structures [13, 14]. Improved resolution of various experimental techniques and improved methods of new materials producing allow direct measurements of surface properties. Therefore, results and predictions of theory and computer simulation of surface characteristics of solid bodies can be successfully tested. And better understanding of surface properties can lead to important new applications.

For elastic nano-scale bodies, the surface tension plays a significant role and influences on the deformation of bodies as a whole. Also, in a number of recent works [15–19], attempts were made to use the accounting methods of surface effects for nano-scale piezoelectric and flexoelectrical bodies. Despite a considerable number of approaches and results, a fairly complete and strict analytic theory of surface effects for nano-scale piezoelectric and ferroelectric metamaterials still does not exist. Furthermore, by taking the defining relations connecting the surface tension and electrical activity, the problems of complete classical and generalized statements encounter serious difficulties. For example, the theory for static problems of piezoelectricity with surface effects, presented in the work [20], see also [17–19, 21, 22], raises a number of problems relating to the formulation of boundary conditions, to the questions of solvability, uniqueness of solutions, etc. Mathematical study of boundary-value problems for elastic solids with surface stresses was performed in [23–27] where existence and uniqueness of week and strong solutions is proved. In particular, in [24] the stiffening effect of surface elasticity was confirmed. In the case of piezoelectric and magnetoelastic solids the similar analysis was given in [28–30].

Therefore, in recent years, methods of computer experiment have become an instrument used in many fields of science. Motivation for their application to the study of physical systems is diverse. One of the main motives is elimination of the limitations inherent in analytical models. Usually, by analytical analysis of the problem (if it is possible in general), different approximations are used. The application of computer simulation techniques gives the opportunity to study complex systems, not investigated analytically previously.

In this paper, for computer simulation of mechanical and electrical properties of metamaterials, we will apply Monte-Carlo method for numerical statistical description of macroscopic systems, widely used for investigations of bulk properties of various materials [31, 32] and a powerful tool in the study of complex molecular systems [15, 33, 34]. In Monte-Carlo method, the solution of dynamic equations for particles is replaced by the generation of some stochastic process. Such technique provides quite simple calculation of average values for various physical quantities within the canonical ensemble. Alongside with the description of bulk characteristics, Monte-Carlo method is used quite successfully for the simulation of crystal growth and studying the properties of emerging surface structures [33, 34]. The properties of crystalline surfaces are often described by means of lattice models. Dynamics of crystal growth is simulated by random processes of adsorption, evaporation and surface transport. The Monte-Carlo method allows directly simulating such dynamic processes.

2 Simulation of the Adsorption Process

Adhesive, strength, optical, isolative and other properties of polymer coatings on solid surfaces are studied at the intersection of various fields of science: metallurgy, thermodynamics, physical chemistry and macromolecular chemistry, solid state physics, classical mechanics, physical and chemical mechanics, mathematical statistics, etc. [14, 35–38].

The formation of polymer coating on substrate surface can be considered to consist of several stages [39]:

- spreading of the adhesive over the surface and wetting (Fig. 2a);
- equilibrium setting of adhesive contact related with molecular properties of the adhesive (Fig. 2b);
- formation of chemical and physical polymer structure during the solidification accompanied by the occurrence of the surface layer with properties different from ones of the bulk (Fig. 2c).

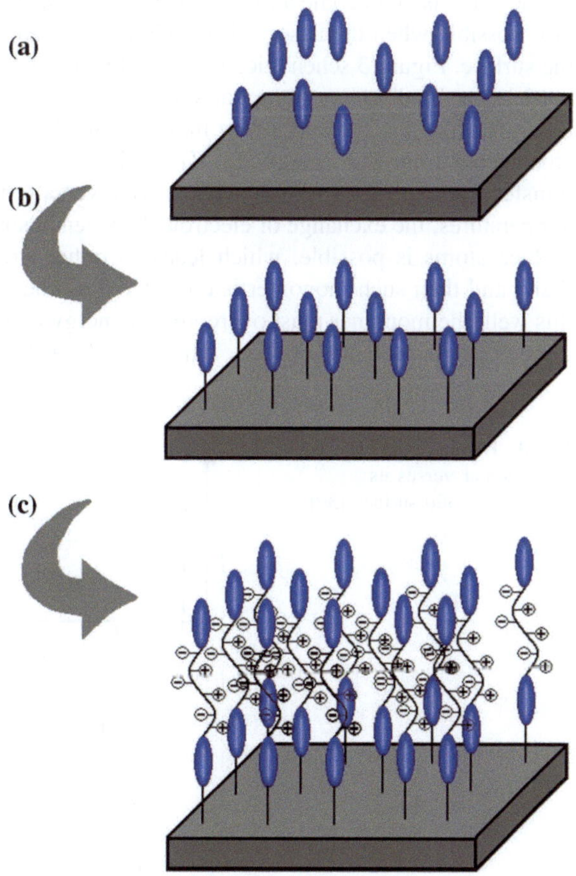

Fig. 2 The formation of polymer coating structure: spreading of the adhesive over the surface and wetting (**a**); equilibrium setting of adhesive contact related with macromolecular properties of the adhesive (**b**); the formation of chemical and physical structure in the polymer during the solidification (**c**)

The last stage includes shrinkage of the adhesive, possible crystallization, allocation of new phases, etc.

Film formation occurs as a result of several physico-chemical processes: evaporation of solvents, latex a stabilization and dehydration, cooling of melt, resulting to formation of the polymer coating with a specific set of properties. Change of aggregate state of polymer is accompanied by jump like change of its specific volume, namely, the minimum of free volume and sharp slow-down of relaxation processes. Simultaneously, the structure formation is no characteristic for highly elastic state of a substance. The properties of such coatings are also affected by the content of compositions and film forming conditions.

Molecular theory treats the adhesion as the result of molecular interaction forces between contacting molecules of adhesive and substrate. Therefore, it is important that the adhesive and the substrate molecules possess polar functional groups capable of interaction [40]. Thermodynamic properties of the coating can be examined if it is considered, for example, as a set of dipoles interacting with a solid surface and with each other. When the particle faces the surface as a result of thermal motion, it can be mirrored from the substrate, or be firmly bound to the surface, that is, to become adsorbed one. The opposite phenomenon of desorption is also possible when the adsorbed particle having sufficient kinetic energy can leave the surface. Figure 3 schematically shows the dependence of potential energy of the particle on the distance to the flat surface [41]. In case of physical adsorption, the monomer binds to the surface by means of weak van der Waals forces, which are characterized by the energy E_p. These bonds are not accompanied by charge transfer from substrate to monomer or vice versa (physical absorption). At higher temperatures, the exchange of electrons between adsorbed monomers and the metal surface atoms is possible, which leads to rather strong chemical bonds between them, and then such monomer is considered as chemisorbed one [36]. To get into this well, the monomer must overcome an energy barrier E_b. Then, the particle is in a much deeper potential well E_c. The value $E_c + E_b$ defines the work function W.

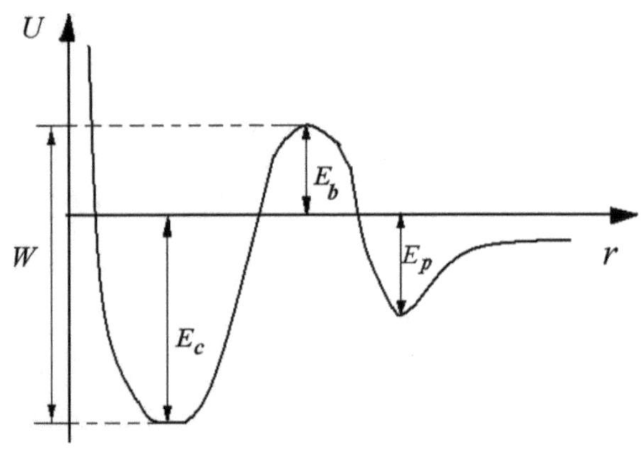

Fig. 3 The potential energy of monomer versus its distance to solid surface [36]

(a) *Simulation algorithm*

The modified Langmuir adsorption model may be used for calculations of the monomer sticking degree to the substrate. In statistical physics, a system of interacting particles placed in a periodic array of equivalent elementary cells is called the lattice gas. Two-dimensional lattice gas model corresponds to the adsorption layer; it is one of the basic models of statistical mechanics. Classical Langmuir model is based on the following assumptions:

- all adsorptive sites of the lattice are equivalent;
- only one particle can occupy an adsorptive site;
- influence of adsorption forces is restricted by thickness of the monolayer coating.

Modified Langmuir model takes into account the interactions of adsorbed molecules both with the substrate surface and between them (side or lateral interactions with the energy $\varepsilon_{lat.}$). Since the particles included in the composition of the adhesive are basically polar molecules [14], the energy of dipole orientational interactions between lattice sites in this model may be represented as:

$$\varepsilon_{lat.} = -K \sum\nolimits_{i,j=1}^{N,M} n_{i,j} n_{i+1,j} - K \sum\nolimits_{i,j=1}^{N,M} n_{i,j} n_{i,j+1}, \quad (1)$$

where K is the constant of interaction between the dipoles. Calculations by the Eq. (1) take into account only the interactions between the nearest lattice sites, since orientational interactions forces decrease sufficiently rapidly with the increase of the distance.

The substrate surface is represented as two-dimensional rectangular lattice containing N sites (adsorption points) along axis X, and M sites—along axis Y. The position of the lattice site is defined by two numbers i and j, and its employment—by the number n_{ij}, which is equal to zero, if the adsorption point is free, and to unity—otherwise [41].

The input parameters for the adsorption simulation program are: the temperature of the system T, the molar mass of the adhesive, the constant of interaction between the dipoles K, the height of the energy barrier E_b, and the potential well depth E_c, the lattice dimensions N and M, and the number of Monte-Carlo steps N_{MC}.

In this model, at the initial moment ($t = 0$), it is assumed that all the values of the occupied lattice sites $n_{ij} = 0$; i.e., there is no adsorption. Further calculations are carried out by Monte-Carlo method with the use of Metropolis algorithm, which is a powerful tool in the study of complex molecular systems [42–44].

- The random selection of a lattice site and the calculation of the energy of the molecule of the adsorbent located near this site:
  - if $n_{i,j} = 0$, the molecule of the adsorbent is free and is not captured by the surface (non-adsorbed one). Therefore, the value of its energy $\varepsilon = \varepsilon_M$ is a random number determined by Maxwell distribution at the given temperature;

  - if $n_{i,j} = 1$, the molecule of the adsorbent is captured by the surface and its energy is defined by the expression: $\varepsilon = \varepsilon_M - E_c + \varepsilon_{lat.}$, where E_c is the depth of the potential well, $\varepsilon_{lat.}$ is the energy (1) of lateral interactions.

- The change of the lattice site state:
  - if $n_{i,j} = 0$ and $\varepsilon < E_b$, the adsorbent molecule cannot overcome the energy barrier, so the lattice site state does not change;
  - if $n_{i,j} = 0$ and $\varepsilon > E_b$, the adsorbent molecule with probability 0.5 is captured by the substrate surface (if the generated random number with uniform distribution at interval [0; 1] is greater than 0.5). In this case, the value $n_{i,j}$ changes and becomes equal to the unity;
  - if $n_{i,j} = 1$ and $\varepsilon < E_b + E_c$, the adsorbent molecule is tightly bound on the solid surface. Then, the value $n_{i,j}$ does not change;
  - if $n_{i,j} = 1$ and $\varepsilon > E_b + E_c$, the desorption is possible with the probability 0.5, i.e., the value $n_{i,j}$ becomes equal zero.

The considered operations are repeated for a certain time proportional to the number of Monte-Carlo steps, which is set by the lattice size and the heating rate of the coating during the solvent evaporation. Relative degree of filling of the metal surface θ (degree of sticking) by dipoles is defined as the ratio of the number of adsorbed sites to the total number of lattice sites:

$$\theta = \frac{1}{NM} \sum_{i,j=1}^{N,M} n_{i,j}. \tag{2}$$

(b) *Simulation results*

Figure 4 shows the results of the calculations of time dependencies of the value θ at the given heating rates $g = \Delta T/\Delta t$ during the evaporation of the solvent from the metal sheet, but at different reduced constants of monomers interaction (K/k_BT_0) and the barrier energy (E_b/k_BT_0), where T_0 is the initial temperature of the system. Calculations were performed on 100×100 lattice at periodic boundary conditions.

It can be seen, that with the increase of monomers interaction, their degree of sticking to metal grows, that is associated with the rise of their cooperativity during the adhesion (Fig. 4a). On the other hand, with the increase of barrier height E_b, the value θ decreases (Fig. 4b), i.e., smaller number of monomers sticks to metal surface.

Figure 5 shows the results of calculation of time dependencies of θ at different heating rates of the metal sheet and at the given constant of monomer interaction and barrier energy. It turns out that under slow heating of sheet, the degree of sticking of monomers to metal surface is low (curve 1). At high speed of heating, there is some sticking to the metal surface, but during further heating the value θ decreases (curve 3). It can be seen that optimal temperature range is determined by curve 2 in Fig. 5, when the adhesion is maximal.

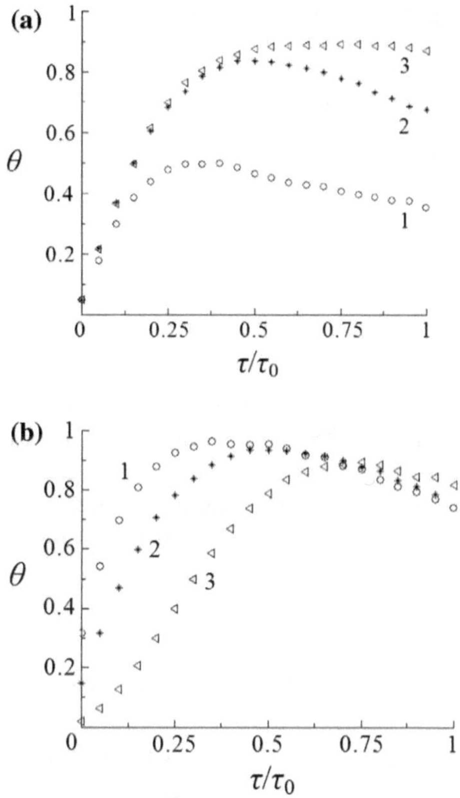

Fig. 4 Time dependencies of the coefficient of sticking of monomers to the metal sheet surface at the heating rates $g = \Delta T/\Delta t$: (**a**) at different reduced constants of monomer interaction $K/k_B T_0 = 0$ (*1*), 1 (*2*), 2 (*3*) and the given barrier energy ($E_b/k_B T_0 = 1$); (**b**) at different reduced barrier energy: $E_b/k_B T_0 = 0, 5$ (*1*), 1 (*2*), 2 (*3*) and the given constant of monomer interaction ($K/k_B T_0 = 1$). T_0 is initial temperature, τ_0 is total simulation time

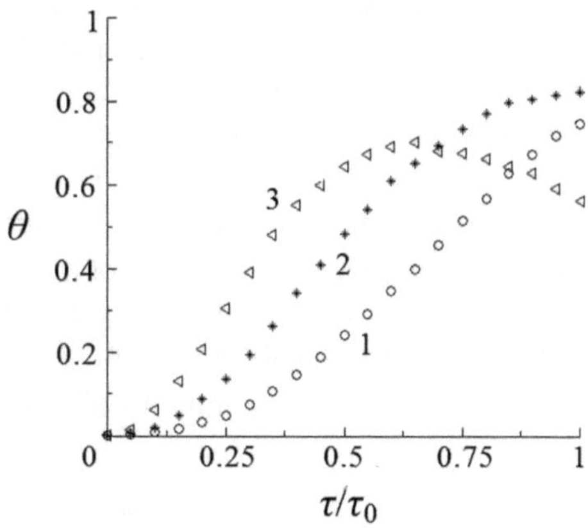

Fig. 5 Time dependencies of the coefficient of monomers sticking θ to the metal sheet surface at different heating rates $g_0 = \Delta T/\Delta t$ (*1*), $2g_0$ (*2*), $3g_0$ (*3*) at the given reduced monomer interaction constant ($K/k_B T_0 = 1$) and the barrier energy ($E_b/k_B T_0 = 3$). T_0 is initial temperature, τ_0 is total simulation time

Increase of sticking degree θ with the rise of temperature in this model can be explained by the increase of kinetic energy of monomers leading to overcoming the energy barrier E_b (Fig. 3), which prevents the formation of the adsorbed state of

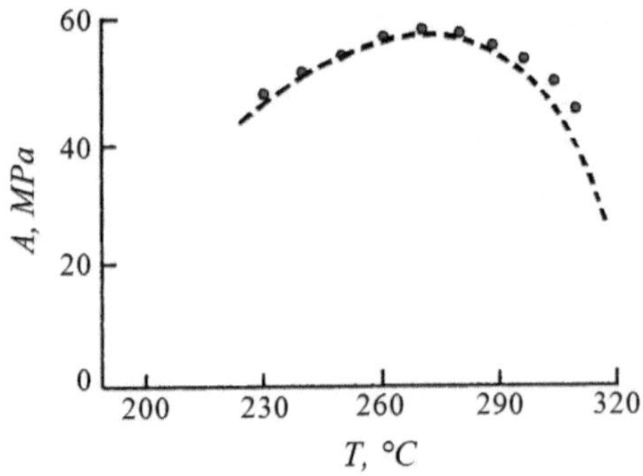

Fig. 6 The adhesive strength of polyvinyl butyral on the steel surface *A* versus the film formation temperature *T*: the dotted line represents the experiment data [87], symbols (o) are simulation results

monomers on metal surface. With further increase of temperature, the degree of sticking of monomers, conversely, decreases due to the growth of thermal fluctuations in the system. The obtained result corresponds to the experimental data [45] on the study of the adhesive strength of the polyvinyl butyral layer depending on the temperature during formation of the polymer coating on the steel sheet surface (Fig. 6).

3 Model "Polymer-Substrate"

Further simulation was associated with the study of systems with an already-formed layer on the substrate surface.

The hybrid discrete-continual model is presented in Fig. 7. This model consists of rotators of length l, connected to the substrate treated as an elastic continuum.

Within the framework of lattice multichain models of polymers [46–52], $N_1N_2N_3$ rigid kinetic chain elements of length l called rotators form the curvilinear "quasi-lattice" $\vec{n} \equiv (n_1, n_2, n_3)$, where indices n_1, n_2, n_3 take the following values: $n_1 = 1, \ldots, N_1$; $n_2 = 1, \ldots, N_2$; $n_3 = 1, \ldots, N_3$.

We assume that the energy of orientational interactions $H^{(i)}_{\vec{n},\vec{m}}$ between chain elements located in nodes $\vec{n}$ and $\vec{m}$ of the "quasi-lattice" depends only on the mutual angle of their orientation. For the potential of dipole type (Keesom's energy), the energy is proportional to cosine of spatial angle $\Phi_{\vec{n},\vec{m}}$ between their axes: $H^{(i)}_{\vec{n},\vec{m}} = -K_i \cos \Phi_{\vec{n},\vec{m}}$, as in the Gotlib-Maksimov multichain models [48], i.e.,

$$H^{(i)}_{\vec{n},\vec{m}} = -K_i \frac{\left(\vec{l}_{\vec{n}}, \vec{l}_{\vec{m}}\right)}{l^2} = -K_i \frac{(u_{\vec{n}}u_{\vec{m}} + v_{\vec{n}}v_{\vec{m}} + w_{\vec{n}}w_{\vec{m}})}{l^2}, \tag{3}$$

where $u_{\vec{n}}$, $v_{\vec{n}}$ and $w_{\vec{n}}$ are projections on the coordinate axis of vector $\vec{l}_{\vec{n}}$, oriented along kinetic unit located in the node $\vec{n}$ ($i = 1, 2, 3$).

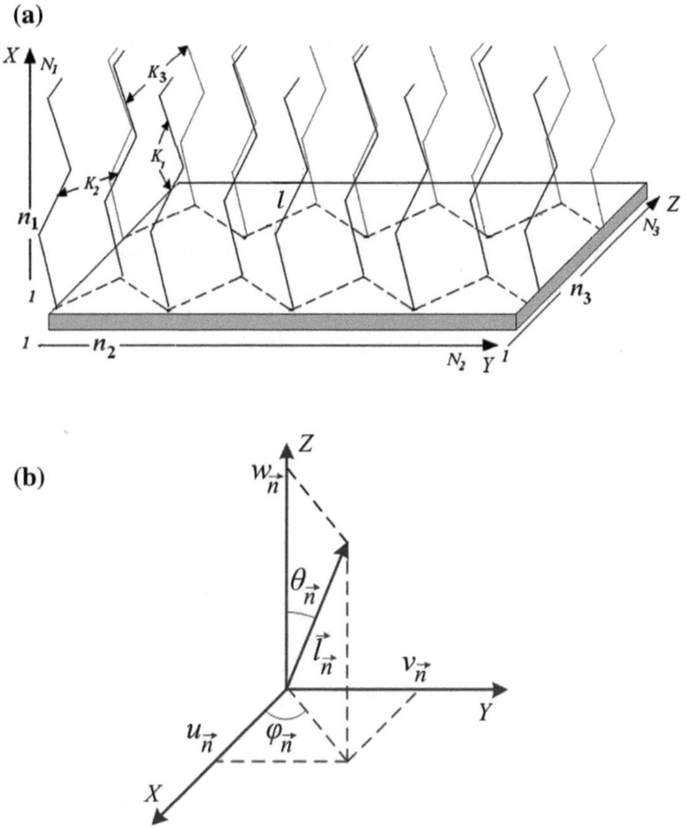

Fig. 7 (**a**) The discrete-continual model of the three-dimensional ordered system of N_2N_3 chains consisting of N_1 rigid kinetic units of length l, so-called rotators, N_2 and N_3 are numbers of chains along directions of the substrate cross-section. (**b**) The orientation vector $\vec{l}_{\vec{n}}$ of chain segment located in the node $\vec{n}$ relatively to the Cartesian (x, y, z) and spherical (φ, θ) coordinate systems

The energetic parameter K_1 along the "longitudinal" curvilinear direction n_1 of the "quasi-lattice" describes the longitudinal orientational interactions and determines the mean cosine of the angle between neighbour kinetic units [48–50]. The energetic parameters K_2 and K_3 characterize orientational interactions of neighbour kinetic units in the transversal directions along n_2 and n_3 of the "quasi-lattice" [49, 50]. The estimation of their characteristic values for rigid as well as for flexible polymer chains is provided in [53].

The potential energy of the dipole orientation interactions of the chains elements, taking into account the interactions of the between nearest rotators, can be represented in the Cartesian coordinate system as:

$$H = -K_1 \sum_{|n_1-m_1|=1} \sum_{n_2=m_2} \sum_{n_3=m_3} \frac{(u_{\vec{n}}u_{\vec{m}} + v_{\vec{n}}v_{\vec{m}} + w_{\vec{n}}w_{\vec{m}})}{l^2}$$
$$- K_2 \sum_{n_1=m_1} \sum_{|n_1-m_1|=1} \sum_{n_3=m_3} \frac{(u_{\vec{n}}u_{\vec{m}} + v_{\vec{n}}v_{\vec{m}} + w_{\vec{n}}w_{\vec{m}})}{l^2} \quad (4)$$
$$- K_3 \sum_{n_1=m_1} \sum_{n_2=m_2} \sum_{|n_3-m_3|=1} \frac{(u_{\vec{n}}u_{\vec{m}} + v_{\vec{n}}v_{\vec{m}} + w_{\vec{n}}w_{\vec{m}})}{l^2},$$

or in the spherical coordinate system: in the terms of polar and azimuthal angles of kinetic units $(0 \le \theta_{\vec{n}} \le \pi, 0 \le \varphi_{\vec{n}} < 2\pi)$:

$$H = -K_1 \sum_{\vec{n}} \left[\begin{array}{c} \cos\theta_{n_1,n_2,n_3} \cos\theta_{n_1-1,n_2,n_3} \\ + \sin\theta_{n_1,n_2,n_3} \sin\theta_{n_1-1,n_2,n_3} \cos(\varphi_{n_1,n_2,n_3} - \varphi_{n_1-1,n_2,n_3}) \end{array} \right]$$
$$- K_2 \sum_{\vec{n}} \left[\begin{array}{c} \cos\theta_{n_1,n_2,n_3} \cos\theta_{n_1,n_2-1,n_3} \\ + \sin\theta_{n_1,n_2,n_3} \sin\theta_{n_1,n_2-1,n_3} \cos(\varphi_{n_1,n_2,n_3} - \varphi_{n_1,n_2-1,n_3}) \end{array} \right] \quad (5)$$
$$- K_3 \sum_{\vec{n}} \left[\begin{array}{c} \cos\theta_{n_1,n_2,n_3} \cos\theta_{n_1,n_2,n_3-1} \\ + \sin\theta_{n_1,n_2,n_3} \sin\theta_{n_1,n_2,n_3-1} \cos(\varphi_{n_1,n_2,n_3} - \varphi_{n_1,n_2,n_3-1}) \end{array} \right].$$

In general, the energy of the interaction between the elements-rotators is described by the Stockmayer potential that is Lennard-Jones potential plus an additional term that takes into account of the energy of dipole interactions (3).

The potential energy of the whole system taking into account also elastic deformations of the substrate $H_{subs.}$ is given by the expression:

$$H = \sum_{\vec{n},\vec{m}} \left(\sum_{i=1}^{3} H_{\vec{n},\vec{m}}^{(i)} + 4\varepsilon \left[\left(\frac{1}{r_{\vec{n},\vec{m}}/\sigma} \right)^{12} - \left(\frac{1}{r_{\vec{n},\vec{m}}/\sigma} \right)^{6} \right] \right) + H_{subs.} \quad (6)$$

In Eq. (6), the parameter ε is the depth of the pit of the Lennard-Jones potential, $r_{\vec{n},\vec{m}}$ is the distance between nearest elements of the polymer chains, σ is the scale distance where the energy of interactions is zero.

Analyzing the expression (6), it should be noted that, firstly, the first term corresponds to the potential energy of orientational interactions; this term is temperature dependent, unlike interactions described by Lennard-Jones potential. With increasing the temperature, the contribution of energy of the dipole orientation interactions becomes less significant in the potential energy of the whole system (Fig. 8). Moreover, this fact has an effect on the density of packaging of the elements-rotators and, respectively, the relative dimensions of the system (Fig. 9). At the same time, both terms depend on the distance between of the interacting elements. With increasing distance between rotators at the deformation, the contribution of both terms in the potential energy (6) is decreased (Fig. 8). Therefore, the substrate plays the dominant role at the large tension or bending of the system, which considered as a continuum and elastic medium with the Hooke's law.

Secondly, this result has a statistical distribution, in the contrast to analytical calculations, because at the simulation of such systems by Monte-Carlo method, the

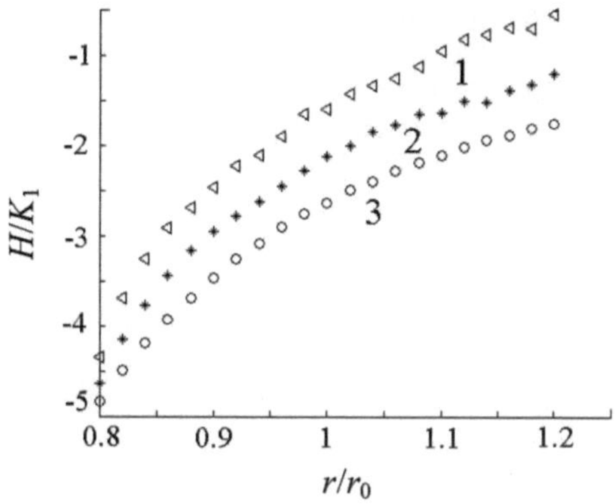

Fig. 8 The average normalized energy of the dipole orientation interactions H/K_1 of the element-rotator versus the relative distance r/r_0 between the nearest neighbors (r_0 is average distance between adjacent rotators in equilibrium configuration of the system) for different normalized temperatures: $T^* = k_BT/K_1 = 1, 2$ (*1*), $0, 8$ (*2*), $0, 4$ (*3*)

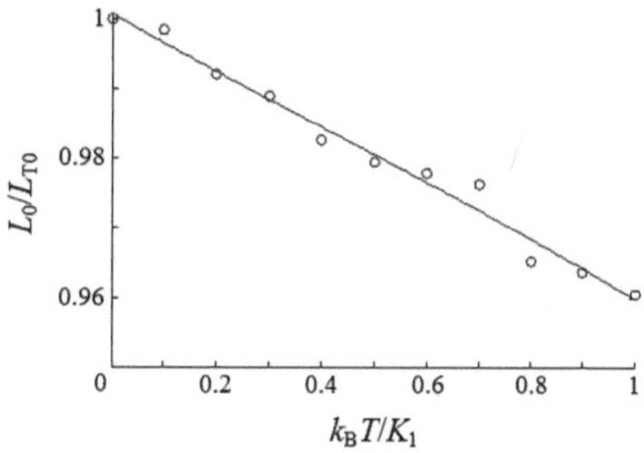

Fig. 9 The relative length L_0/L_{T0} of the system in the equilibrium initial non-deformed state versus the normalized temperature $T^* = k_BT/K_1$ (L_{T0} is the length of the system for $T^* = 0$)

random orientations of the rotators are used. Therefore, the use of the Monte-Carlo method requires a detailed analysis of the influence of such values as the number of Monte-Carlo steps, lattice size and type of boundary conditions etc. on the simulation results.

In Fig. 10, we present the dependence of the average energy of a rotator H/K_1 on the number of Monte-Carlo steps N_{MC} at different reduced temperatures $T^* = k_BT/K_1$ and lattice sizes for periodic boundary conditions. As can be seen from this figure, the average value of the segment's energy at sufficient value N_{MC} does not depend on the lattice dimensions. The dependence $H/K_1(\lg(N_{MC}))$ reaches quickly a saturation by small number of particles, but the energies have greater dispersion. It can be concluded that the simulation of the system with the particle number less than 10^3 does not allow achieving satisfactory results, even at large number of Monte-Carlo steps N_{MC}. An increase in the number of particles requires an increase of the value N_{MC} and calculations on high-performance computers.

Third, to determine the required characteristics of the system under study, it is necessary to specify its initial configuration. A wrong choice of initial conditions can lead to incorrect results when the Monte-Carlo method is used. In Fig. 10, we present the calculated average energy of the rotator H/K_1 on the number of Monte-Carlo steps N_{MC} at a fixed temperature and different initial state of the

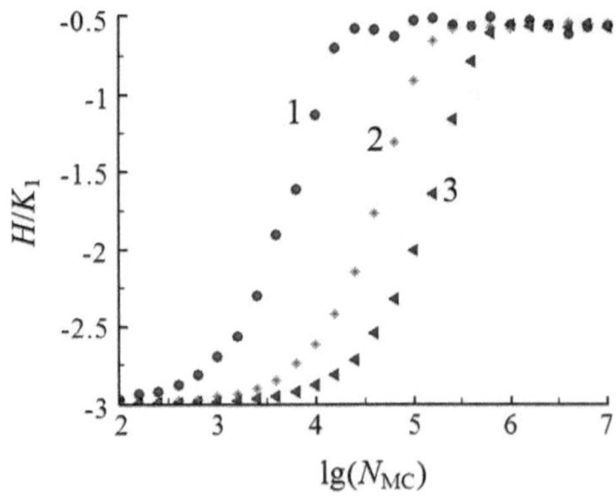

Fig. 10 The average normalized energy H/K_1 of the element-rotator versus the number of Monte-Carlo steps N_{MC} for the given reduced temperature $T^* = k_BT/K_1 = 2$. The simulation was made on the lattices: $10 \times 10 \times 10$ (*1*), $20 \times 20 \times 20$ (*2*), $30 \times 30 \times 30$ (*3*)

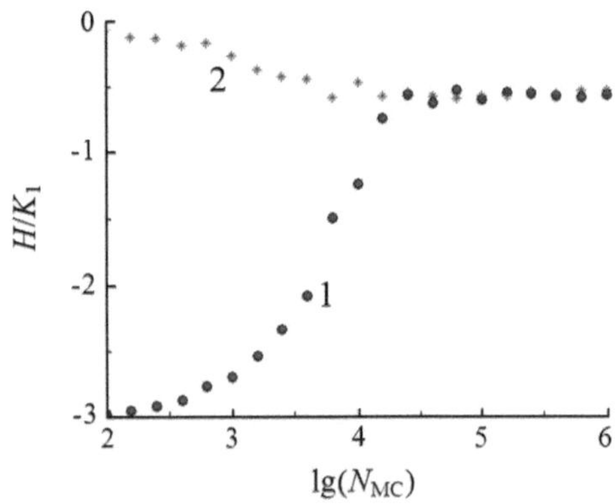

Fig. 11 The average normalized energy H/K_1 of the element-rotator versus the number of Monte-Carlo steps N_{MC} for the given reduced temperature $T^* = k_BT/K_1 = 2$. The simulation was made on the lattice $20 \times 20 \times 20$ at initial parallel (*1*) and chaotic (*2*) orientation of the rotators

system: rotators are arranged in parallel configuration (curve 1) or chaotic one (curve 2). The minimum number of Monte-Carlo steps N_{MC} required to obtain the correct simulation results can be determined from the lowest abscissa of the intersection point for these curves ($\sim 10^4 \div 10^5$ steps for the lattice $20 \times 20 \times 20$) (Fig. 11).

4 Simulation of the Simplest Types of Mechanical Deformation

In computer simulation of the simplest types of deformation, it assumed that the system is subjected to tensile force F (Fig. 12a) or bending moment M (Fig. 12b), simultaneously applied to the substrate and the polymer coating.

For polymer chains with polar groups, the parameters K_2 and K_3 can be evaluated from the multipolar expansion [54] of the interaction energy of two identical dipoles with electric dipole moment p placed in neighboring chains:

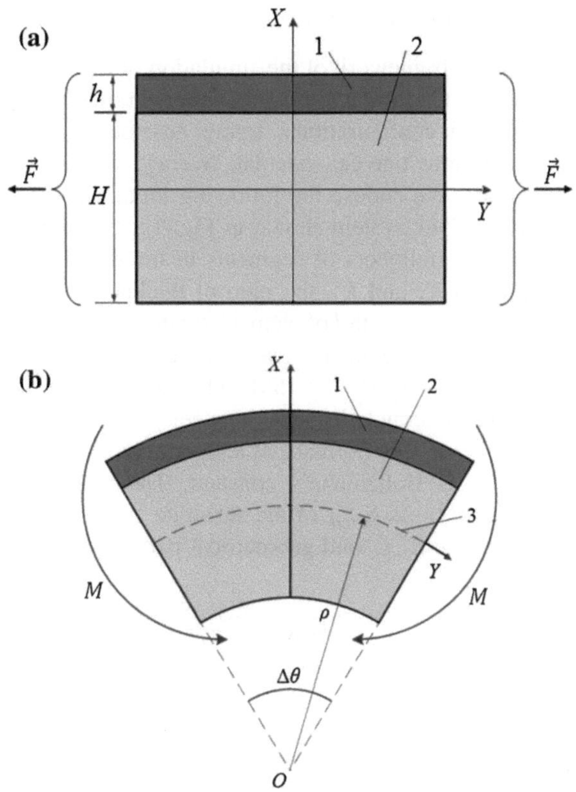

Fig. 12 The schemes of deformation of the discrete-continuous system: the tension under the influence of the force *F* (*a*) and the bending under the influence of the moment *M*(*b*). *1*—the polymer coating with the thickness *h*, *2*—the substrate with the thickness *H*, *3*—the position of the neutral layer, where there no tension/compression; ρ is the radius of curvature of the neutral layer corresponding to the angular deformation $\Delta\theta$, *O* is the center of curvature

$$K_2 \cong K_3 = \frac{1}{4\pi\varepsilon_0} \cdot \frac{p^2}{r^3}, \tag{7}$$

where r is the mean distance between the neighboring chains, and ε_0 is the dielectric constant.

Therefore, during the system deformation, i.e., for stretching and bending, parameter K_2 was calculated from the dependence interactions energy of chains on the relative displacements of chain elements:

$$K_2 = K_3 \cdot \frac{1}{(r/r_0)^3}. \tag{8}$$

In Eq. (8), the parameter r_0 is the mean distance between neighbour chain elements in the direction of force action (along n_2) at equilibrium configuration of the system, $r = r_0 + \Delta r$, where Δr characterizes the change of element position during the deformation. The energetic parameter K_3 keeps constant value during the deformation.

(a) *Simulation algorithm*

Within the framework of the simulation of the presented above system by means of the Monte-Carlo method, we generate a random process consisting of sequence of the system configurations. Using ensembles with the large enough number of configurations one can calculate average values of almost all equilibrium physical quantities. We choose the following input parameters for the simulation of deformations of the system shown in Fig. 12: the dimensions of the "quasi-lattice" N_1, N_2 and N_3 (numbers of segments in the directions n_1, n_2 and n_3), the interaction parameters K_1 and K_3, the ratio of thickness of the polymer coating and the substrate h/H, the length l of element-rotator, and the number of Monte-Carlo steps. All quantities having energetically dimension were normalized with respect to the parameter $K_1 \sim 10^{-20}$ J [53], whereas length-scale parameters were normalized using the average distance between neighbour chains of the polymer coating $a \sim 10^{-10}$ m. The normalized temperature of the system is defined as $T^* = k_B T / K_1$, where k_B is Boltzmann's constant. The effective tensile stiffness $k_{tens.}$ and the bending stiffness $k_{bend.}$ of the substrate are defined through the Young modulus of the substrate $E_{subs.}$ and geometrical parameters of the system:

$$k_{tens.} = \frac{E_{subs.} S_{subs.}}{L_0}, \, k_{bend.} = \frac{E_{subs.} I_z^{subs.}}{L_0}. \quad (9)$$

In (9), the quantity $S_{subs.}$ is the area of the transverse substrate cross-section in the plane X–Z, L_0 is the length of the system in the non-deformable state along the axes Y and Z, $I_z^{subs.} = \int x^2 dS$ is the inertia moment of the substrate cross-section with respect to the axis Z. If the axis X and Z are the principal central inertia axes, then, $I_z = ba^3/12$ for the rectangular cross-section with the height a and the width b. Since the centres of mass of the substrate and coating are not coincide we find that $I_z^{subs.} = bH(H^2 + 3h^2)/12$, where H and h are correspondingly the thickness of the substrate and the polymer coating.

For the simulation of the deformations of compression/tension and bending of the system, we use the following algorithm:

- We assume the following initial state of the system. Chain elements located at the interface coating-substrate as well at ends of the system along the axis Y (in the direction n_2) are fixed. At the top boundary along axis X (in the direction n_1) segments are free, and boundary conditions along the axis Z (in the direction n_3) are periodic ones since in this direction, the system is infinite. In order to achieve faster the equilibrium state at low temperatures, all segments are parallel each other, whereas for high temperatures, their orientation is given randomly.
- For determination of the parameter L_0 in non-deformable state, we simulate its uniform compression/extension along Y axis and find equilibrium configuration of the system for which the energy of the polymer part (i.e., without the substrate), is minimal at the given temperature. For this purpose, we imply cycle of the Monte-Carlo steps with Metropolis algorithm and the self-consistency

condition. If the energy of the system decreases with the change of its linear dimensions along the Y axis, we replace the value of r_0 in Eq. (8) using current value of the distance between elements in the direction n_2. If the energy does not decrease, we keep the initial value of r_0.

- We simulate the tension or the plane pure bending of the system (from the equilibrium value L_0). At each step of the bending deformation, we decrease the curvature radius ρ of the neutral layer through the increase of the angle $\Delta\theta = L_0/\rho$, whereas the distance n_2 along the axis Y between segments of the polymer coating is determined by the formula:

$$r = r_0 + \left(\frac{(H-h)}{2} + l\left(n_1 - \frac{1}{2}\right)\right)\frac{L_0}{(N_2-1)\rho}. \tag{10}$$

In Eq. (10), the quantities H and h are correspondingly the substrate and polymer coating thickness (Fig. 12), n_1 is the number of element-rotator along the axis X. The distance between neighbour segments in Eq. (10) is defined as the distance between their geometrical centres. Since we consider thin coatings, the ratio $h/H < 1$, and the polymer layer at the simulation is always under tension ($r > r_0$).

- At each step of bending deformation, we calculate the energy of equilibrium state of the system. We take the energy of elastic substrate as $H_{subs.} = k_{tens.}(\Delta x)^2/2$ at the stretching deformation and $H_{subs.} = k_{bend.}(\Delta\theta)^2/2$ at the bending deformation.
- According of Hooke's law, the force F applied to the system is proportional to the displacement of chain elements from their initial positions, in other words $F \sim \Delta x$. The work of external force during the deformations transforms to the energy change of the system: $A = -\Delta H$, where

$$A = \frac{F\Delta x}{2}. \tag{11}$$

Considering that $\Delta H = k_{syst.}(\Delta x)^2/2$ and using the relations similar to Eq. (9) but for whole system, we obtain that:

$$A = \frac{ES_{syst.}}{L_0}\frac{(\Delta x)^2}{2}. \tag{12}$$

In Eq. (12), E is the effective Young modulus of the system, the quantity $S_{syst.}$ is the area of the transverse cross-section of the system in the plane X–Z. As a result, using Eqs. (11) and (12) it is possible to calculate, the value of F and Young's modulus E of the system in dependence on the deformation Δx.

- According to Hooke's law the bending moment M applied to the system is proportional to the angular displacements of the particles of the system that is $M = k_{syst.}\Delta\theta$, where the quantity $k_{syst.}$ is the effective bending stiffness of the system as whole. Since the value of angular strain $\Delta\theta$ is related with the curvature radius of the neutral curve $\Delta\theta = L_0/\rho$, we find that $M \sim 1/\rho$. Note, that the work of external loads A for an elastic system stores in the change of its potential energy, i.e., $A = \Delta H$, where

$$A = \frac{M\Delta\theta}{2}. \qquad (13)$$

Considering that $\Delta H = k_{syst.}(\Delta\theta)^2/2$ and using the relations similar to Eq. (9) but for the system as whole, we obtain that:

$$A = \frac{EI_z^{syst.}}{L_0}\frac{(\Delta\theta)^2}{2}. \qquad (14)$$

In Eq. (14), $I_z^{syst.} = b(H+h)^3/12$ is the inertia moment of the cross-section of the system with respect to the axis Z. As a result, using Eqs. (13) and (14), we can calculate the dependencies of the moment M and Young's modulus E of the system on the angular deformation $\Delta\theta$, or on the curvature radius ρ of the neutral line of the system.

(b) *The simulation results*

In Figs. 13, 14 and 15, we present dependencies of the force, bending moment and Young's modulus depending on the type of system deformation (stretching/bending) for different interactions of the chains and temperatures. It is seen, that with the increase of the deformations, the influence of the coating becomes more and more negligible. With the decrease of the parameter K_3 describing interactions between chains (and, therefore, with the decrease of K_2), the influence of the coating disappears faster (Figs. 13 and 14).

In Figs. 13a and b, the regions placed after peaks can be explained by the break of the bonds between neighbour chains of polymer coating and the increase of a role of the substrate during the system deformation. Similar stress-strain dependencies have been observed during the stretching of thin films from.

crystalline polymers such as caoutchouc and gutta-percha [55], for which such behaviour was explained by the structural transformations on material from ordered microstructure to amorphous one. For polymers with polar groups, it is transition from ferroelectric phase to paraelectric one. Further asymptotically linear behaviour of the given curves demonstrates negligible influence of the coating and the dominant role of the substrate which deformation follows Hooke's law.

The bending stiffness of the system consisting of the substrate and the coating (Fig. 12) with different elastic properties is given as a sum of stiffness parameters of the system parts, i.e., it is given by

$$ES_{syst.} = E_{subs.}S_{subs.} + E_{pol.}S_{pol.} \text{ (for the compression/tension);} \qquad (15a)$$

Fig. 13 (**a**) The normalized force Fa/K_1 applied to the system versus the relative strain $\Delta x/L_0$ for the given temperature $T^* = k_BT/K_1 = 0, 1$ and renormalized intrachain interaction parameter ($K_1 = 1$) and interchain one: $K_3 = 0.1$ (*1*), 0.05 (*2*), 0.01 (*3*); force Fa/K_1 versus the strain $\Delta x/L_0$ for the substrate (*4*) (Hooke's law). (**b**) The renormalized bending moment applied to the system M/K_1 versus the angular deformation $\Delta\theta$ for the given renormalized temperature$T^* = k_BT/K_1 = 0, 1$ and normalized intrachain interaction parameter ($K_1 = 1$) and interchain one: $K_3 = 0, 15$ (*1*), 0, 10 (*2*), 0, 05 (*3*); the bending moment M/K_1 versus the angular deformation $\Delta\theta$ for the substrate (*4*) (Hooke's law)

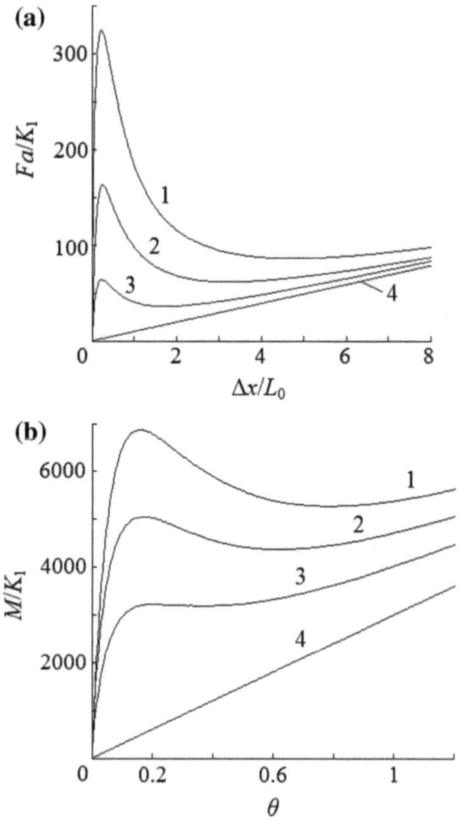

$$EI_z^{syst.} = E_{subs.}I_z^{subs.} + E_{pol.}I_z^{pol.} \text{ (for the bending)}. \quad (15b)$$

In Eqs. (15a, 15b), the quantity $E_{pol.}$ is the elastic modulus, $S_{pol.}$ is the area of the transverse cross-section in the plane X–Z, and $I_z^{pol.} = bh(3H^2 + h^2)/12$ is the inertia moment of the cross-section of the polymer coating with respect the axis Z. With the increase of tension deformations and decrease of interactions between chains the influence of the polymer layer on the stiffness tends to zero, i.e., $E_{pol.} \to 0$. Therefore, from Eqs. (15a, 15b), it follows that the relative Young's modulus of the system "substrate-coating" $E/E_{subs.}$ tends to the limit value which depends on the ratio of the coating and substrate thicknesses h/H:

$$\frac{E}{E_{subs.}} \to \frac{S_{subs.}}{S_{syst.}} = \frac{1}{1 + h/H} \text{ (for the tension)}; \quad (16a)$$

$$\frac{E}{E_{subs.}} \to \frac{I_z^{subs.}}{I_z^{syst.}} = \frac{1 + 3(h/H)^2}{[1 + (h/H)]^3} \text{ (for the bending)}. \quad (16b)$$

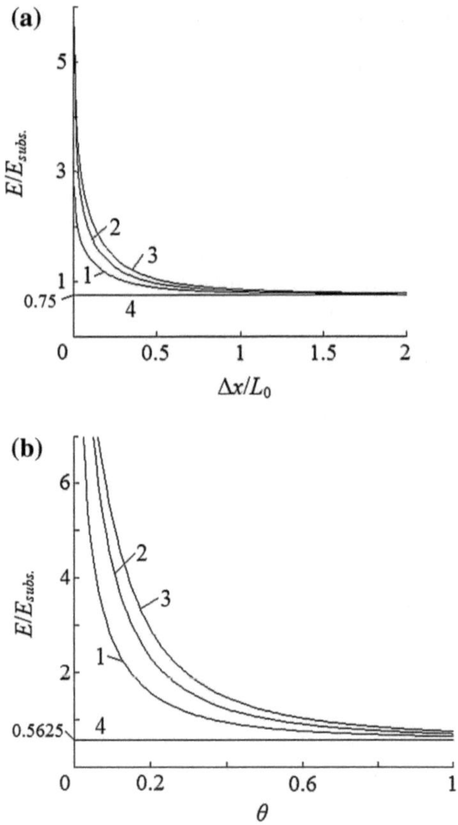

Fig. 14 (*a*) The relative effective Young's modulus of the system $E/E_{subs.}$ versus the relative strain $\Delta x/L_0$ for the given temperature $T^* = k_BT/K_1 = 0, 1$ and renormalized intrachain interaction parameter ($K_1 = 1$) and interchain one: $K_3 = 0.4$ (*1*), 0.7 (*2*), 1 (*3*); the limit value of the relative Young's modulus for large deformations of system (*4*). (*b*) The relative effective Young's modulus of the system $E/E_{subs.}$ versus the angular deformation $\Delta\theta$ for the given temperature $T^* = k_BT/K_1 = 0, 1$ and renormalized intrachain interactions parameter ($K_1 = 1$) and interchain one: $K_3 = 0, 01$ (*1*), 0, 03 (*2*), 0, 05 (*3*); the limit value of the relative Young's modulus for large deformations of the system (*4*)

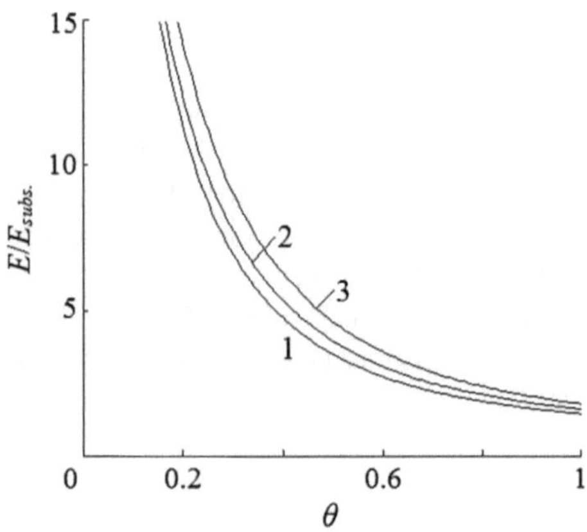

Fig. 15 The relative effective Young's modulus of the system $E/E_{subs.}$ versus the angular deformation $\Delta\theta$ for various renormalized temperature: $T^* = 0$ (*1*), 0, 5 (*2*), 1 (*3*) and renormalized intrachain interaction parameter ($K_1 = 1$) and interchain one ($K_3 = 0, 1$)

Within this simulation, we take that the ratio $h/H = 1/3$, thusthe value $E/E_{subs.} \rightarrow 0, 75$ (for the tension of the system, Fig. 14a), or the value $E/E_{subs.} \rightarrow 0, 5625$ (for the bending of the system, Fig. 14b).

In Fig. 15, the dependencies of the effective Young's modulus of the system on the deformations level for various temperatures are shown. It is seen that with increasing temperature, the system becomes stiffer. This stiffening can be explained through the decrease of its dimensions (L_0) along the axes Y and Z in the equilibrium non-deformable state (Fig. 9) and, therefore, through the increase of interactions between chains.

Varying the length of attached to the substrate chains and their density one can analyse the dependencies of the elastic parameters of the system on its dimensions along one direction or on number of polymer chains per unit surface area.

5 Simulation of Ferroelectric Properties of Thin Films

(a) *structural transitions in thin films*

In the absence of external electric field, ferroelectrics possess a spontaneous electric polarization below the Curie temperature ($T < T_C$). The transition to polarized state is associated with ordering of certain structure elements, which were in disordered state before transition (at $T > T_C$). For the study of the effects of ordering in ferroelectric films, we will use the three-dimensional lattice model, consisting of $N_1N_2N_3$ rigid kinetic units—rotators (Fig. 7a).

The long-range orientational order parameter is calculated as the mean cosine of the angle of rotator orientation relative to the axis of the preferred direction:

$$\mu = \langle \cos(\theta) \rangle. \tag{17}$$

For ferroelectric systems, we assume that the potential energy consists of two components:

$$H_{total} = H + H_{ext}. \tag{18}$$

The first term in Eq. (18) is the potential energy of interactions between rotators (see Eq. 6); the second term describes the presence of external field.

$$H_{ext} = p \cdot E \sum\nolimits_{i,j,k} \cos\left(\varphi_{i,j,k}\right), \tag{19}$$

where p is the module of the vector of the rotator dipole moment, E is the external electric field strength, $\varphi_{i,j,k}$ the angle between the directions of vectors $\vec{E}$ and $\vec{p}$.

Periodical boundary conditions are used, it is assumed that the upper and the lower boundaries of the film (respectively for the lattice planes with $n_1 = 1$ and $n_1 = N_1$) can be freely oriented or fixed.

The calculation results show that the long-range order is absent ($\mu = 0$) in one monolayer, even at low temperatures. However, the long-range order exists on the surface due to bulk effects. Figure 16 shows the dependences of the long-range

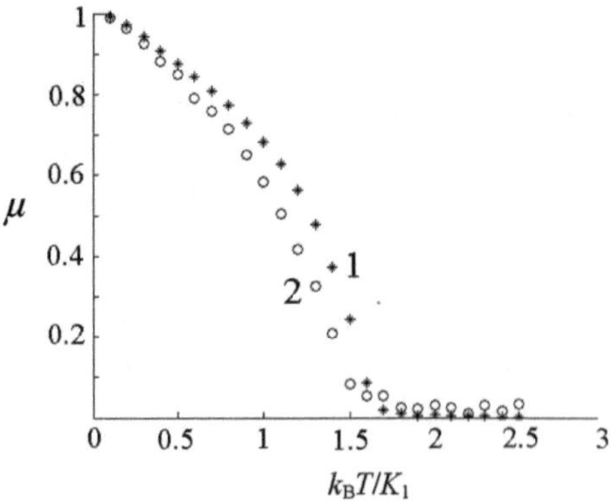

Fig. 16 The long-range orientational order parameter versus the reduced temperature k_BT/K_1 inside the ferroelectric film (*1*) and on the surface at free (2) boundary conditions. $K_1 = K_2 = k_BT$

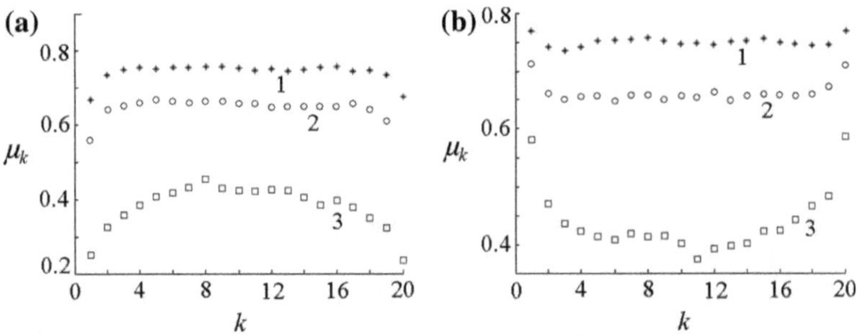

Fig. 17 The order parameter versus the number of layers k in the ferroelectric film for free (**a**) and fixed (**b**) boundary conditions at different reduced temperatures $k_BT/K_1 = 0, 9$ (*1*), 1 (*2*), 1, 4 (*3*). $K_1 = K_2 = K_3$

orientational order on the reduced temperature k_BT/K_1 in the bulk and on the surface with free boundaries. It is evident, that the orientational order on the surface of ferroelectric film is smaller than inside the bulk, and this difference is the greatest nearby the phase transition. The difference between the values of the long-range orientational order in the bulk and the film at free boundaries also depends on its thickness. Thus, the order parameter in thicker films is larger, than in thin films (Fig. 16).

Mutual influence of surface and bulk properties occurs in the surface layer of the film. The authors of the work [56] suggest that the surface layer has a certain thickness. The calculated dependencies of long-range order parameter $\mu_k = \langle \cos(\varphi_{n_1 = k, n_2, n_3}) \rangle$ in different layers at free and fixed boundary conditions are shown in Fig. 17. As can be seen, the effective thickness of the surface layer depends on the temperature. Nearby the phase transition point, its thickness increases to infinity.

Depolarizing fields in ferroelectrics are related both with the induced dynamic polarization and with the existence of spontaneous polarization occurring below the temperature phase transition in ferroelectric state.

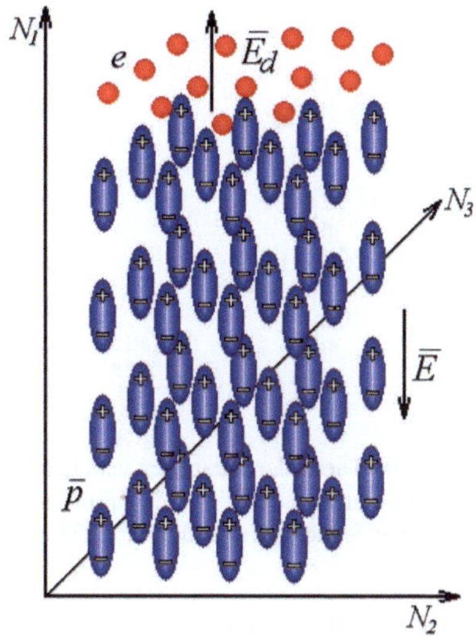

Fig. 18 Three-dimensional lattice model consisting of $N_1N_2N_3$ rigid kinetic units

The polarization of the film results in the existence of the internal electric field $\vec{E}$. Under the influence of this field, the free electrons move to the external boundary of the film ($n_1 = N_1$) and create the depolarizing field E_d (Fig. 18), which value is determined by the equation:

$$E_d = E_{d0}\mu e^{-\lambda(N_1 - n_1)}, \tag{20}$$

where E_{do} and λ are the constants determined by the number of free carriers and μ is the orientational order parameter in ferroelectric film.

By the simulation, it is assumed that free electrons are located on the upper free boundary ($n_1 = N_1$), and create the depolarizing field (Eq. 20). To calculate the order parameter μ_k and the value of the depolarizing field E_d it is necessary to solve the self-consistency problem. The polarization vector at free and periodic boundary conditions can rotate freely without an external field. To eliminate this effect, at the lower boundary of the film ($n_1 = 1$), the fixed boundary conditions are introduced.

Figure 19a shows the dependence of the order parameter on the layer number at different values of the constant E_{dO} in Eq. (20). It is evident that if we choose a sufficiently large value E_{dO}, the areas appear in the film, in which the rotators are oriented oppositely to the direction of the internal electric field $\vec{E}$. Thus, in this work, the value of the constant E_{d0} is chosen so that the polarization is zero at the external boundary. Figure 19b shows the dependence of the order parameter on the layer number at different values of the reduced temperature k_BT/K_1. The effective thickness of the surface layer (where the surface order parameter differs from the order parameter in the bulk), as in the case of the absence of depolarizing field, depends on the temperature.

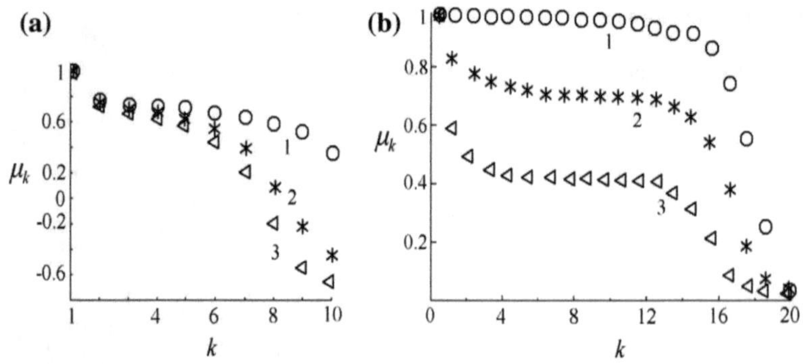

Fig. 19 The order parameter versus the number of layers k in ferroelectric film at the presence of the depolarizing field: **a** at different constants $pE_{d0} = K_1$ (*1*), 2 K_1 (2), 2, 5 K_1 (*3*), $k_BT = K_1$; **b** at different reduced temperatures k_BT/K_1 = 0, 2 (*1*), 1 (2), 1, 5 (*3*)

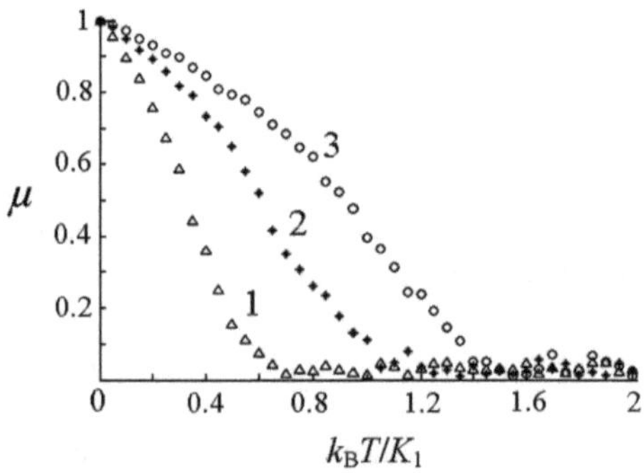

Fig. 20 The long-range dipole order parameter μ versus the reduced temperature k_BT/K_1 at different renormalized field parameter $h = pE/K_1 = 0$ (*1*), 1 (2), 3 (*3*)

(b) *structural transitions in thin films in external electric field*

From the calculated temperature dependences of the long-range order parameter and the susceptibility (Figs. 20 and 21) by the change of the renormalized field parameter $h = E\,p/K_1$ it is evident that the inclusion of the electric field leads to the increase of temperature of transition and to its "smearing", what is consistent with classical representations of the theory of phase transitions [57].

The dielectric susceptibility is calculated by the following formula:

$$\chi = \frac{1}{k_B T}\left[\langle\cos\theta\rangle^2 - \langle\cos^2\theta\rangle\right]. \quad (21)$$

(c) *structural transitions in thin films in alternating external electric field*

The interaction potential (18) with a continuous symmetry group has an infinite number of energy minima, and in general, the potential relief can be represented by a surface of rotation. Therefore, within the framework of three-dimensional model, the repolarization process can take place continuously, without energy costs due to

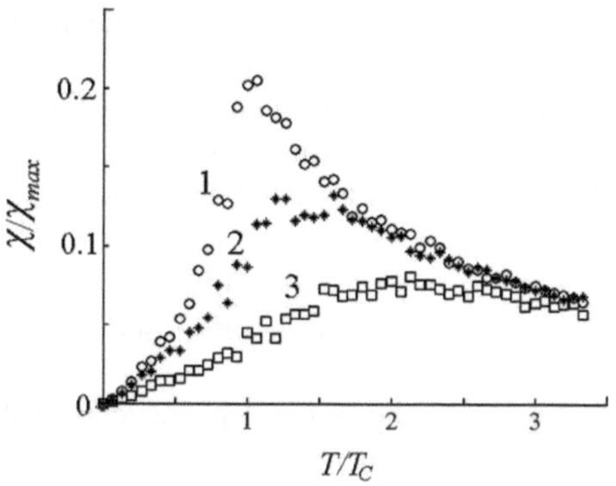

Fig. 21 The susceptibility χ versus the reduced temperature T/T_c at different renormalized external field parameter $h = pE/K_1 = 0$ (*1*), 0, 5 (*2*), 2 (*3*) for systems of rotators with dipole interactions

rotator motions near the potential minimum and without crossing the barrier. So, the dependence of the vector of polarization on electric field strength in an ideal ferroelectric must be reversible (without hysteresis) [58]. However, nevertheless, there is a hysteresis loop in real ferroelectric samples due to relaxation effects at sufficiently rapid change of electric field, when a mutual orientation of rotators in the system does not have time to change due to rotator motions near the potential minimum without crossing the barrier.

At the harmonic time dependence of the field, the module of polarization vector changes as:

$$\mathrm{P(t)} = \varepsilon_0 \chi_0 E_0 \cos(\omega t - \delta), \tag{22}$$

where χ_0 is the susceptibility of the system in constant field, δ is the angle of retard. In the complex form, the expression (22) has the following counterpart:

$$\dot{P}(t) = \varepsilon_0 \chi_0 E_0 \mathrm{e}^{i\omega t} \mathrm{e}^{-i\delta} = \varepsilon_0 \chi_0 E(\cos\delta - i\,\sin\delta). \tag{23}$$

Thus, the inertness of polarization process results in the appearance of real component ($\chi = \chi_0\cos\delta$) and imaginary one ($\chi = \chi_0\sin\delta$) of dielectric susceptibility.

The simplest model of the polarization process of the medium has been proposed by Debye [59], who suggested that interaction forces between particles impede the turn of the dipole moments of the particles of the medium in the direction towards the external electric field. These forces are proportional to the speed of the turn, i.e., the rotation (orientation) of the particles occurs with some "friction". At high enough frequencies, the susceptibility is equal to zero ($\chi = 0$).

The real and imaginary parts of the dielectric susceptibility in Debye model are equal accordingly:

$$\chi' = \frac{\chi_0}{1 + (\omega\tau)^2}, \qquad \chi'' = \frac{\chi_0 \omega\tau}{1 + (\omega\tau)^2}. \tag{24}$$

Another well-known model of the medium is the Drude-Lorentz model, which is based on the assumption that under the influence of external field, charged particles of the medium make damping harmonic oscillations with their own cyclic frequency ω_0 and the constant of damping time τ. In Drude-Lorentz model the real and the imaginary parts of the dielectric susceptibility are equal:

$$\chi' = \frac{\chi_0 \omega_0^2 (\omega_0^2 - \omega^2)}{(\omega_0^2 - \omega^2) + 4\omega^2/\tau^2}, \quad \chi'' = \frac{2\chi_0 \omega \omega_0^2/\tau^2}{(\omega_0^2 - \omega^2) + 4\omega^2/\tau^2}. \tag{25}$$

Figure 22 shows the dependencies of the real part of complex susceptibility on the reduced frequency (in the log- scale) at different temperatures and constants of intermolecular interactions, calculated for 3*d*-model of rotators by Monte-Carlo method. The obtained dependences are similar to the curves obtained for the Debye model: susceptibility increases with the rise of temperature below the phase transition point ($T < T_c$) (Fig. 22a), and, on the contrary, susceptibility decreases above the transition point ($T > T_c$) (Fig. 22b).

The imaginary part of the complex dielectric susceptibility χ'' determines the amount of absorption of energy of the electric field, which is proportional to the

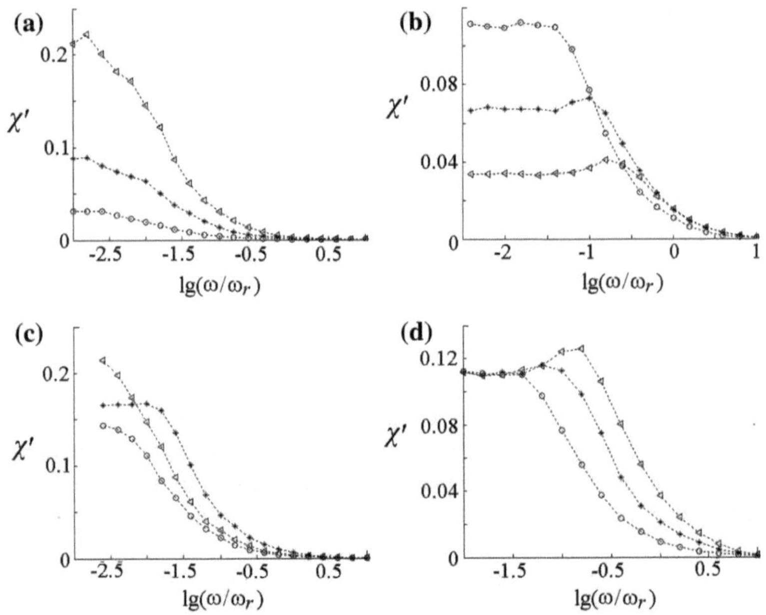

Fig. 22 The real part of complex susceptibility χ, versus the reduced frequency ω/ω_r of the external field (in the log-scale) for systems with isotropic (**a, b, c**), and anisotropic interactions (**d**) at different reduced temperatures k_BT/K_1 in the low-temperature region **a** $k_BT/K_1 = 0.5$ (o), 1 (*), 1.5 (Δ); in the high-temperature region **b** $k_BT/K_1 = 3$ (o), 5 (*), 10 (Δ); in the point of phase transition **c** $k_BT/K_1 = 1.3$ (o), 1.5 (Δ), 2 (*); **d** $k_BT/K_1 = 3$. The constants of orientational interactions $K_2 = K_3 = K_1$ (o), $K_2 = K_3 = 0.5K_1$ (*), $K_2 = K_3 = 0.2K_1$ (Δ)

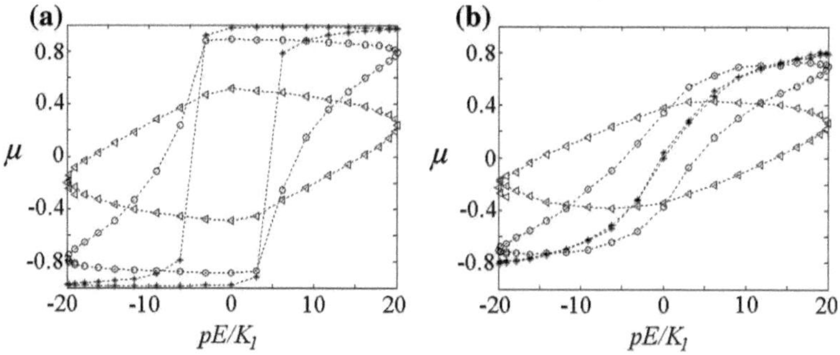

Fig. 23 The hysteresis curves for ferroelectric systems in subcritical area at the reduced temperature $k_BT/K_1 = 0.2$ (**a**) and postcritical area at reduced temperature $k_BT/K_1 = 5$ (**b**) at different frequency of the external electric field ω: $0.1\omega_r$ (*), ω_r (o), $5\omega_r$ (Δ), where ω_r corresponds to the number of steps $N_{MC} = 10^3$

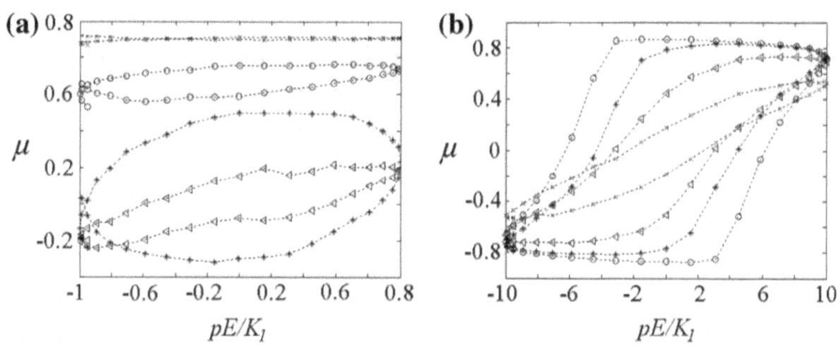

Fig. 24 The hysteresis curves for ferroelectric systems in subcritical area (**a**) and postcritical area (**b**) at different reduced temperatures k_BT/K_1: **a** $pE_0/K_1 = 1$, $k_BT/K_1 = 0.5$ (x), 1 (o), 3 (Δ); **b** $pE_0/K_1 = 10$, $k_BT/K_1 = 0.2$ (o), 1.5 (*), 3 (Δ), 6 (x)

area of the hysteresis curve. Figure 23 shows the change of hysteresis curve shape depending on the external field frequency, so-called dispersion, explained by the fact that the process of polarization of the medium requires a certain finite time.

The complex part of dielectric susceptibility depends on the amplitude of the external field and the reduced temperature k_BT/K_1 (Fig. 24). The sample has its own polarization at low temperatures, and weak external field is not able to destroy it (Fig. 24a). At high values of the external field amplitude (Fig. 24b), the width of the hysteresis curve only decreases with the increase of the reduced temperature k_BT/K_1.

The kind of hysteresis curve depends on the temperature at which the process in question occurs, namely, above or below the phase transition point. At low temperatures (in subcritical area), the system has polarization in the absence of the field. Change of the external field at low frequencies primarily affects the direction of polarization. With the increasing field frequency due to inertness of the process of

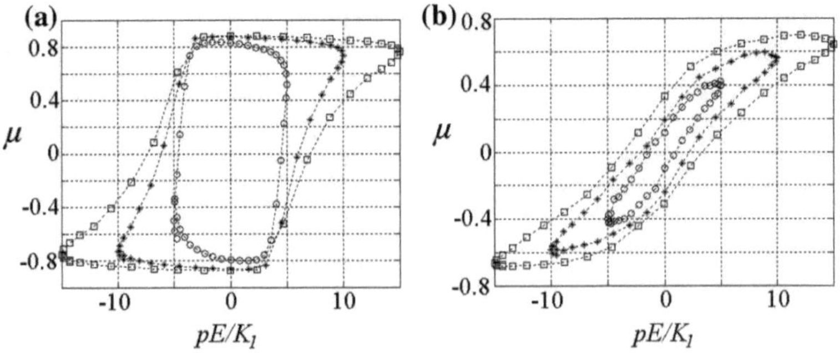

Fig. 25 The hysteresis dependences (pE/K_1) for ferroelectric systems in the subcritical area at the reduced temperature $k_BT/K_1 = 0.2$ (**a**) and the postcritical area at the reduced temperature $k_BT/K_1 = 4$ (**b**) at different reduced amplitudes of the external electric field strength: $pE_{max}/K_1 = 5$ (o), 10 (*), 15 (•)

reorientation of rotators, the maximum value of the order parameter decreases and becomes less than in the absence of the field. At temperatures above the phase transition point (in the postcritical area) and low frequencies, there is a hysteresis loop, since it is necessary a some time to orient the rotators in a particular direction. With increasing the frequency, the hysteresis curve becomes wider, but the maximum value of the order parameter decreases, since at very rapid changes of the field, rotators due to their inertness do not have time to change their orientation and therefore the polarization vector of the medium becomes zero.

Figure 25 shows the dependencies of the long-range orientational order parameter μ for ferroelectric system, calculated for the 3*d*-model of rotators on the electric field strength, alternating according to the harmonic law $E = E_{max}$•$\sin\omega t$ (ω is the cyclic frequency of the field). Form of hysteresis curves obtained depends both on the frequency and amplitude of the external electric field, temperature, and interactions constants. The account of the frequency of the external electric field is as follows: the value ω must be inversely proportional to the number of Monte-Carlo steps N_{MC}. For example, at high frequency of the field, when the system does not have time to restructure, a relatively small value N_{MC} should be taken. As the standard frequency in this work, value ω_r is adopted, corresponding to the number of Monte-Carlo steps $N_{MC} = 10^3$ at the simulation of the system consisting of $N = 8000$ rotators under periodic boundary conditions.

Figure 26 shows the dependences of the areas (Ω) of hysteresis curves $\mu(pE/K_1)$ on the logarithm of the reduced frequency of the external field, obtained for ferroelectric systems at different values of temperature and interaction constants. Therefore, at a certain frequency of the dependence $\chi'\{\lg(\omega/\omega_0)\}$ in critical area: near the phase transition point of the external field (ω_0), the resonance occurs, and a maximum of the imaginary part of the dielectric susceptibility (the area of hysteresis curve) is observed. The position of the maximum is weakly dependent on the

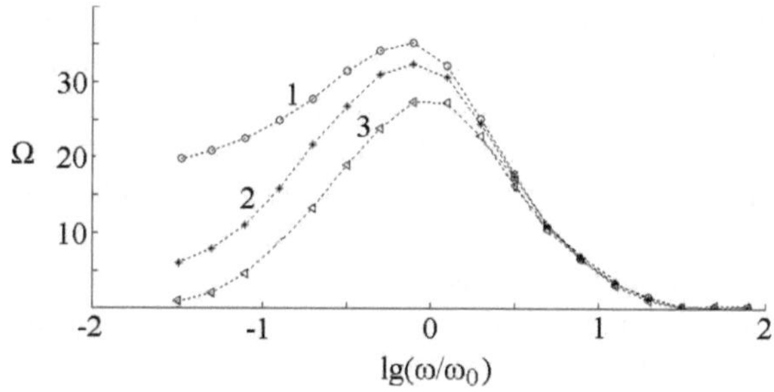

Fig. 26 The area of hysteresis curves Ω versus the logarithm of the reduced frequency of the external field calculated for ferroelectric systems at different reduced temperatures: $k_B T/K_1 = 0.2$ (*1*); 2 (*2*); 5 (*3*)

temperature and interaction constants as in Drude-Lorentz model, and the peak height increases with the decrease of temperature and the increase of the energy of interaction between the molecules.

6 Application to Surface Elasticity

The presented above detailed analysis of mechanical properties of polymer coatings based on the multichain model of polymer has direct relation to the models of surface elasticity. By the term of surface elasticity we mean the enhanced models of the elasticity which take into account surface tension or surface stresses. Among various models of the surface elasticity the most popular is the Gurtin-Murdoch model [60, 61]. From the mechanical point view the Gurtin-Murdoch model describes the deformation of a solid with attached on its surface an elastic membrane. The mechanical properties of the membrane should be introduced explicitly independently on the material behavior in the bulk. The model found various applications in nanomechanics, see [16, 34, 62–67] and the reference therein. In particular, it was shown that the surface enhancements leads to stiffening at the nanoscale, that is to the positive size-effect [68–70]. For example, the apparent Young modulus becomes dependent on the specimen size. Analysis of corresponding boundary-value problems is given in [27, 62–64] where in addition to stiffening it was shown that surface elasticity changes the smoothness of the solutions [5, 23, 25, 26, 71–74]. The surface strain energy density for an isotropic material within the Gurtin-Murdoch model is

$$U = \frac{1}{2}\alpha\epsilon_{nn}^2 + \beta\epsilon_{mn}\epsilon_{mn}, \tag{26}$$

where α and β are elastic moduli called surface Lame parameters, ϵ_{mn} is the surface strains. As in the classic elasticity, instead of α and β the surface engineering moduli can be introduced, such as surface Young's modulus and Poisson's ratio.

The crucial point for the model is the determination of the parameters α and β of the model. The direct measurements of these moduli are difficult, in general, and can be performed for example using three-point bending tests of nanobeams. More efficient calculation is based on the molecular dynamics simulations [75–78] where the crystal lattice dynamics is considered. Mechanics of three-layered plates with surface stresses was used in [67]. The possible scheme of mechanical tests bases on tension and bending as shown in Fig. 12. As result, for example, for β the following formula can be used: $\beta = \mu h$, where μ is the shear modulus in the bulk, and h is the coating thickness [65].

Steigmann and Ogden [79, 80] extended the Gurtin-Murdoch

$$U = \frac{1}{2}\alpha\epsilon_{nn}^2 + \beta\epsilon_{mn}\epsilon_{mn} + \frac{1}{2}\gamma k_{nn}^2 + \tau k_{mn}k_{mn}, \tag{27}$$

where k_{mn} is the tensor of curvature change. In (27) we have introduced new elastic moduli γ and τ related to bending stiffness parameters. As a result, in the Steigmann-Ogden model there more material parameters. It can be shown that the Steigmann and Ogden model leads to more stiffening effects than the Gurtin-Murdoch model [26].

In addition to the Gurtin-Murdoch and Steigmann-Ogden models there are other generalizations of the surface elasticity, considering the changes of the thickness during deformations, orientational interactions, etc., see [65, 81, 82] and reference therein. Let us note that from mechanical point of view the considered here multichain model is quite similar to the continuum models based on Cosserat approach. Indeed, the orientational interactions between dipoles bring us rotations as independent kinematical variable as in the Cosserat continuum. Moreover, the multichain model can be treated as more general model, since its kinematics is described by more kinematic variables, so we come to multipolar models of continuum. Another similarity there is between multichain model and the recently discovered structured based on woven fabrics called pantographic lattices [83–86]. In fact, nearly inextensible fibers in woven fabrics can be modelled as chains as shown in Fig. 7. Thus, the models of polymers used in the polymer science can be also treated as a foundation of models generalized continua that are continua with additional internal degrees of freedom.

Presented here calculations of mechanical properties of polymeric coatings extend the previous calculations [75–78] in order to take into account the real microstructure of polymeric brushes. Similar idea for modelling of coatings with complex inner structure was discussed in [65]. Now, having in hands the mechanical properties of the polymeric coating we can use these parameters for any model of surface elasticity. The simplest case is the Gurtin-Murdoch model for which we can directly use obtained parameters such as Young's modulus.

7 Conclusion

In this work, by means of analytical methods and computer simulation (by Monte-Carlo method), we have investigated the processes of forming of the polymer coating on the solid substrate, namely, on the metal surface, and some mechanical and electrical properties of metamaterials.

The results of the study of the polymer coating formation and the adsorption process by means of computer simulation show that at the increase of monomers interaction, their degree of sticking to metal grows, that is associated with the rise of their cooperativity during the adhesion process. It is proved that under slow heating of metal sheet, the degree of sticking of monomers to its surface is low. At high speed of heating, there is some sticking to the metal surface, but during further heating the sticking coefficient decreases due to the increase of thermal fluctuations in the system. Optimal temperature range is found, in which the coating adhesion is maximal.

Hybrid discrete-continuum model has been developed to study the system consisting of flexible substrate and polymer coating in depending on intermolecular interactions and temperature. It was obtained that at the increase of deformation degree, the polymer coating surface gradually ceases to affect the investigated dependences. When the interchain interaction parameter decreases, the influence of the coating disappears more quickly. The dependencies of the effective Young modulus for the system on the strain at different temperatures shown that with the increase of temperature, the system becomes more rigid.

In this study we also have investigated ferroelectric properties of thin films depending on the frequency and amplitude of the external electric field, temperature and interchain interactions. We found that the presence of the electric field leads to the increase in the transition temperature and to its “smearing”, what is consistent with classical concepts in the theory of phase transitions. The form of obtained hysteresis curves depend on both the frequency and amplitude of the external electric field, and on temperature and of interaction constants. The analysis of hysteresis curves shows that the polarization of the medium requires a finite time.

Thus, the results of theoretical study and computer simulation of characteristics of metamaterials obtained in this study, allow to come to better understanding of their surface properties and can lead to identification of new important and interesting properties of such materials.

Acknowledgements The work is performed within the framework of the project “Methods of microstructural nonlinear analysis, wave dynamics and mechanics of composites for research and design of modern metamaterials and elements of structures made on its base” (grant №15-19-10008 of by the Russian Science Foundation).

References

1. Bhushan, B., Jung, Y.: Natural and biomimetic artificial surfaces for superhydrophobicity, self-cleaning, low adhesion, and drag reduction. Prog. Mater Sci. **56**, 1–108 (2011)
2. Bico, J., Thiele, U., Quere, D.: Wetting of textured surfaces. Colloid Surf. A Physicochem. Eng. Asp. **206**(1–3), 41–46 (2002)
3. Boinovich, L.B., Emelyanenko, A.M.: Russ. Chem. Rev. **77**(7), 583 (2008)
4. Callies, M., Quere, D.: On water repellency. Soft Matter **1**(1), 55–61 (2005)
5. Kushch, V.I., Chernobai, V.S., Mishuris, G.S.: Longitudinal shear of a composite with elliptic nanofibers: local stresses and effective stiffness. Int. J. Eng. Sci. **84**, 79–94 (2014)
6. Zheng, L., Wu, X., Lou, Z., et al.: Superhydrophobicity from microstructured surface. Chin. Sci. Bull. **49**, 1779–1787 (2004)
7. Lisichkina G.V.: Chemistry of grafted of the surface compounds. FIZMATLIT, Moscow (2003)
8. Shirtcliffe, N., McHale, G., Atherton, S., et al.: An introduction to superhydrophobicity. Adv. Colloid Interface Sci. **161**, 124–138 (2010)
9. Zhang, X., Shi, F., Niu, J., et al.: Superhydrophobic surfaces: from structural control to functional application. J. Mater. Chem. **18**, 621–633 (2008)
10. Ageev, A.A., Aksenova, I.V., Volkov, V.A., Eleev, A.F.: About fluorine-containing surface-active substances capable to modify polymer fabrics' fibers. Fluorine Notes **4**(83), 5–6 (2012)
11. Rigoberto, C.A., et al.: Polymer Brushes: Synthesis, Characterization, Applications. WILEY-VCH Verlag GmbH & Co. KGaA, Weinheim (2004)
12. Birstein, T.M.: The polymer brushes. Soros Educ. J. **5**, 42–47 (1999)
13. Mamonova, M.V., Prudnikov, V.V., Prudnikova, V.V., Prudnikova, I.A.: The theoretical and experimental methods in surface physics, Omsk (2009)
14. Oura, K., Lifschic, V.G., Sarznin, A.A., Zotov, A.V., Katayama, M.: Introduction to Surface Physics, Moscow, Nauka, p. 496 (2006)
15. Huang, G.Y., Yu, S.W.: Effect of surface piezoelectricity on the electromechanical behavior of a piezoelectric ring. Phys. Status Solid. B **243**(4), 22–24 (2006)
16. Wang, G.F., Feng, X.Q.: Effect of surface stresses on the vibration and buckling of piezoelectric nanowires. EPL **91**, 56007 (2010)
17. Yan, Z., Jiang, L.Y.: Surface effects on the electromechanical coupling and bending behaviors of piezoelectric nanowires. J. Phys. D Appl. Phys. **44**, 075404 (2011)
18. Yan, Z., Jiang, L.Y.: Vibration and buckling analysis of a piezoelectric nanoplate considering surface effects and in-plane constraints. Proc. R. Soc. A **468**, 3458–3475 (2012)
19. Zhang, J., Wang, Ch., Adhikari, S.: Surface effect on the buckling of piezoelectric nanofilms. J. Phys. D Appl. Phys. **45**, 285301 (2012)
20. Pan, X.H., Yu, S.W., Feng, X.Q.: A continuum theory of surface piezoelectricity for nanodielectrics. Sci. China: Phys., Mech. Astron. **54**(4), 564–573 (2011)
21. Yan, Z., Jiang, L.: Electromechanical response of a curved piezoelectric nanobeam with the consideration of surface effects. J. Phys. D Appl. Phys. **44**, 365301 (2011)
22. Yan, Z., Jiang, L.: Surface effects on the electroelastic responses of a thin piezoelectric plate with nanoscale thickness. J. Phys. D Appl. Phys. **45**, 255401 (2012)
23. Altenbach, H., Eremeyev, V.A., Lebedev, L.P.: On the existence of solution in the linear elasticity with surface stresses. ZAMM **90**(3), 231–240 (2010)
24. Altenbach, H., Eremeyev, V.A., Lebedev, L.P.: On the spectrum and stiffness of an elastic body with surface stresses. ZAMM. **91**(9), 699–710 (2011)
25. Eremeyev, V.A., Lebedev, L.P.: Existence of weak solutions in elasticity. Math. Mech. Solids **18**(2), 204–217 (2013)
26. Eremeyev, V.A., Lebedev, L.P.: Mathematical study of boundary-value problems within the framework of Steigmann-Ogden model of surface elasticity. Continuum Mech. Thermodyn. **28**(1–2), 407–422 (2016)

27. Schiavone, P., Ru, C.Q.: Solvability of boundary value problems in a theory of plane–strain elasticity with boundary reinforcement. Int. J. Eng. Sci. **47**, 1331–1338 (2009)
28. Nasedkin, A.V., Eremeyev, V.A.: Spectral properties of piezoelectric bodies with surface effects. In: Surface Effects in Solid Mechanics, pp. 105–121. Springer (2013)
29. Nasedkin, A.V., Eremeyev, V.A.: Harmonic vibrations of nanosized piezoelectric bodies with surface effects. ZAMM **94**(10), 878–892 (2014)
30. Nasedkin, A.V., Eremeyev, V.A.: Some models for nanosized magnetoelectric bodies with surface effects. In: Advanced Materials, pp. 373–391. Springer (2016)
31. Binder, K.: Monte-Carlo Methods in Statistical Physics. Mir, Moscow (1982)
32. Binder, K., Heermann, D.W.: Monte-Carlo Simulation in Statistical Physics, 5th edn. Springer Heidelberg, Dordrecht, London, NY (2010)
33. Multer-Krumbhaar, H., Burkhardt, T.W., Kroll, D.: A generalized kinetic equation for crystal growth. J. Cryst. Growth. **38** (1977)
34. Wang, J., Huang, Z., Duan, H., Yu, S., Feng, X., Wang, G., Zhang, W., Wang, T.: Surface stress effect in mechanics of nanostructured materials. Acta Mech. Solida Sin. **24**, 52–82 (2011)
35. Bogdanova, YuG: The Adhesion and its Role in the Strength of Polymer Composites. Moscow University Press, Moscow (2010)
36. Pratton, M.: Introduction to Surface Physics, p. 256. Izhevsk, NIC RHD (2000)
37. Roldugin, V.I.: Physical Chemistry of Surface, p. 586. Dolgoprudnyy, Intellect (2008)
38. Vaz, C.A.F., Bland, J.A.C., Lauhoff, G.: Magnetism in ultrathin film structures. Rep. Prog. Phys. **71** (2008)
39. Lipatov, Y.S.: The Colloidal Polymer Physics, p. 344. Kiev, Nauka, Dumka (1984)
40. Berlin, A.A., Basin, V.E.: Fundamentals of Polymer Adhesion, p. 319. AM, Chemistry (1969)
41. Proutorov, E.V., Maksimova, O.G., Maksimov, A.V.: Simulation of the adhesion contact on the interface of polymer-metal. J. Phys: Conf. Ser. **633**, 012044 (2015)
42. Allen, M.P., Tildesley, D.J.: Computer Simulation of Liquids. Clarendon, Oxford (1987)
43. Binder, K., Levek, D., Weiss, J., et al.: Monte-Carlo Methods in Statistical Physics. Springer, Berlin (1979)
44. Metropolis, N., Rosenbluth, A.W., Rosenbluth, M.N., Teller, A.H., Teller, E.: Equations of state calculations by fast computing machines. J. Chem. Phys. **21**, 1087–1091 (1953)
45. Yakovlev, A.D.: Chemicals and Coatings Technology, p. 448. SPb., HIMIZDAT (2008)
46. Gerasimov, R.A., Maksimov, A.V., Petrova, T.O., Maksimova, O.G.: Ordering and the relaxation properties of macromolecules in ferroelectric polymer films. Phys. Solid State **54** (5), 1002–1004 (2012)
47. Gerasimov, R.A., Eremeyev, V.A., Petrova, T.O., Egorov, V.I., Maksimova, O.G., Maksimov, A.V.: Computer simulation of the mechanical properties of metamaterials. J. Phys: Conf. Ser. **738**(1), 012100 (2016)
48. Gotlib, Y.Y., Maximov, A.V.: A theory of orientational ordering in two-dimemsional multichain polymer systems with dipole interaction. Polym. Sci. A **34**(11), 902–907 (1992)
49. Maksimov, A.V., Pavlov, G.M.: The molecular orientational order in surface layers of polymer films. Polym. Sci. A **49**(7), 828–836 (2007)
50. Maksimov, A.V., Gerasimov, R.A.: Anisotropic models of polymer ferroelectrics. Phys. Solid State **51**(7), 1365–1369 (2009)
51. Maksimov, A.V., Gerasimov, R.A., Maksimova, O.G.: Ordering and large-scale relaxation properties of macromolecules in ferroelectric polymer films. Ferroelectrics **432**(1), 32–40 (2012)
52. Petrova, T.O., Maksimova, O.G., Gerasimov, R.A., Maksimov, A.V.: Application of analytical and numerical methods to simulation of systems with orientation interactions. Phys. Solid State **54**(5), 937–939 (2012)
53. Maksimova, O.G., Maksimov, A.V.: Orientational order in two-dimensional polymer systems as described in terms of the Vaks-Larkin model. Polym. Sci. A **45**(9), 1476–1485 (2003)

54. Bakingem, E.: Basic theory of intermolecular forces. Application to the small molecules. In: Pullman, B. (ed.) Intermolecular interactions: from diatomic molecules to biopolymers, pp. 9–98. Moscow, Mir (1981)
55. Askadskii, A.A.: Deformation of Polymers. Khimiya, Moscow (1973)
56. Cui Lian, XuXu, JiXin, Che, Zelong, He, Huijie, Xue, Tianquan, Lv: Properties of phase transformation of ferroelectric thin films with surface layers. J. Mod. Phys. **2**, 1037–1040 (2011)
57. Landau, L.D.: Theoretical Physics. Nauka. VIII, Moscow (1982)
58. Sonin, A.S.: Introduction to the Physics of Liquid Crystals. Nauka, Moscow (1983)
59. Landau, L.D., Lifschic, E.M.: Statistical Physics. Nauka. V, Moscow (1976)
60. Gurtin, M.E., Murdoch, A.I.: A continuum theory of elastic material surfaces. Arch. Ration. Mech. Anal. **57**, 291–323 (1975)
61. Gurtin, M.E., Murdoch, A.I.: Surface stress in solids. Int. J. Solid. Struct. **14**(6), 431–440 (1978)
62. Duan, H.L., Wang, J., Huang, Z.P., Karihaloo, B.L.: Size-dependent effective elastic constants of solids containing nanoinhomogeneities with interface stress. J. Mech. Phys. Solids **53**, 1574–1596 (2005)
63. Duan, H.L., Wang, J., Karihaloo, B.L.: Theory of elasticity at the nanoscale. In: Aref, H., Van der Giessen, E. (eds.) Advances in Applied Mechanics, vol. 42, pp. 1–68. Elsevier, Amsterdam (2008)
64. Duan, H.L., Wang, J., Karihaloo, B.L., Huang, Z.P.: Nanoporous materials can be made stiffer than non-porous counterpartsby surface modification. Acta Mater. **54**, 2983–2990 (2006)
65. Eremeyev, V.A.: On effective properties of materials at the nano- and microscalesconsidering surface effects. Acta Mech. **227**, 29–42 (2016)
66. Javili, A., McBride, A., Steinmann, P.: Thermomechanics of solids with lower-dimensional energetics: on the importanceof surface, interface, and curve structures at the nanoscale. A unifying review. Appl. Mech. Rev. **65**, 010802-1–31 (2012)
67. Zhu, H.X., Wang, J.X., Karihaloo, B.L.: Effects of surface and initial stresses on the bending stiffness of trilayer plates and nanofilms. J. Mech. Mater. Struct. **4**, 589–604 (2009)
68. Chen, C., Shi, Y., Zhang, Y., Zhu, J., Yan, Y.: Size dependence of Young's modulus in ZnO nanowires. Phys. Rev. Lett. **96**, 075505 (2006)
69. Cuenot, S., Frétigny, C., Demoustier-Champagne, S., Nysten, B.: Surface tension effect on the mechanical properties of nanomaterials measured by atomic force microscopy. Phys. Rev. B **69**, 165410 (2004)
70. Jing, G.Y., Duan, H.L., Sun, X.M., Zhang, Z.S., Xu, J., Li, Y.D., Wang, J.X., Yu, D.P.: Surface effects on elastic properties of silver nanowires: contact atomic-force microscopy. Phys. Rev. B **73**, 235409–6 (2006)
71. Kim, C., Ru, C., Schiavone, P.: A clarification of the role of crack-tip conditions in linear elasticity with surface effects. Math. Mech. Solids **18**, 59–66 (2013)
72. Kim, C.I., Schiavone, P., Ru, C.Q.: Effect of surface elasticity on an interface crack in plane deformations. Proc. R. Soc. A **467**, 3530–3549 (2011)
73. Mishuris, G.S.: Interface crack and nonideal interface concept (Mode III). Int. J. Fract. **107**, 279–296 (2001)
74. Mishuris, G.S., Kuhn, G.: Asymptotic behaviour of the elastic solution near the tip of a crack situated at a nonidealinterface. ZAMM **81**, 811–826 (2001)
75. Arroyo, M., Belytschko, T.: An atomistic-based finite deformation membrane for single layer crystalline films. J. Mech. Phys. Solids **50**, 1941–1977 (2002)
76. Miller, R.E., Shenoy, V.B.: Size-dependent elastic properties of nanosized structural elements. Nanotechnology **11**, 139 (2000)
77. Sfyris, D., Sfyris, G., Galiotis, C.: Curvature dependent surface energy for a free standing monolayer graphene: someclosed form solutions of the non-linear theory. Int. J. Non-Linear Mech. **67**, 186–197 (2014)
78. Shenoy, V.B.: Atomistic calculations of elastic properties of metallic fcc crystal surfaces. Phys. Rev. B **71**, 094104 (2005)

79. Steigmann, D.J., Ogden, R.W.: Plane deformations of elastic solids with intrinsic boundary elasticity. Proc. R. Soc. A **453**, 853–877 (1997)
80. Steigmann, D.J., Ogden, R.W.: Elastic surface–substrate interactions. Proc. R. Soc. A **455**, 437–474 (1999)
81. Povstenko, Y.: Mathematical modeling of phenomena caused by surface stresses in solids. In: Altenbach, H., Morozov, N.F. (eds.) Surface Effects in Solid Mechanics, pp. 135–153. Springer, Berlin (2013)
82. Rubin, M., Benveniste, Y.: A Cosserat shell model for interphases in elastic media. J. Mech. Phys. Solids **52**, 1023–1052 (2004)
83. dell'Isola, F., Steigmann, D., Della Corte, A.: Synthesis of fibrous complex structures: designing microstructure to deliver targeted macroscale response. Appl. Mech. Rev. **67**(6), 060804-060804-21 (2016)
84. dell'Isola, F., Giorgio, I., Pawlikowski, M., Rizzi, N.L.: Large deformations of planar extensible beams and pantographic lattices: heuristic homogenization, experimental and numerical examples of equilibrium. Proc. Roy. Soc. A **472**(2185) (2016). doi:10.1098/rspa.2015.0790
85. Scerrato, D., Giorgio, I., Rizzi, N.L.: Three-dimensional instabilities of pantographic sheets with parabolic lattices: numerical investigations. Z. Angew. Math. Phys. **67**(3), 1–19 (2016)
86. Scerrato, D., Zhurba Eremeeva, I. A., Lekszycki, T., and Rizzi, N. L.: On the effect of shear stiffness on the plane deformation of linear second gradient pantographic sheets, Z. Angew. Math. Mech. (2016). doi:10.1002/zamm.201600066
87. Leonova, T.M., Kastro, R.A.: The dielectric properties of the MIS structures based on aluminum oxide. Materials of the XII-th International Conference "Physics of Dielectrics", St. Petersburg (2011)

Identification of Arrays of Cracks in the Elastic Medium by the Ultrasonic Scanning

M.A. Sumbatyan and M. Brigante

Abstract In the present paper we study the problem on image identification for clusters of linear cracks located inside an unbounded elastic medium, by using a circular Ultrasonic echo-method. The parameters to be reconstructed are the number of cracks, their size, location and the slope of each defect. The scanning is performed by an ultrasonic transducer of a fixed frequency placed at a certain distance in a far-zone, which can generate an ultrasonic wave incident to the system of cracks at arbitrary angle. The input data, used for the reconstruction algorithm, is taken as the back-scattered amplitudes measured in the echo method for the full circular interval of the scanning angle. The diffraction of the elastic waves is studied in the scalar approximation. The proposed numerical algorithm is tested on some examples with clusters of cracks whose position and geometry are known a priori.

Keywords Crack identification · Array of cracks · Ultrasonic circular scanning · Inverse problem · Optimization · Boundary Integral Equation

1 Introduction

Recently, many published works have been devoted to inverse identification problems. This interest is caused by the importance of such investigations to many practical applications, in particular—in the Ultrasonic (US) Nondestructive Testing (NDT). Chronologically, first recognition methods were based upon an approximation of weak wave interaction, namely upon the theories of Ramm [1]. The well

M.A. Sumbatyan (✉)
Institute of Mathematics, Mechanics and Computer Science, Southern Federal University, Milchakova Street 8a, Rostov-on-Don 344090, Russia
e-mail: masumbatyan@sfedu.ru

M. Brigante
Department of Structures for Engineering and Architecture, University of Naples - Federico II, Via Claudio 21, 84125 Naples, Italy
e-mail: brigante@unina.it

M.A. Sumbatyan (ed.), *Wave Dynamics and Composite Mechanics for Microstructured Materials and Metamaterials*,
Advanced Structured Materials 59, DOI 10.1007/978-981-10-3797-9_4

developed methods of the acoustical tomography based on the Radon transformation are also related to the class of theories using the hypothesis of weak scattering [2].

Unfortunately, the defects in solids cannot be considered as weak scatterers. That is why many recent methods of identification are based on the strict diffraction theories aimed at both direct and inverse diffraction problems [3–5].

In the ambit of the NDT the acoustic and electromagnetic methods are based on wave properties of damaged materials and their difference from analogous properties of perfect (undamaged) bodies. The problem of identification of the near-surface texture and the defects arising in masonry structures is being successfully studied by using both acoustic and electromagnetic waves [6–8]. In particular, the US methods [9] demonstrate high efficiency in defect identification. Other acoustic methods are based on the measurement of the wave field over some parts of the sample's boundary, when the latter is loaded by a certain time-harmonic outer force. The theoretical results show that in many cases the shape of the boundary uniquely determines the geometry of the internal defects [10–12]. A contiguous method is founded upon measurements of the natural mode frequencies of the sample. This permits the reconstruction on the basis of alternative ideas, taking into account that the geometry of the internal defects influences significantly the measurable set of the natural frequencies [13].

It should be noted that from the mathematical point of view the image identification problem is a typical inverse problem. This is intensively developing nowadays [3, 4, 14]. This theory is connected with the so-called ill-posed problems [15], which generate instability when one applies standard direct numerical methods to construct the solution. The present authors have published a series of works on the image identification of defects and other related problems of the US scanning in various theoretical and practical aspects [16–19]. The US methods demonstrate high efficiency even in the case of multiple scattering [20, 21].

The main goal of the present work is to develop an efficient algorithm to solve the identification problem for a set of linear cracks located in the unbounded elastic medium. The investigation is performed under the assumption of scalar US wave propagation [9]. This means that in the case of solid medium, where both longitudinal and transverse types of waves may propagate through the medium, we assume that the transverse wave is predominant in the considered anti-plane problem, so that the longitudinal one can be neglected. We first reduce the wave problem in the damaged medium to a Boundary Integral Equation (BIE), which by using the standard "displacement discontinuity method" [22] further reduces the problem to a system of linear algebraic equations (SLAE). Then we formulate the inverse reconstruction problem. We assume that an US transducer is placed in a far zone to work in the echo-scanning regime. It is also assumed that this sensor can record the back-scattered US signal in the whole angular interval $\alpha \in (0, 2\pi)$, with the measured amplitude of the reflected signal forming the data base for the formulated inverse identification problem. We do not pay much attention to such fundamental questions as existence and uniqueness. Instead we develop an efficient algorithm for practical recognition. We demonstrate some examples on cracks' identification problem, for various dimension of the cluster. In the identification algorithm we do not assume that the number of cracks is known a priori, and cluster's dimension is an additional unknown parameter, among other unknown quantities, to be determined.

It should also be noted that the analogous problem for a cluster of linear cracks, located inside a rectangular elastic specimen, has recently been studied by the present authors in [23].

2 Basic Equations of the Elasticity Theory for Anti-plane Mode of Deformation

Let a finite array of linear cracks be located in a linear isotropic homogeneous elastic medium, inside a certain bounded domain, l_n $(n = 1, \dots, N_c)$ denotes the surface of n-th crack, and $l = \bigcup_{n=1}^{N_c} l_m$ denotes the full set of cracks. Let us choose the origin of the Cartesian coordinate system somewhere in this domain, close to the cluster of cracks. In order to identify the geometry of each crack in this array, we apply a circular scanning by an US sensor which works in the echo-regime, see Fig. 1. The current position of the sensor is given by the Cartesian coordinates $(R\cos\alpha, R\sin\alpha)$, and it is assumed that the amplitude of the back-scattered impulse is known for the full angular scanning: $\alpha \in (0, 2\pi)$.

Let us cite the governing equations to the formulated problem. In the case of the so-called "anti-plane" (or "shear-stress", SH) problem, this stress is directed perpendicularly to the considered plane (x, y), so that the mode of deformation is identical for all cross-sections $z = \text{const}$ [24]. Then in a fixed rectangular Cartesian coordinate system $Oxyz$ the components of the displacement vector are $\bar{u} = \{0,\ 0,\ u_z(x, y)\}$, this determines the two non-trivial components of the stress tensor as follows:

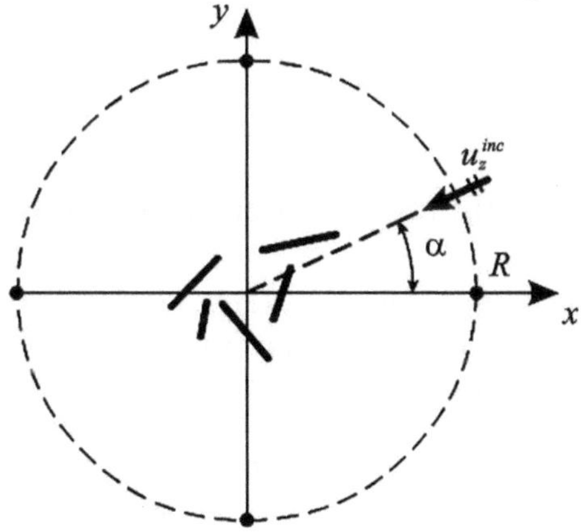

Fig. 1 The circular scanning of the array of linear cracks by an US transducer in the echo-regime

$$\tau_{xz}(x,y) = \mu \frac{\partial u_z(x,y)}{\partial x}, \qquad \tau_{yz}(x,y) = \mu \frac{\partial u_z(x,y)}{\partial y}, \tag{1}$$

where μ is the shear elastic modulus. Under such conditions the equations of motion reduce to the single scalar Helmholtz equation regarding function $u_z(x,y)$ [9, 24]:

$$\frac{\partial^2 u_z(x,y)}{\partial x^2} + \frac{\partial^2 u_z(x,y)}{\partial y^2} + k^2 u_z(x,y) = 0, \qquad k = \frac{\omega}{c_s}, \tag{2}$$

where k is the wave number related to the transverse wave speed, the time-dependent factor $e^{-i\omega t}$ is omitted in all formulas, and c_s is the transverse wave speed.

The internal stress vector $\bar{T} = \{T_x, T_y, T_z\}$ over arbitrary elemental area with the normal $\bar{n} = \{n_x, n_y, 0\}$ in the considered case of anti-plane deformation has the only non-trivial component:

$$T_x = T_y = 0, \qquad T_z = \tau_{xz} n_x + \tau_{yz} n_y = \mu \left(\frac{\partial u_z}{\partial x} n_x + \frac{\partial u_z}{\partial y} n_y \right) = \mu \frac{\partial u_z}{\partial n}. \tag{3}$$

Then the boundary conditions over the set of cracks l located in the medium should be satisfied over the faces of all defects which are free of load. Let us represent the full wave field as a sum of the incident and the scattered ones:

$$\begin{aligned} u_z &= u_z^{inc} + u_z^{sc}(x,y), \quad u_z^{inc}(x,y) = e^{-ik(x\cos\alpha + y\sin\alpha)}, \quad w(x,y) = u_z^{sc}(x,y), \\ T_z &= T_z^{inc} + T_z^{sc}, \qquad T_z^{inc} = -ik\mu(n_x \cos\alpha + n_y \sin\alpha)\, e^{-ik(x\cos\alpha + y\sin\alpha)}, \end{aligned} \tag{4}$$

where α is the angle of incidence.

It is obvious that the boundary condition for the scattered wave field implies the cracks' faces to be loaded by the tangential stress:

$$T_z\big|_l = 0, \sim T_z^{sc}\Big|_l = -\, T_z^{inc}\Big|_l, \implies \frac{\partial u_z}{\partial n}\bigg|_l = 0, \sim \frac{\partial w}{\partial n}\bigg|_l = -\frac{T_z^{inc}}{\mu}\bigg|_l. \tag{5}$$

3 Mutual Influence of the Elementary Cracks

In the forthcoming investigation of the array of cracks we follow the so-called "displacement discontinuity method" [22]. This method is allied to the standard BEMs, see for example [25]. Let us begin with a single linear elementary crack Γ of small length ε located horizontally in the chosen Cartesian coordinate system (x,y), see Fig. 2. If the length of the elementary crack is small, then the stress T_z may be set constant over such a short interval $(-\varepsilon/2, \varepsilon/2)$.

By using the well known Green's function for the Helmholtz equation in the two-dimensional space, $\Phi(\xi,\eta,x,y) = (i/4)H_0^{(1)}(kr)$, $r = \left[(\xi - x)^2 + (\eta - y)^2\right]^{1/2}$, one

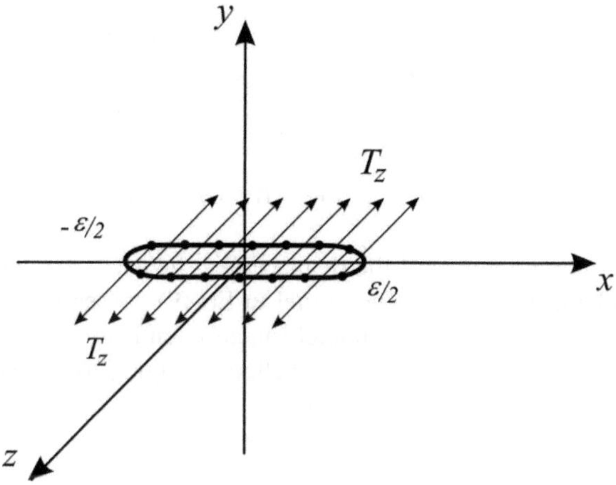

Fig. 2 The elementary crack of the length ε under the anti-plane stress field

can write out the basic integral representation of the classical potential theory [25]:

$$w(x, y) = \int_{\Gamma} \left[u_z(\xi, \eta) \frac{\partial \Phi(\xi, \eta, x, y)}{\partial n} - \Phi(\xi, \eta, x, y) \frac{\partial u_z(\xi, \eta)}{\partial n} \right] dl, \tag{6}$$

where both the outer unit normal vector $\mathbf{n}(\xi, \eta)$ and the elementary arc of length $dl(\xi, \eta)$ are connected with the point (ξ, η), not (x, y). Besides, we imply that contour Γ represents itself a smooth closed contour surrounding the elementary crack under consideration. Note that the basic integral representation (6) is valid for a smooth closed surrounding contour, but the latter can be arranged arbitrarily close to the elementary linear crack [25]. It should also be noted that the second integral in (6) is equal to zero, due to boundary condition (5), therefore

$$w(x, y) = \int_{\Gamma} u_z(\xi, \eta) \frac{\partial \Phi(\xi, \eta, x, y)}{\partial n(\xi, \eta)} dl(\xi, \eta), \tag{7}$$

Now, by bringing nearer the surrounding smooth contour Γ close to the faces of the crack under consideration, it can easily be seen that with this limit the following relations hold:

$$\left[\frac{\partial \Phi(\xi, \eta, x, y)}{\partial n(\xi, \eta)} \right]_+ = \left. \frac{\partial \Phi}{\partial \eta} \right|_{\eta=0}, \quad \left[\frac{\partial \Phi(\xi, \eta, x, y)}{\partial n(\xi, \eta)} \right]_- = - \left. \frac{\partial \Phi}{\partial \eta} \right|_{\eta=0}, \tag{8}$$

where the subscript $+$ is related to the upper side $y = +0$ of the elementary crack, and the subscript $-$ is related to the lower side $y = -0$ of the crack, see Fig. 2. As a result, the displacement field at arbitrary point (x, y) caused by the elementary displacement discontinuity g, see Fig. 2, can directly be extracted from (7), in the following form:

$$w(x,y) = \int_{-\varepsilon/2}^{\varepsilon/2} g(\xi) \left.\frac{\partial \Phi}{\partial \eta}\right|_{\eta=0} d\xi = \frac{iky g}{4} \int_{-\varepsilon/2}^{\varepsilon/2} \frac{H_1^{(1)}\left[k\sqrt{(\xi-x)^2+y^2}\right]}{\sqrt{(\xi-x)^2+y^2}} d\xi, \quad (9)$$

where

$$g(\xi) = u_z(\xi,+0) - u_z(\xi,-0) = w(\xi,+0) - w(\xi,-0) \quad (10)$$

is the difference of the displacement between the upper and the lower faces of the elementary crack. The kernel in Eq. (9) possesses a hyper-singular behavior when the argument of the Hankel function tends to zero [5]. By extracting explicitly the singularity of the Hankel function at origin [26], one can develop the more efficient representation for the kernel:

$$w(x,y) = \frac{iy g}{4} \int_{-\varepsilon/2}^{\varepsilon/2} K(\xi - x, y)\, d\xi, \quad K(x,y) = \frac{kH_1^{(1)}\left(k\sqrt{x^2+y^2}\right)}{\sqrt{x^2+y^2}} =$$

$$= -\frac{2i}{\pi(x^2+y^2)} + Q(x,y), \quad Q(x,y) = \frac{kH_1^{(1)}\left(k\sqrt{x^2+y^2}\right)}{\sqrt{x^2+y^2}} + \frac{2i}{\pi(x^2+y^2)}. \quad (11)$$

Now, since $Q(x,y)$ is a regular function when compared with $K(x,y)$, the integration in Eq. (11) can be performed explicitly:

$$w(x,y) = \frac{ig}{4}\left\{\frac{\varepsilon y}{2}\left[Q\left(\frac{\varepsilon}{2} - x, y\right) + Q\left(\frac{\varepsilon}{2} + x, y\right)\right] + \right.$$
$$\left. + \frac{2i}{\pi}\left(\arctan\frac{\varepsilon/2 - x}{y} + \arctan\frac{\varepsilon/2 + x}{y}\right)\right\}. \quad (12)$$

The integration of the hyper-singular, first function in Eq. (11), has been performed here by using a tabulated integral, while the integration of the regular function Q is performed accepting that this function is almost constant on the short interval of length ε. Besides, instead of to take the value of respective integrand at the central point $\xi = 0$ multiplied by ε, we apply an arithmetic average of the integrand at the end-points. Such a treatment can guarantee that the arising expressions have no singular values which may occur in the case $x^2 + y^2 = 0$.

Expressions (11) and (12) permit calculation of the components of the stress tensor, on the basis of Eq. (1):

$$\frac{\tau_{xz}^{sc}(x,y)}{\mu} = \frac{\partial w(x,y)}{\partial x} = \frac{iy g}{4} \int_{-\varepsilon/2}^{\varepsilon/2} \frac{\partial}{\partial x} K(\xi - x, y)\, d\xi =$$

$$= -\frac{iy g}{4} \int_{-\varepsilon/2}^{\varepsilon/2} \frac{\partial}{\partial \xi} K(\xi - x, y)\, d\xi = -\frac{iy g}{4}\left[K\left(\frac{\varepsilon}{2} - x, y\right) - K\left(-\frac{\varepsilon}{2} - x, y\right)\right],$$

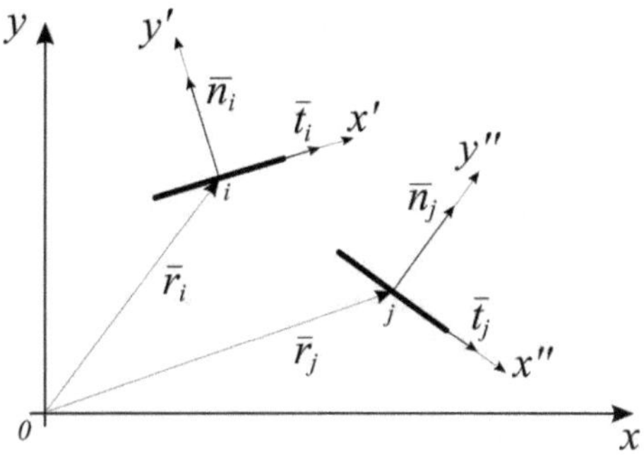

Fig. 3 Mutual influence of i-th and j-th elementary cracks

$$\frac{\tau_{yz}^{sc}(x,y)}{\mu} = \frac{\partial w(x,y)}{\partial y} = \frac{i\,g}{4}\left\{\frac{\varepsilon}{2}\left[Q_y\left(\frac{\varepsilon}{2}-x,y\right)+Q_y\left(\frac{\varepsilon}{2}+x,y\right)\right]+\right.$$

$$\left.+\frac{2i}{\pi}\left[\frac{x-\varepsilon/2}{(x-\varepsilon/2)^2+y^2}+\frac{x+\varepsilon/2}{(x+\varepsilon/2)^2+y^2}\right]\right\},\; Q_y(x,y)=\frac{\partial}{\partial y}\left[y\,Q(x,y)\right]=$$

$$=k\left[\frac{ky^2}{r^2}H_0^{(1)}(kr)+\frac{x^2-y^2}{r^3}H_1^{(1)}(kr)\right]+\frac{2i(x^2-y^2)}{\pi r^4},\quad r=\sqrt{x^2+y^2}. \quad (13)$$

Let us consider the pair of elementary displacement discontinuities which are described, respectively, by the data $\{x_i, y_i, \bar{n}_i, \bar{t}_i\}$ and $\{x_j, y_j, \bar{n}_j, \bar{t}_j\}$ in a fixed Cartesian coordinate system (x, y). Here the first two quantities designate coordinates of the central point, and the quantities $\bar{t}$, $\bar{n}$ are directed along the elementary crack and in direction to its normal, see Fig. 3. Let us study the contribution of j-th element to the tangential stress at i-th element. For this aim we couple with i-th crack the local Cartesian axes (x', y'), and with j-th point the local axes (x'', y''). It is obvious from Fig. 3 that if the radius-vector of point i in the coordinate system (x'', y'') is denoted by $\bar{r}_i'' = \{x_i'', y_i''\}$, then

$$\bar{r}_i'' = \bar{r}_j - \bar{r}_i, \quad x_i'' = \bar{r}_i'' \cdot \bar{t}_j = (\bar{r}_j - \bar{r}_i)\cdot \bar{t}_j, \quad y_i'' = \bar{r}_i'' \cdot \bar{n}_j = (\bar{r}_j - \bar{r}_i)\cdot \bar{n}_j. \quad (14)$$

Therefore, in the coordinate system (x'', y'') one obtains, see Eq. (13):

$$\frac{\tau_{xz}^{''sc}(x_i,y_i)}{\mu} = -\frac{i\,y_i''\,g_j}{4}\left[K\left(\frac{\varepsilon}{2}-x_i'',y_i''\right)-K\left(-\frac{\varepsilon}{2}-x_i'',y_i''\right)\right],$$

$$\frac{\tau_{yz}^{''sc}(x_i,y_i)}{\mu} = \frac{i\,g_j}{4}\left\langle\frac{\varepsilon}{2}Q_y\left(\frac{\varepsilon}{2}-x_i'',y_i''\right)+Q_y\left(\frac{\varepsilon}{2}+x_i'',y_i''\right)+\right.$$

$$\left.+\frac{2i}{\pi}\left[\frac{x_i''-\varepsilon/2}{(x_i''-\varepsilon/2)^2+(y_i'')^2}+\frac{x_i''+\varepsilon/2}{(x_i''+\varepsilon/2)^2+(y_i'')^2}\right]\right\rangle. \quad (15)$$

Then in the coordinate system (x'', y'') the stress vector on i-th elementary area is

$$T_z^{sc}(x_i, y_i) = \tau''^{sc}_{xz} n''_{ix} + \tau''^{sc}_{yz} n''_{iy} = \tau''^{sc}_{xz}\, \bar{n}_i \cdot \bar{t}_j + \tau''^{sc}_{yz}\, \bar{n}_i \cdot \bar{n}_j, \tag{16}$$

where quantities τ''^{sc}_{xz}, τ''^{sc}_{yz} are defined by expressions (15). This form is very close to a certain one containing only the quantities expressed in the basic coordinate system (x, y). The remaining local variables in Eq. (15) can obviously be determined from Eq. (14) as follows:

$$x''_i = (x_j - x_i)t_{jx} + (y_j - y_i)t_{jy}, \qquad y''_i = (x_j - x_i)n_{jx} + (y_j - y_i)n_{jy}. \tag{17}$$

Expressions (15)–(17) determine the influence of the displacement discontinuity g_j located at j-th point to the shear stress $T_z^{sc}(x_i, y_i)$ acting on the elementary area at i-th point, and all quantities are now defined in the basic coordinate system. If the total number of elements is I, then the complete contribution of all elements to $T_z^{sc}(x_i, y_i)$ is a superposition of elementary expressions taken in the form (16). This can be written symbolically as follows:

$$\frac{T_z^{sc}(x_i, y_i)}{\mu} = \sum_{j=1}^{I} K_{ij}\, g_j, \qquad i = 1, \dots, I. \tag{18}$$

with some known quantities K_{ij}. By taking into account boundary condition (5), with quantity T_z^{inc} defined in (4), the problem is finally reduced to the following system of linear algebraic equations (SLAE):

$$\sum_{j=1}^{I} K_{ij}\, g_j = i\, k(n_{ix} \cos\alpha + n_{iy} \sin\alpha)\, e^{-i\, k(x_i \cos\alpha + y_i \sin\alpha)}, \qquad i = 1, 2, \dots, I. \tag{19}$$

where the unit vector $\{-cos\alpha, -sin\alpha\}$ defines the direction of the incidence.

Once this system is numerically solved, the value of the scattered stress field at arbitrary point (x, y) inside the medium can directly be calculated by formula coinciding with (18), applied at such interior point (x, y) instead of the element-location point (x_i, y_i). We thus can conclude that the solution of the direct diffraction problem for the array of linear cracks is reduced to the SLAE (19), with further summation in Eq. (18). The imaginary unit in the argument of the exponential function cannot be confused with the subscript i in the arising elements of vectors and matrices.

It should be noted that every elementary crack considered in the present section may be a small part of a certain crack from the full array of cracks under diffraction. If n-th elementary crack from the system of N_c real cracks at hand is divided to J_n small elementary cracks, then it is obvious that the total number of elementary cracks is $I = \sum_{n=1}^{N_c} J_n$.

4 Identification Problem and Some Details of the Numerical Algorithm

As indicated in the final part of the previous section, when integral equation (19) is solved, i.e. all quantities g_j, $(j = 1, \ldots, I)$ are defined, the acoustic wave field at any point in the medium can be calculated from Eq. (9), as a superposition of all particular contributions with $g = g_j$. Let us calculate the diffracted wave field in the far zone. Operating in the local coordinate system connected with the i-th elementary crack, as shown in Fig. 2, let us estimate the quantity $\sqrt{(\xi - x)^2 + y^2}$ in the far zone, i.e. when $R = \sqrt{x^2 + y^2} \to \infty$:

$$\sqrt{(\xi - x)^2 + y^2} = \sqrt{x^2 + y^2 - 2x\xi + \xi^2} = R\sqrt{1 - \frac{2x\xi}{R^2} + \frac{\xi^2}{R^2}} \sim$$

$$\sim R\left[1 + \frac{1}{2}\left(-\frac{2x\xi}{R^2} + \frac{\xi^2}{R^2}\right) - \frac{1}{8}\left(-\frac{2x\xi}{R^2} + \frac{\xi^2}{R^2}\right)^2\right] \sim$$

$$\sim R - \frac{x\xi}{R} + \frac{\xi^2}{2R} - \frac{x^2\xi^2}{2R^3} \sim R - \xi\cos\beta, \quad (x = R\cos\beta,\ y = R\sin\beta), \qquad (20)$$

if rejecting infinitesimal terms.

After that, by applying the far-field asymptotic of the Hankel function, one obtains from (9):

$$w(x, y) = w(R, \beta) = \frac{i k y g_i}{4} \int_{-\varepsilon/2}^{\varepsilon/2} \frac{H_1^{(1)}\left[k\sqrt{(\xi - x)^2 + y^2}\right]}{\sqrt{(\xi - x)^2 + y^2}} d\xi \sim A y g_i \times$$

$$\times \int_{-\varepsilon/2}^{\varepsilon/2} \frac{\exp\left[k\sqrt{(\xi - x)^2 + y^2}\right]}{\left[(\xi - x)^2 + y^2\right]^{3/4}} d\xi \sim B y g_i \int_{-\varepsilon/2}^{\varepsilon/2} \frac{\exp[i k(R - \xi\cos\beta)]}{R^{3/2}} d\xi =$$

$$= \frac{B g_i \sin\beta}{\sqrt{R}} e^{ikR} \int_{-\varepsilon/2}^{\varepsilon/2} e^{-ik\xi\cos\beta} d\xi = D g_i \tan\beta \cdot \sin(k\cos\beta \cdot \varepsilon/2), \quad (R \to \infty), (21)$$

where the quantities B and D are certain constants if the wave number k and the distance R are fixed. Since we apply the circular US scanning with a certain large fixed value of R at a fixed scanning frequency of the transducer in the echo-method, hence in our case the quantity B in the final representation in (21) is constant indeed.

Expression (21) is written in the coordinate system coupled with the i-th elementary crack. Let us pass to the main Cartesian coordinate system. It is obvious that angle β in Eq. (21) is the angle between the unit tangential vector $\bar{t}_i$ and the direction of the incidence. Therefore,

$$\cos\beta = t_{ix}\cos\alpha + t_{iy}\sin\alpha, \tag{22}$$

which should be substituted into Eq. (21). After that the full amplitude of the back-scattered far-field wave can be calculated as a sum of all elementary contributions given by Eq. (21):

$$A(\alpha) = \left| D\sum_{i=1}^{I} g_i \tan[\arccos(t_{ix}\cos\alpha + t_{iy}\sin\alpha)]\sin[k(t_{ix}\cos\alpha + t_{iy}\sin\alpha)\varepsilon/2] \right|. \tag{23}$$

Further we give the details of a reconstruction algorithm, which reduces the identification problem to an optimization problem for a certain strongly nonlinear objective functional. A special stochastic numerical technique is applied to solve this optimization problem.

If the geometry of the array of linear cracks is known then, in the discretization, the length of the elementary cracks depends upon the wave length connected with the frequency of oscillations, as well as upon the length of the current crack. In any way, one should take at least ten nodes along the wave length. In order to be more specific, let us take into account the fact well known in the US detection: it is not realistic to detect cracks whose length is smaller than the wave length λ. We thus assume, when developing our algorithm, that the length of all cracks under detection is greater than the wave length (it is admitted that the two quantities may be equal too). Starting from this assumption, we accept each elementary crack to be of the length which is less than $\lambda/10$. This guarantees that in the worst case, when the length of the crack is equal to λ, the minimum number of grid nodes over the current crack from the system at hand is 10. Once again, the total number of elementary cracks is $I = \sum_{n=1}^{N_c} J_n$.

Let the system under identification consist of array of linear cracks l_n, $n = 1, \ldots, N_c$. Each crack is defined by its central point with its Cartesian coordinates (a_n, b_n), by its length ζ_n, and by the angle of slope θ_n ($|\theta_n| \leq \pi/2$) with respect to x-axis. If n-th crack contains J_n elementary cracks of the length $h_n = \zeta_n/J_n$, then the full set of elementary cracks is a union of ones when running over the given system of linear cracks:

$$\begin{cases} x_i = a_n + [h_n(q-0.5) - \zeta_n/2]\cos\theta_n, \\ y_i = b_n + [h_n(q-0.5) - \zeta_n/2]\sin\theta_n, \end{cases} \quad (q = 1, \ldots, J_n;\ n = 1, \ldots, N_c). \tag{24}$$

It should be noted that the grid nodes over any given linear crack are distributed so that they are situated just at the half-way point between the end-points of a current elementary segment. Such an arrangement provides a symmetry so desired for any algorithm.

Coming to the formulated identification problem, let us estimate the total number of the unknown parameters to be reconstructed. If the number of linear cracks is N_c then for each of them one has four unknown parameters: $a_n, b_n, \zeta_n, \theta_n$, $(n = 1, \dots, N_c)$. Therefore, the total number of the unknown parameters, to be reconstructed, is $4N_c$.

In order to find all these unknowns, we construct an objective functional, and reduce the reconstruction to an optimization problem for this functional. For this purpose let us represent system (19) in the operator form:

$$\mathbf{K} g = f, \ \mathbf{K} = \{K_{ij}\}, \ g = \{g_i\}, \ f = \{f_i\}, \quad (i,j = 1, \dots, I), \tag{25}$$

whose solution can be expressed in terms of the inverse matrix, as follows

$$g = \mathbf{K}^{-1} f, \quad \Longrightarrow \quad g_i = \left(\mathbf{K}^{-1} f\right)_i . \tag{26}$$

Obviously, the right-hand side in Eq. (26) depends upon all $4N_c$ unknown parameters, as well as upon angle of incidence α: $\mathbf{K}^{-1} f = (\mathbf{K}^{-1} f)(a_n, b_n, \zeta_n, \theta_n, \alpha)$.

Now, let us pass to the question what is the measured information which can be used as input data for the inverse identification problem. We assume that the scanning US sensor can measure the amplitude of the echo-impulse at a fixed distance R, in a far zone, for M positions of the sensor corresponding to the values of the irradiation angle $\alpha = \alpha_m$, $(m = 1, \dots, M)$. For the sake of brevity, we assume that the values α_m are uniformly distributed over the full circular interval $(0, 2\pi)$. The registered values of the back-scattered amplitude form the array of the input data F_m, $(m = 1, \dots, M)$. In our numerical experiments the quantities F_m may be taken as the amplitude $A(\alpha)$ calculated from the solution to respective direct problem: $F_m = A(\alpha_m)$, see Eq. (23). In order to simulate the more realistic noisy input experimental data, the calculated quantities F_m are artificially subjected to some noisy perturbations which form the noisy input data F_m^*. Then, by substituting (26) into (23), one comes to the system of nonlinear equations for parameters $a_n, b_n, \zeta_n, \theta_n$ written in the following discrete form:

$$\left| D \sum_{i=1}^{I} (\mathbf{K}^{-1} f)(a_n, b_n, \zeta_n, \theta_n, \alpha_m) \tan[\arccos(t_{ix} \cos\alpha_m + t_{iy} \sin\alpha_m)] \times \right.$$
$$\left. \times \sin\left[\frac{k(t_{ix}\cos\alpha_m + t_{iy}\sin\alpha_m)\varepsilon}{2}\right] \right| = F_m^*, \qquad (m = 1, \dots, M). \tag{27}$$

It is interesting to compare the number of the unknown parameters and the number of equations. As indicated above, the first quantity is $4N_c$. Obviously, the second one is equal to the number of sensor positions, M. Subject to which quantity is greater among these two ones, system (27) may be underdetermined, overdetermined, or well determined. In every case it is unclear how one can solve this system directly. Intuitively, one could suppose the more sensor measurement points the higher precision of the reconstruction. The method proposed here works well for any input data. It is thus indifferent to which case from the three ones described above takes place indeed and what is the real number of the sensor points. Obviously, the number of the sensor points may affect the precision of the reconstruction. However, the technique itself is universal with respect to this number.

The system of equations (27) can be resolved by a minimization of the discrepancy functional:

$$\min[\Omega(a_n, b_n, \zeta_n, \theta_n)], \qquad \Omega(a_n, b_n, \zeta_n, \theta_n) =$$

$$= \left\| \left| D \sum_{i=1}^{I} (\mathbf{K}^{-1} f)(a_n, b_n, \zeta_n, \theta_n, \alpha_m) \tan[\arccos(t_{ix} \cos\alpha_m + t_{iy} \sin\alpha_m)] \times \right.\right.$$

$$\left.\left. \times \sin\left[\frac{k(t_{ix}\cos\alpha_m + t_{iy}\sin\alpha_m)\varepsilon}{2}\right] \right| - F_m^* \right\|^2 =$$

$$= \sum_{m=1}^{M} \left\{ \left| D \sum_{i=1}^{I} (\mathbf{K}^{-1} f)(a_n, b_n, \zeta_n, \theta_n, \alpha_m) \tan[\arccos(t_{ix} \cos\alpha_m + t_{iy} \sin\alpha_m)] \times \right.\right.$$

$$\left.\left. \times \sin\left[\frac{k(t_{ix}\cos\alpha_m + t_{iy}\sin\alpha_m)\varepsilon}{2}\right] \right| - F_m^* \right\}^2 . \quad (28)$$

It should be noted that in the case of absolute precision of the input data the true geometry of the cracks cluster returns zero minimum value to functional Ω. However, there arises the problem of uniqueness since we cannot prove that only true geometry of the cracks makes this functional trivial.

The minimization of functional (28) can be attained by any classical method of optimization [27, 28]. However, the main restriction of regular iterative schemes is that they give a local minimum of respective functional only. Another difficulty is connected with a non-uniqueness of the solution, the question already discussed above but to be concerned again in a different aspect. Namely, it is not evident that a local minimum is at the same time the global minimum of the functional. In fact, it is well known that for nonlinear equations such values of local minima may be too far from the desired value $\Omega = 0$.

For this reason, we used in our numerical experiments a version of the method of global random search [29] contiguous to the one described in detail in [30]. This algorithm is developed to seek maxima, but it can be applied to minima too. It is constructed so that it moves both up and downhill and as the optimization process proceeds, it focuses on the most promising area. As a first step, it randomly chooses a trial point within the step of the user selected starting point. The function is evaluated at this trial point and its value is compared to its value at the initial point. In the minimization problem, all downhill moves are accepted and the algorithm continues from that trial point. The relationship between the initial value of F and the resulting step length is function dependent. This algorithm shows perfect convergence for many problems, also for our inverse identification problem, but unfortunately sometimes it arrives at a local extremum, instead of the global one, in the cases when there are a lot of global minima of the objective function.

The algorithm applied is a slight modification of this idea. It possesses the two following specific features: (1) random sampling of values in the neighborhood of the points, for which the values of the functional are smaller, happens more frequently than in the neighborhood of worse points, and (2) the domains, in which random values of variables are chosen, are gradually contracted to the small neighborhoods of the points with smaller values of the functional. This technique demonstrates remarkable convergence for all considered examples.

Let us test the proposed method on some examples of the simultaneous identification of several cracks. If one applies US probes with the cyclic frequency $f = 1$ mHz and the wave speed in the medium is 6 km/s, then the wave length is $\lambda = 6 \times 10^3/1 \times 10^6 = 6 \times 10^{-3}$ m = 6 mm.

For all examples listed below the input data for the reconstruction is taken from the solution of respective direct problem, with some stochastic perturbations of the so obtained data, in order to model the input data with a certain error. In our simulation we always used $M = 72$ points of measurements, uniformly distributed over the total circular interval of the incident polar angle $\alpha_m \in (0, 2\pi)$, $m = 1, \dots, M$, with the step $360°/72 = 5°$. For all examples demonstrated below the maximum possible number of cracks in the cluster is taken $N_c = 4$. The real number of cracks is the additional, fifth parameter to be identified. It should be noted that the intersection of cracks is also permissible.

Let us estimate the efficiency of the proposed algorithm, in the case related to typical examples considered below and applied with 500 iterations in the described stochastic search, with 30 trials on each iteration step, for every of N_c combinations of the number of cracks. The algorithm thus uses $30 \times 500 \times 4 = 60{,}000$ trials of the objective functional Ω that takes near one hour of calculations when implemented on PC with AMD Athlon Core2 processor of 6.0 HGz CPU clock (recall that each trial requires solution of a certain SLAE of dimension $I \times I$, see Eq. (19)). If anyone applies a direct random search, without any acceleration technique, then one should calculate functional Ω depending upon desired scale in the precision of the identification of all geometric parameters of the cracks. Let with a rough treatment one seek the system of cracks in a 40×40 mm square domain, of dimension $(-20, 20) \times (-20, 20)$ mm, as shown in Fig. 4. Then with the scale step 1 mm one should take 40 values for each of the two parameters a and b. Let then the maximum crack length be 20 mm and the minimum one 6 mm (coinciding with the wave length), with the same scale 1 mm, and at least 10 possible values for the slope angle θ. Then for each crack one should apply $40 \times 40 \times 15 \times 10 = 240{,}000$ trials. In the total, one should apply $(240{,}000)^{N_c}$ trials, where $N_c = 4$ is the maximum possible number of cracks. The reader can easily understand that the calculation time for such “direct” numerical experiments exceeds the capabilities of any existing computer.

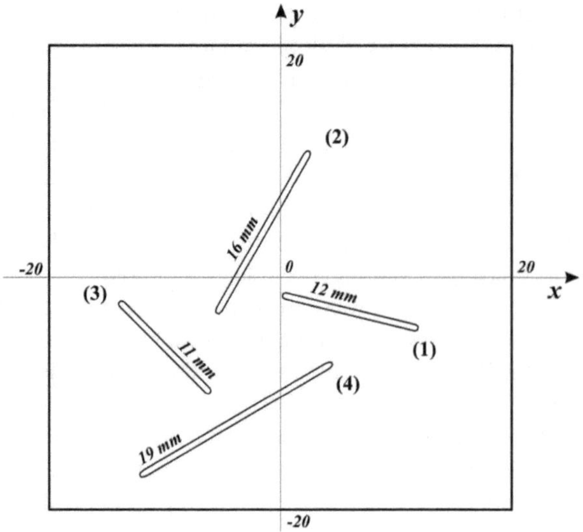

Fig. 4 The array of four cracks to be identified in the $(-20, 20) \times (-20, 20)$ mm quadratic domain

Table 1 Identification of the single crack (1), see Fig. 4, noise level 10%

Configuration	a (mm)	b (mm)	ζ (mm)	θ
Exact	6.000	−3.000	12.000	$-\pi/9 = -0.349$
One crack	**6.292**	**−2.179**	**12.063**	**−0.344**
Two cracks	7.184	−1.441	11.985	−0.344
	19.101	−16.204	19.049	−0.768
Three cracks	5.685	−3.798	12.108	−0.344
	15.497	11.536	12.200	0.303
	−2.863	−12.732	9.715	1.170
Four cracks	6.630	−2.995	11.989	−0.344
	8.342	8.657	18.003	−1.401
	−16.966	−11.689	14.631	0.832
	14.160	17.624	6.851	−0.249

It is obvious that the proposed algorithm, in the particular case when the admitted maximum number of cracks is $N_c = 1$, can be used for any single crack reconstruction. Some examples on the reconstruction of a single crack and arrays of cracks are presented in Tables 1, 2, 3 and 4. All sizes are given in *mm*. The configuration of the system of cracks is shown in Fig. 4.

For multiple cracks array the algorithm works so that this selects the most likely geometry, sequentially for $N_c = 1, 2, 3, 4$ number of cracks. Then the algorithm makes a choice on the best value of functional Ω, among these 4 geometries, to come to the true solution, including the recognition of the true number of cracks by itself.

Table 2 Identification of the two cracks (1) and (2), see Fig. 4, noise level 0%

Configuration	a (mm)	b (mm)	ζ (mm)	θ
Exact	6.000	−3.000	12.000	$-\pi/9 = -0.349$
	−1.500	4.000	16.000	$\pi/3 = 1.047$
One crack	6.872	6.404	15.952	1.048
Two cracks	**11.363**	**0.367**	**12.065**	**−0.351**
	3.981	**7.435**	**16.033**	**1.047**
Three cracks	16.228	1.016	11.987	−0.357
	5.661	13.488	15.804	1.050
	−17.602	−0.689	10.338	1.353
Four cracks	17.085	−9.321	10.858	−0.349
	−5.497	−2.791	16.021	1.050
	12.574	−14.159	10.004	0.447
	4.851	12.337	18.869	0.085

Table 3 Identification of the three cracks (1)–(3), see Fig. 4, noise level 10%

Configuration	a (mm)	b (mm)	ζ (mm)	θ
Exact	6.000	−3.000	12.000	$-\pi/9 = -0.349$
	−1.500	4.000	16.000	$\pi/3 = 1.047$
	−10.000	−6.000	11.000	$-\pi/4 = -0.785$
One crack	15.973	15.093	16.176	1.050
Two cracks	−1.743	−0.588	12.917	−0.526
	−8.586	−2.220	16.895	1.017
Three cracks	**5.302**	**−9.656**	**12.020**	**−0.370**
	−5.303	**−8.662**	**17.214**	**1.050**
	−14.496	**−17.069**	**12.818**	**−0.788**
Four cracks	−11.336	−6.188	13.504	−0.374
	−16.966	−11.689	14.631	0.832
	−4.815	−11.485	10.936	−0.798
	14.160	17.624	6.851	−0.249

5 Conclusions

It is clear from the tables that, for any configuration of N_c cracks to be identified, the algorithm tries to find an appropriate system of 1,2,3,4 cracks sequentially. Physically, such a problem is solvable only for the true number of cracks N_c, the cells which are marked in the tables by the bold font. Mathematically, this means that the minimum value of the functional Ω close to 0 can be attained only if N_c is true. Of course, for arbitrary number of cracks chosen, the algorithm approaches to a certain

Table 4 Identification of the four cracks (1)–(4), see Fig. 4, noise level 0%

Configuration	a (mm)	b (mm)	ζ (mm)	θ
Exact	6.000	−3.000	12.000	$-\pi/9 = -0.349$
	−1.500	4.000	16.000	$\pi/3 = 1.047$
	−10.000	−6.000	11.000	$-\pi/4 = -0.785$
	−4.000	−12.200	19.000	$\pi/6 = 0.524$
One crack	−7.592	−6.829	9.331	0.766
Two cracks	−8.465	10.833	18.015	1.112
	−5.137	12.156	19.529	0.560
Three cracks	4.627	0.840	8.123	−0.356
	0.755	−12.263	17.862	0.991
	9.958	−4.747	16.079	0.498
Four cracks	**−10.940**	**19.407**	**10.587**	**−0.337**
	12.259	**0.785**	**16.449**	**1.048**
	−4.169	**13.786**	**9.401**	**−0.816**
	−6.472	**−6.611**	**19.298**	**0.521**

minimum value of the functional. However, our calculations show that the attained minimum for the true value of N_c is at least by one order smaller than for other three cases.

From the presented tables as well as from other numerous calculations performed the following conclusions can be extracted:

1. The main conclusion is that the position of the cracks forming the array is never identified correctly. This is easily explained by the fact that the far-field scattered wave field, expression (23), is free of coordinates x_i, y_i, being dependent upon the angle of the incidence only. It is obvious that one cannot hope to identify the real position of a reflector if only the information from the far-field input data is used.
2. The number of cracks N_c for the particular configuration, as a rule, is reconstructed correctly. Below it is discussed that this question is resolved successfully not always; this strongly depends on the geometry of the configuration, see point 7 of the Conclusions.
3. Two geometrical parameters, the length of the crack ζ and the slope angle θ, can typically be identified quite precisely. Physically, this can be explained by the fact that for a short-wave diffraction in the echo-regime the reflected amplitude diagram has a sharp maximum just when the incidence is orthogonal to the faces of the crack. This predetermines a sufficiently precise identification of the angles of slope for all cracks from the array, except those cases described in point 7, see below.
4. In the reconstruction of single crack the behavior of the convergence is typically monotonic, in the sense that with the increasing number of iterations the two parameters ζ and θ become closer to their exact values, and so is the value of the

functional Ω. Recall that exact values of the parameters in our test examples are known a priori since they are constructed from respective direct problems.

5. If we construct the input data from a respective direct problem, without any random perturbation, then it is obvious that theoretically the value of the discrepancy functional is zero for the exact solution. In practice, when implementing the proposed numerical algorithm, the attained minimal value is small but not trivial. However, the behavior of Ω versus number of iteration steps is typically monotonic, even in the case of multiple cracks.
6. The convergence of the reconstructed parameters versus iteration number is typically not monotonic. This means that after more iteration steps some parameters may approach closer to their exact values, however some other parameters sometimes may be slightly more distant from respective exact values. This is quite natural in multidimensional optimization, since a smaller value of the discrepancy functional does not always mean closer values for all variables to their exact quantities.
7. One more interesting conclusion is that the identified arrays of cracks, found with an incorrect number of cracks, contain, as a rule, some correct cracks. This can be explained by the fact that the circular scanning cannot reconstruct correctly those cracks which are hidden in a "shadow" of some other long ones. In fact, imagine a pair of cracks parallel and close to each other, like those located on the two opposite sides of a narrow rectangle. It is very hard to believe that the circular scanning can recognize this pair of cracks separately, not as only a single crack. Such "hidden" configuration can be imagined for arbitrary number of cracks, for example every time when a system of large cracks obstructs a short crack; then the latter is practically invisible with the US scanning. In such cases the number of cracks in the array are identified with error, even if some cracks from the array are identified correctly. The typical examples are given by the case "Three cracks" in Table 1, by the case "Three cracks" in Table 2, by the case "One crack" in Table 3, and many others.
8. Typically, the error in the identification of the inclination angle θ does not exceeds few percents in the case of exact input data, and does not exceed the error level of 10% in the case of the input data subjected to the noisy perturbation. The error in the identification of cracks' length ζ is typically from few percents for the single crack to the value of 10–15% for multiple cracks. In all cases, the error of the identification grows with the increasing of the number of cracks.
9. To improve the precision of the identification, we should resign the fixed frequency hypothesis and the time-harmonic regime of oscillations. In fact, by using the well known "time-of-flight" information, one may hope to improve the precision of the reconstruction operating with the real transient US impulse. Apparently, the precision of the identification can also be improved by using several carrier frequencies, instead of only one fixed frequency. The time-of-flight information can considerably improve the identification of cracks' position, two first parameters a and b.

Acknowledgements The first author is thankful to the Russian Science Foundation (RSCF), for its support by Project 15-19-10008.

References

1. Ramm, A.G.: Scattering by Obstacles. Reidel Publish, Dordrecht (1986)
2. Natterer, F.: The Mathematics of Computerized Tomography. Wiley, Stuttgart (1986)
3. Colton, D., Kress, R.: Integral Equation Methods in Scattering Theory. Wiley, New York (1983)
4. Colton, D., Kress, R.: Inverse Acoustic and Electromagnetic Scattering Theory. Springer, Berlin (1992)
5. Sumbatyan, M.A., Scalia, A.: Equations of Mathematical Diffraction Theory. CRC Press, Boca Raton (2005)
6. Binda, L., Lenzi, G., Saisi, A.: NDE of masonry structures: use of radar test for the characterization of stone masonries. J. NDT E Int. **31**, 411–419 (1998)
7. Daniels, J.: Surface Penetrating Radar. Institute of Electrical Engineering, London (1996)
8. Maierhofer, C., Leipold, S.: Radar investigation of masonry structures. J. NDT E Int. **34**, 139–147 (2001)
9. Krautkrämer, J., Krautkrämer, H.: Ultrasonic Testing of Materials. Springer, Berlin (1983)
10. Abda, A.B., et al.: Line segment crack recovery from incomplete boundary data. Inverse Prob. **18**, 1057–1077 (2002)
11. Alessandrini, G., Beretta, E., Vessella, S.: Determining linear cracks by boundary measurements: Lipschitz stability. SIAM J. Math. Anal. **27**, 361–375 (1996)
12. Friedman, A., Vogelius, M.: Determining cracks by boundary measurements. Indiana Univ. Math. J. **38**, 527–556 (1989)
13. Rucevskis, S., Sumbatyan, M.A., Akishin, P., Chate, A.: Tikhonov's regularization approach in mode shape curvature analysis applied to damage detection. Mech. Res. Commun. **65**, 9–16 (2015)
14. Potthast, R.: Point Sources and Multipoles in Inverse Scattering Theory. Chapman and Hall, London (2001)
15. Tikhonov, A.N., Arsenin, V.Y.: Solutions of Ill-Posed Problems. Winston, Washington (1977)
16. Brigante, M., Sumbatyan, M.A.: Application of the inverse diffraction technique to the problem of reconstruction of complexly shaped flaws by ultrasonic methods. Russ. J. NDT **46**, 98–111 (2010)
17. Brigante, M., Sumbatyan, M.A.: An efficient numerical algorithm for crack reconstruction in elastic media by the circular US scanning. Inverse Prob. Sci. Eng. **18**, 361–379 (2010)
18. Brigante, M.: Numerical algorithm for defect reconstruction in elastic media with a circular ultrasonic scanning. Eng. Anal. Bound. Elem. **37**, 551–557 (2013)
19. Scalia, A., Sumbatyan, M.A.: On efficient quantitative analysis in real-time ultrasonic detection of cracks. Ultrasonics **37**, 239–245 (1999)
20. Brigante, M.: On multiple scattering in acoustic media: a deterministic ray tracing method for random structures. Ultrasonics **53**, 652–657 (2013)
21. Sumbatyan, M.A., Brigante, M.: Analysis of strength and wave velocity for micro-damaged elastic media. Eng. Fract. Mech. **145**, 43–43 (2015)
22. Crouch, S.L., Starfield, A.M.: Boundary Element Methods in Solid Mechanics. George Allen and Unwin, London (1983)
23. Brigante, M., Sumbatyan, M.A.: Reconstruction of crack clusters in the rectangular domain by ultrasonic waves. Res. NDE **21**, 193–212 (2010)
24. Achenbach, J.: Wave Propagation in Elastic Solids. North-Holland, Amsterdam (1987)
25. Bonnet, M.: Boundary Integral Equations Methods for Solids and Fluids. Wiley, New York (1999)

26. Abramowitz, M., Stegun, I.: Handbook of Mathematical Functions. Dover, New York (1965)
27. Gill, P.E., Murray, W., Wright, M.H.: Practical Optimization. Academic Press, London (1981)
28. Kantorovich, L.V., Akilov, G.P.: Functional Analysis. Pergamon Press, Oxford (1982)
29. Zhigljavsky, A.A.: Theory of Global Random Search. Kluwer, Dordrecht (1991)
30. Corana, M., et al.: Minimizing multimodal functions of continuous variables with the simulated annealing algorithm. ACM Trans. Math. Softw. **13**, 262–280 (1987)

Short-Wave Diffraction of Elastic Waves by Voids in an Elastic Medium with Double Reflections and Transformations

Nikolay V. Boyev

Abstract The classical diffraction problem of high-frequency waves emitted from a point source in an elastic medium with a void or a system of voids is considered. The voids are bounded by arbitrary smooth surfaces. Single reflection cases of longitudinal and transverse waves are studied by taking into account their transformations on the boundary surface. Double reflection cases are investigated for two different possibilities of transformations of elastic waves: the longitudinal wave transformed to the transverse one, and vice versa. The developed method of the research is based on the evaluation of diffraction integrals by means of multidimensional stationary-phase method. The novel approach allows us to obtain the leading asymptotic term of the diffracted displacement field as a closed-form expression in the cases of single and double reflections, which corresponds to the geometrical theory of diffraction (GTD).

Keywords Short-wave diffraction · Elastic waves · Double reflection and transformations · Defects

1 Introduction

The classical problem of the diffraction by the boundaries in continuous media has important technical applications. In the acoustic media the reflected waves contain the necessary information about the shape of the obstacle. In the ultrasonic non-destructive testing this information serves as a basis for the reconstruction of the characteristic size and the shape of the defects. In the case of single reflection of elastic waves, the solution to such a problem in a two-dimensional case is attained by different methods in [2, 4, 22]. For a three-dimensional case in [3, 11] the short-wavelength approximation is obtained in a closed form for the pressure of acoustic wave in the case of a single reflection, and in [3] for multiple reflections

N.V. Boyev (✉)
Southern Federal University, 105/42 Bolshaya Sadovaya Str., Rostov-on-Don 344006, Russian Federation
e-mail: boyev@math.sfedu.ru

M.A. Sumbatyan (ed.), *Wave Dynamics and Composite Mechanics for Microstructured Materials and Metamaterials*,
Advanced Structured Materials 59, DOI 10.1007/978-981-10-3797-9_5

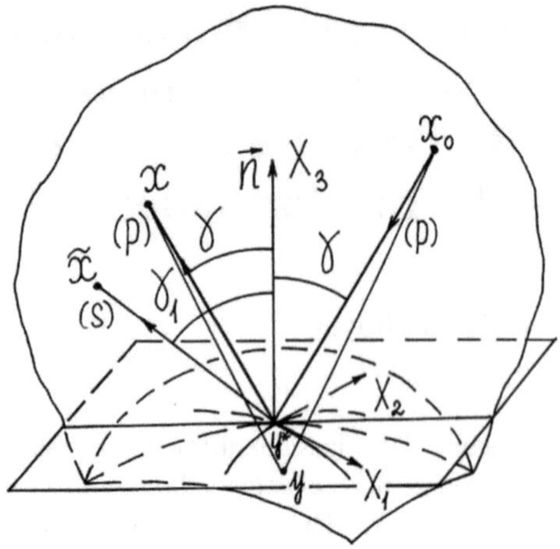

Fig. 1 Scattering of the high-frequency longitudinal wave in the longitudinal reflection (p-p reflection) and transformation of the longitudinal wave to the transverse one (p-s transformation), on the free boundary of the cavity located inside the elastic medium

from arbitrary smooth boundaries. The short-wavelength diffraction approaches developed in the present chapter can be efficiently applied to the study of elastic media with concentration of solid inclusions, namely, metamaterials. Such materials are known to be used as wave filters [10]. Some analytical and numerical approaches for two-dimensional problems of acoustic and elastics waves propagation, mainly in low-frequency range, are developed in [12, 14–17, 21] and in a number of other works. Similar three-dimensional problems are discussed in [18, 19, 23].

2 Problem Statement

Suppose the spherical monochromatic high-frequency wave falls from the point x_0 of an infinite elastic medium on the boundary of the flaw located in this medium (Fig. 1). The wave is generated by the local force $Qe^{-i\omega t}$ concentrated at this point x_0, where ω is the oscillation frequency.

At the same time, the displacements of the point y in the elastic medium are defined by the Kupradze matrix [9]:

$$
\begin{aligned}
U_j^{(k)}(y, x_0) &= U_{jp}^{(k)}(y, x_0) + U_{js}^{(k)}(y, x_0), \quad k, j = 1, 2, 3 \\
U_{jp}^{(k)}(y, x_0) &= -\frac{1}{4\pi\rho\omega^2}\frac{\partial^2}{\partial y_k \partial y_j}\left(\frac{e^{ik_p R_0}}{R_0}\right), \quad R_0 = |x_0 - y|, \\
U_{js}^{(k)}(y, x_0) &= \frac{1}{4\pi\rho\omega^2}\left[k_s^2\delta_{kj}\left(\frac{e^{ik_s R_0}}{R_0}\right) + \frac{\partial^2}{\partial y_k \partial y_j}\left(\frac{e^{ik_s R_0}}{R_0}\right)\right],
\end{aligned}
\tag{1}
$$

where ρ is density, λ, μ are Lamè coefficients, $k_p = \omega/c_p$, $k_s = \omega/c_s$, c_p, c_s are the wave numbers and the wave speed of the longitudinal and the transverse waves, respectively, δ_{kj} are the Kronecker symbols. The goal of the present study is to find the amplitude characteristics of the scattered field on the stress-free cavity surface.

3 Method of Solution

The time-dependence parameter of the problem is monochromatic, in particular, for the displacement in an elastic media it can be written as: $u(x_1, x_2, x_3, t) = Re[u(x_1, x_2, x_3)\exp(-i\omega t)]$. The Kupradze matrix defines the displacements of the longitudinal (p-wave) and transverse (s-wave) waves at the point y, in the direction of the radial axis: $\mathbf{q} = \frac{\mathbf{x}_0 y}{|x_0 y|}$.

$$\mathbf{u}_{\mathbf{q}}^{(p)}(y) = Q_{\mathbf{q}}\mathbf{q}\frac{k_p^2}{4\pi\rho\omega^2}\left(1 + i\frac{2}{k_p R_0} - \frac{2}{(k_p R_0)^2}\right)\frac{e^{ik_p R_0}}{R_0}, \quad Q_{\mathbf{q}} = (\mathrm{Q}, \mathrm{q}) \tag{2}$$

$$\mathbf{u}_{\mathbf{q}_1}^{(s)}(y) = Q_{\mathbf{q}_1}\mathbf{q}_1\frac{k_s^2}{4\pi\rho\omega^2}\left(1 - i\frac{1}{k_s R_0} - \frac{1}{(k_s R_0)^2}\right)\frac{e^{ik_s R_0}}{R_0}, \quad Q_{\mathbf{q}_1} = (\mathrm{Q}, \mathrm{q}_1) \tag{3}$$

The tangential direction $\mathbf{q}_1$ is perpendicular to $\mathbf{q}$, Q_q and Q_{q_1} are the projections of the force $\mathbf{Q}$ in the directions $\mathbf{q}$ and $\mathbf{q}_1$. The displacement vector components at the point x of the wave scattered by the stress-free boundary surface are defined by the following the Somigliana integral [13]:

$$u_k(x) = \iint_S \mathbf{T}_y[\mathbf{U}^{(k)}(y, x)] \cdot \mathbf{u}(y) dS_y, \tag{4}$$

$$\mathbf{T}_y\left[\mathbf{U}^{(k)}(y, x)\right] = 2\mu\frac{\partial \mathbf{U}^{(k)}}{\partial n} + \lambda \mathbf{n}\mathrm{div}\left(\mathbf{U}^{(k)}\right) + \mu\left(\mathbf{n} \times rot\left(\mathbf{U}^{(k)}\right)\right) \tag{5}$$

where the Kupradze matrix $\mathbf{U}^k(y, x)$ is obtained from the matrix $\mathbf{U}^k(y, x_0)$ (1), by substituting x_0 for x and R_0 for $R = |y - x|$, $\mathbf{T}_y$ is a force vector at the point y, $\mathbf{u}(y)$ is a vector of the full displacement field on the boundary surface, $\mathbf{n}$ is normal to the surface S. Let us extract the terms defined by the longitudinal (p) and the transverse (s) waves in the full displacement vector (4) on the boundary surface and in the force vector at the point y (5).

$$\begin{aligned} u_k(x) = &\iint_S \mathbf{T}_y[\mathbf{U}_p^{(k)}(y, x)] \cdot \mathbf{u}(y; p) dS_y + \iint_S \mathbf{T}_y[\mathbf{U}_s^{(k)}(y, x)] \cdot \mathbf{u}(y; p) dS_y + \\ &+ \iint_S \mathbf{T}_y[\mathbf{U}_p^{(k)}(y, x)] \cdot \mathbf{u}(y; s) dS_y + \iint_S \mathbf{T}_y[\mathbf{U}_s^{(k)}(y, x)] \cdot \mathbf{u}(y; s) dS_y \end{aligned} \tag{6}$$

here $\mathbf{u}(y; p)$ ($\mathbf{u}(y; s)$) is the full displacement vector on the boundary S produced by the incidence of the longitudinal (transverse) wave on the boundary S. The first and the last terms in (6) describe $p-p$ and $s-s$ reflections, where the second and the third $p-s$ and $s-p$ reflections are transformations.

As in the classical geometric theory of diffraction developed in the scalar acoustic problems [5] and extended to the problems of dynamical theory of elasticity [1], one should distinguish between high-frequency asymptotics in the local and the global senses. In the global formulation, the problem of computing the full field at the point x is being put forward. In such a case, applying the Eq. (6), the full field at the point x of the elastic medium is constructed by the four terms in (6) and the incident spherical wave. The asymptotic solution presented below is local in character and provides the leading asymptotic term of the amplitude of the diffracted field in a small neighborhood of any ray emitted from the point x_0, reflected by the surface at the point y^* and arriving at the point x. Evidently, such rays can only exist if both point y^* and x lie in the illuminated area.

4 Single Reflection of the Longitudinal Wave by the Cavity Surface

Let us analyze in detail the problem of the propagation of the longitudinal (p) term (2) of the spherical wave (1) in the fixed direction q in an elastic media, when p-wave interacts with a free boundary of the cavity and reflects from it. In such a case the longitudinal wave is reflected to the p-wave ($p-p$ reflection) and will be transformed to the transversal s-wave ($p-s$ transformation). The receiving points x in the case of $p-p$ reflection and $\tilde{x}$ in the case of $p-s$ transformation is different being located on the rays, along which the reflected longitudinal and transverse waves propagate. In this case the point y^* of the intersection of the smooth cavity surface with the direction $\mathbf{q} = \frac{\mathbf{x}_0 y^*}{|x_0 y^*|}$ is the specular reflection point of the incident p-wave (2). Below, we obtain the expressions for the amplitudes of the reflected waves at the receiving points x and $\tilde{x}$.

The case of the $p-p$ reflection. Let us refer the incident direction of wave $\mathbf{q} = \{-\cos\alpha, -\cos\beta, -\cos\gamma\}$ to the right Cartesian coordinate system $OX_1X_2X_3$ at the point y^* , where axis OX_3 coincides with the outer normal vector $\mathbf{n}$ towards the cavity boundary oriented in the direction of the elastic medium, and axes OX_1 and OX_2 coincide with tangent lines to the curvature lines of the boundary at the point y^* (Fig. 1). In such a coordinate system vector $\mathbf{q}_1$ has the following coordinates $= \{-\mathrm{ctg}\gamma\cos\alpha, -\mathrm{ctg}\alpha\cos\beta, -\sin\gamma\}$ and normal $\mathbf{n} = \{0, 0, 1\}$.

The coordinates of the displacement vector in the reflected wave are defined by the first term of the Eq. (6) which we write out here in the expanded form:

$$u_k^{(p)}(x) = \iint_S \left[\mu \sum_{m=1}^{2} \left(\frac{\partial U_{mp}^{(k)}}{\partial y_3} + \frac{\partial U_{3p}^{(k)}}{\partial y_m} \right) u_m(y;p) + \right. \\ \left. + \left(2\mu \frac{\partial U_{3p}^{(k)}}{\partial y_3} + \lambda \operatorname{div} \mathbf{U}_p^{(k)} \right) u_3(y;p) \right] dS_y \tag{7}$$

To perform an asymptotic estimation of the integrals (7) at $k_p \to \infty$, the following asymptotic representation is used:

$$\operatorname{div} \mathbf{U}_p^{(k)}(y,x) = i \frac{k_p^3}{4\pi\rho\omega^2} \frac{e^{ik_pR}}{R} \frac{\partial R}{\partial y_k} \left[1 + \mathrm{O}\left((k_pR)^{-1} \right) \right] \tag{8}$$

$$\frac{\partial U_j^{(k)}}{\partial y_m} = i \frac{k_p^3}{4\pi\rho\omega^2} \frac{e^{ik_pR}}{R} \frac{\partial R}{\partial y_m} \frac{\partial R}{\partial y_k} \frac{\partial R}{\partial y_j} \left[1 + \mathrm{O}\left((k_pR)^{-1} \right) \right], k,j,m = 1,2,3 \\ y(y_1,y_2,y_3); \quad x(x_1,x_2,x_3); \quad y \in S, \\ \frac{\partial R}{\partial y_1} = \frac{y_1 - x_1}{R} = -\cos\alpha, \frac{\partial R}{\partial y_2} = \frac{y_2 - x_2}{R} = -\cos\beta, \frac{\partial R}{\partial y_3} = \frac{y_3 - x_3}{R} = \cos\gamma. \tag{9}$$

Here $\{-\cos\alpha, -\cos\beta, \cos\gamma\}$ are the direction cosines of the vector **yx**. After substitution (8) and (9) in (7), we obtain:

$$u_k^{(p)}(x) = i \frac{k_p^3}{4\pi\rho\omega^2} \iint_S \Phi(y) \frac{\partial R}{\partial y_k} \frac{e^{ik_pR}}{R} dS_y \\ \Phi(y) = 2\mu \left[\frac{\partial R}{\partial y_1} u_1(y;p) + \frac{\partial R}{\partial y_2} u_2(y;p) \right] \frac{\partial R}{\partial y_3} + \left[2\mu \left(\frac{\partial R}{\partial y_2} \right)^2 + \lambda \right] u_3(y;p)$$

We now pass on to the spherical coordinate system r, θ, ψ at the point y^*. The displacement vector components are reduced to the form

$$u_r^{(p)}(x) = i \frac{k_p^3}{4\pi\rho\omega^2} \iint_S \Phi(y) \frac{e^{ik_pR}}{R} dS_y, u_\theta^{(p)}(x) = 0, u_\psi^{(p)}(x) = 0 \\ \Phi(y) = -2\mu[\cos\alpha u_1(y;p) + \cos\beta u_2(y;p)] \cos\gamma + [2\mu \cos^2\gamma + \lambda] u_3(y;p) \tag{10}$$

When an asymptotic estimate of the Kirchhoff integral in the Eq. (10) is performed, the components of the full displacement field $u_k(y;p)$, $k = 1,2,3$ under the integration sign should be taken as a solution of a local diffraction problem of the plane incident p-wave scattered by the flat boundary in the elastic half-space [6, 8].

$$u_m(y;p)=\left(1+V_{pp}(y)-\frac{k_s}{k_p\sin\gamma}\sqrt{1-\frac{k_p^2}{k_s^2}\sin^2\gamma V_{ps}(y)}\right)u_{mq}^{(p)}(y),m=1,2$$
$$u_3(y;p)=\left(1-V_{pp}(y)-tg\gamma V_{ps}(y)\right)u_{3q}^{(p)}(y) \quad (11)$$

where V_{pp} and V_{ps} are the coefficients of $p-p$ reflection and $p-s$ transformation [6, 8].

$$V_{pp}=\frac{1}{z}\left[4ctg\gamma ctg\gamma_1-\left(1-ctg^2\gamma_1\right)^2\right],V_{ps}=\frac{4}{z}ctg\gamma\left(1-ctg^2\gamma_1\right)$$
$$z=4ctg\gamma ctg\gamma_1+\left(1-ctg^2\gamma_1\right)^2 \quad (12)$$

The substitution of the relations (12) to (11) and (2) in the expression under the integral sign (10), and taking the non-oscillating functions out of the integral sign in the high-frequency approximation, leads to the following main integral representation of the radial displacement $u_r^{(p)}(x)$

$$u_r^{(p)}(x)=\frac{Q_{\mathbf{q}}}{4\pi\mu}i\frac{k_p^2}{k_s^2}\cdot\frac{k_p}{2\pi}\frac{\cos\gamma}{L_0L}V_{pp}\left(y^*\right)\iint_S e^{ik_p\varphi_{pp}}dS$$
$$\varphi_{pp}=|x_0-y|+|y-x|,L_0=\left|x_0-y^*\right|,L=\left|y^*-x\right|. \quad (13)$$

The asymptotic estimation of the integral in the representation (13) by the two-dimensional stationary phase method [7] is given in [3] for the scattering problem of acoustic wave by the boundary of an obstacle. Using this estimation, we can write out the leading term of the radial displacement amplitude in the reflected p-wave:

$$u_r^{(p)}(x)=\frac{Q_{\mathbf{q}}}{4\pi\mu}\frac{k_p^2}{k_s^2}V_{pp}\left(y^*\right)\cos\gamma\frac{\exp\left\{i\left[k_p(L_0+L)+\frac{\pi}{4}\left(\delta_2^{(pp)}-2\right)\right]\right\}}{L_0L\sqrt{\left|\det\left(D_2^{(pp)}\right)\right|}} \quad (14)$$

where $D_2^{(pp)}$ is the Hessian matrix with symmetric structure $(d_{ij}=d_{ji};i,j=1,2)$ and $\delta_2^{(pp)}=sign\,D_2^{(pp)}$ is the difference between the number of positive and negative eigenvalues of matrix $D_2^{(pp)}$. By calculating the determinant in the denominator and taking into account that $d_{12}=d_{21}$, we can write out the Eq. (14) in the explicit form:

$$u_r^{(p)}(x)=\frac{B\times V_{pp}(y^*)\exp\left\{i\left[k_p(L_0+L)+\frac{\pi}{4}\left(\delta_2^{(pp)}-2\right)\right]\right\}}{\sqrt{\left|(L_0+L)^2+2L_0L(L_0+L)\left(k_2\sin^2\alpha+k_1\sin^2\beta\right)\cos^{-1}\gamma+4L_0^2L^2K\right|}}$$

$$B = \frac{Q_{\mathbf{q}}}{4\pi\mu} \frac{k_p^2}{k_s^2} \tag{15}$$

Here $K = k_1 k_2$ is the Gaussian curvature, $k_1,\ k_2$ are principal curvatures of the boundary at the point y^*. Equation (15) provides the leading asymptotic term of pressure at $k_p L_0 \gg 1,\ k_p L \gg 1,\ k_p R_1 \gg 1,\ k_p R_2 \gg 1$. The structure of the function in the denominator of the Eq. (15) for the reflected longitudinal wave is the same as in the case of a single reflection of acoustic wave [3]. The presence of the coefficient $V_{pp}(y^*)$ in the numerator of the Eq. (15) indicates that the qualitative characteristics of the reflection of the longitudinal—to longitudinal wave remain the same as in the case of the wave scattered from the tangential plane towards the point of the specular reflection. Let us extract two extreme cases for the Eq. (15). If $k_1 = k_2 = 0$, then the well-known result follows from (15) for the radial displacement in a wave scattered by the free boundary plane:

$$u_r^{(p)}(x) = \frac{Q_{\mathbf{q}}}{4\pi\mu} \frac{k_p^2}{k_s^2} V_{pp}(y^*)(L_0 + L)^{-1} \exp\left[ik_p(L_0 + L)\right] \tag{16}$$

In the case of back scattering, $(V_{pp} = -1)$ in a far field Eq. (15) for radial displacement coincides with the relation for the pressure in the scalar case [20].

$$u_r^{(p)}(x) = -0.5 Q_{\mathbf{q}} i L_0^{-2} \sqrt{R_1 R_2} \exp\left[i\left(2k_p L_0 + \frac{\pi}{4}\delta_2^{(pp)}\right)\right] \tag{17}$$

The relation (15) is obtained for the case when the high-frequency longitudinal wave falls on the convex part of the boundary surface of the elastic medium. If a wave is incident on a concave surface, then the principal curvatures k_1 and k_2 should be taken negative. The Eq. (15) differs from the analogous one in [3, 11] for the pressure $p(x)$ in a scattered high-frequency acoustic wave in the scalar case only by the presence of the reflection coefficient V_{pp} in the case of elastic medium. The expression for the pressure in a wave scattered just once by the acoustically hard boundary is obtained in monography [11] in the light of the geometrical theory of diffraction (GTD) and in [3] making use of the estimation of the diffraction integral by the two-dimensional stationary phase method. It means, that the leading asymptotic term t in (15) of the diffraction integral agrees with the calculations of the pressure in the scattered wave using GTD.

The case of $p - s$ transformation. In this case the Cartesian components of the displacement vector u_k^s, $k = 1, 2, 3$ are defined by the second term of the integral representation (6)

$$u_k^{(s)}(\tilde{x}) = \iint_S \mathbf{T}_y \left[\mathbf{U}_s^{(k)}(y, \tilde{x})\right] \cdot \mathbf{u}(y; p) dS_y$$

By implementing the transformations in the same manner as in the previous case, the displacement $u_\theta^s(\tilde{x})$ is reduced to the form (after taking in high-frequency approximation non-oscillating functions behind the integral sign)

$$u_\theta^{(s)}(\tilde{x}) = -\frac{Q_{\mathbf{q}}}{4\pi\mu} i\left(\frac{k_p}{2\pi}\right)\frac{\cos\gamma_1}{L_0 L_1} V_{ps}\left(y^*\right)\iint_S e^{ik_p\phi_{ps}} dS \tag{18}$$

$$\varphi_{ps} = |x_0 - y| + \frac{k_s}{k_p}|y - \tilde{x}|, L_0 = \left|x_0 - y^*\right|, L_1 = \left|y^* - \tilde{x}\right|. \tag{19}$$

The leading asymptotic term of the integral (18) can be obtained by applying the two-dimensional stationary phase method [7]

$$u_\theta^{(s)}(\tilde{x}) = -\frac{Q_{\mathbf{q}}}{4\pi\mu}\cos\gamma_1 V_{ps}\left(y^*\right)\frac{\exp\left\{i\left[k_p L_0 + k_s L_1 + \frac{\pi}{4}\left(\delta_2^{(ps)} - 2\right)\right]\right\}}{L_0 L_1\sqrt{\left|\det\left(D_2^{(ps)}\right)\right|}} \tag{20}$$

where the elements of the symmetric ($d_{12} = d_{21}$) Hessian matrix $D_2^{(ps)} = d_{ij}$, $i, j = 1, 2$ are defined by the Eqs. (21), $\delta_2^{(ps)} = sign\, D_2^{(ps)}$;

$$\begin{aligned}
d_{11} &= L_0^{-1}\sin^2\alpha + \frac{k_s}{k_p}L_1^{-1}\sin^2\alpha_1 - k_1\left(\cos\gamma - \frac{k_s}{k_p}\cos\gamma_1\right)\\
d_{12} &= -\left(L_0^{-1}\cos\alpha\,\cos\beta + \frac{k_s}{k_p}L_1^{-1}\cos\alpha_1\cos\beta_1\right)\\
d_{22} &= L_0^{-1}\sin^2\beta + \frac{k_s}{k_p}L_1^{-1}\sin^2\beta_1 + k_2\left(\cos\gamma - \frac{k_s}{k_p}\cos\gamma_1\right)
\end{aligned} \tag{21}$$

The resulting expression (21) for the tangential displacements $u_\theta^{(s)}(\tilde{x})$ at $p-s$ reflection contains the determinant of the Hessian matrix $\det(D_2^{(ps)})$, whose elements contain a larger number of parameters than in the case of $p-p$ scattering. The leading asymptotic term of the displacement in the reflected s-wave at $p-s$ transformation can be written out in the explicit form, however, in order to attain the result for a specific case, it is reasonable to use direct calculation of the determinant $\det(D_2^{(ps)})$. The presence of the transformation coefficient $V_{ps}(y^*)$ in the Eq. (20) indicate that the qualitative characteristics at $p-s$ transformation in the case of the surface remain the same just as in the case of the wave reflection by the tangential plane to the surface at the point of the specular reflection.

5 Single Reflection of the Transverse Wave by the Cavity Surface

The calculation of the full diffracted field on the boundary of an obstacle in general case (6) also includes the waves scattered at the incidence of the tangential component (3) of a spherical wave (1). As it was done for the previous case, let us study the amplitude characteristics of the reflected s-waves and p-waves with the incidence of the tangential s-wave (3). This analysis is performed in the same coordinate system as in the case of p-wave reflection. Let us present the final expressions for displacements.

The case of $s-s$ reflection.

$$u_{\theta}^{(s)}(x) = -\frac{Q_{q_1}}{4\pi\mu}V_{ss}\left(y^{*}\right)\frac{\exp\left\{i\left[k_s(L_0+L_1)+\frac{\pi}{4}\left(\delta_2^{(ss)}-2\right)\right]\right\}}{L_0L_1\sqrt{\left|\det(D_2^{(ss)})\right|}} \tag{22}$$

where $D_2^{(ss)}$ is the Hessian matrix of a symmetric structure $(d_{ij}=d_{ji}; i,j=1,2)$, and $\delta_2^{(ss)}=sign\, D_2^{(ss)}$. By calculating the determinant in the denominator and taking into account that $d_{12}=d_{21}$, we write out the Eq. (22) in the explicit form:

$$u_{\theta}^{(s)}(x) = \frac{B_1\times V_{ss}(y^{*})\exp\left\{i\left[k_s(L_0+L_1)+\frac{\pi}{4}\left(\delta_2^{(ss)}-2\right)\right]\right\}}{\sqrt{\left|(L_0+L_1)^2+2L_0L_1(L_0+L_1)(k_2\sin^2\alpha_1+k_1\sin^2\beta_1)\cos^{-1}\gamma_1+4L_0^2L_1^2K\right|}}$$

$$B_1 = -\frac{Q_{q_1}}{4\pi\mu} \tag{23}$$

Here $\{-cos\,\alpha_1,\ -\cos\beta_1,\ -\cos\gamma_1\}$ is the vector defining the incident direction of ray x_0-y^{*} in the chosen coordinate system.

The case of $s-p$ transformation.

$$u_r^{(p)}(x) = \frac{Q_{q_1}}{4\pi\mu}\frac{k_p}{k_s}V_{sp}\left(y^{*}\right)\cos\gamma\frac{\exp\left\{i\left[k_sL_0+k_pL+\frac{\pi}{4}\left(\delta_2^{(sp)}-2\right)\right]\right\}}{L_0L\sqrt{\left|\det\left(D_2^{(sp)}\right)\right|}} \tag{24}$$

where the elements of the symmetric Hessian matrix $D_2^{(sp)}=d_{ij}$, $i,j=1,2$ are defined by the expressions (25), $\delta_2^{(sp)}=sign\, D_2^{(sp)}$.

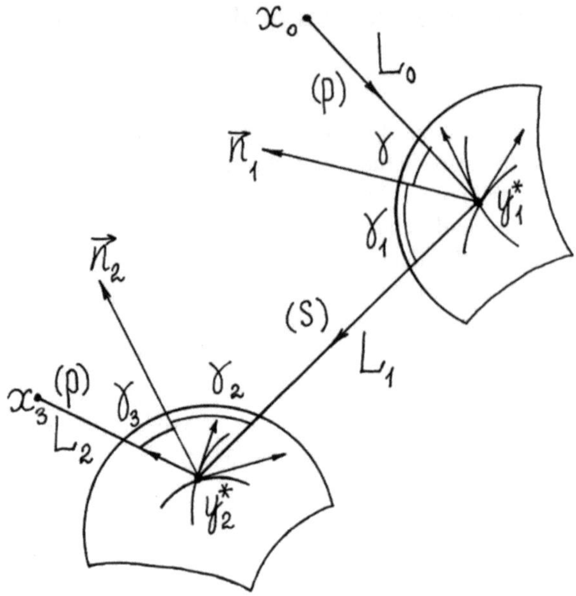

Fig. 2 Double $p-s-p$ transformation of a high-frequency longitudinal wave along the ray $x_0-y_1^*-y_2^*-x_3$ on the free boundary surfaces of two cavities placed in the elastic medium

$$
\begin{aligned}
d_{11} &= \frac{k_s}{k_p} L_0^{-1} \sin^2\alpha_1 + L^{-1}\sin^2\alpha + k_1\left(\frac{k_s}{k_p}\cos\gamma_1 - \cos\gamma\right) \\
d_{12} &= -\left(\frac{k_s}{k_p} L_0^{-1}\cos\alpha_1\cos\beta_1 + L^{-1}\cos\alpha\,\cos\beta\right) \\
d_{22} &= \frac{k_s}{k_p} L_0^{-1} \sin^2\beta_1 + L^{-1}\sin^2\beta + k_2\left(\frac{k_s}{k_p}\cos\gamma_1 - \cos\gamma\right)
\end{aligned}
\tag{25}
$$

Thus, in the first two cases all possible single reflections and transformations of the longitudinal and the transverse waves on the free boundary of elastic media have been studied.

6 The Double Re-Reflection of Elastic Waves in View of All Possible Transformations

The application of GTD, which is based on the use of divergence coefficients, even in the case of double wave reflection is sufficiently lengthy. In our opinion, studying the problem of the re-reflection of a high-frequency wave by the boundary of an elastic body, with all possible resulting wave transformations of arbitrary number N of times, it is more convenient to start with the evaluation of 2 N-dimensional diffraction integral by the multiple stationary-phase method. The basis for studying a general case of an arbitrary number of re-reflections is the problem of double scattering (Fig. 2) which is considered below.

A direct application of the integral representation (6) on all area of "light" for the reflected waves is impossible [5], since it is not suitable for the case of wave multiple reflections. If we take the Somigliana's formula (6) and substitute $\mathbf{u}(y)$ with the values (2) and (3) of the primary field, then the integral Eq. (6) gives only a primary reflection.

A double-reflected wave can be attained only when the values $\mathbf{u}(y)$ include both primary field and its primary reflection. To solve the problem of a double rereflection we rely on the modification [5] of the integral formula (6). Following this modification the doubly reflected waves can be obtained by the integration of rays received from a single reflection by the neighborhood S_1 of the first specular reflection point y_1^* over the neighborhood S_2 of the second specular reflection point y_2^*. Such a modification means that when finding a leading asymptotic term of the quadruple diffraction integral, we deal with the GTD calculations of the displacement amplitude in a doubly reflected wave.

Further we investigate the re-reflection of the high-frequency wave by using as an example the reechoing of the ray $x_0 - y_1^* - y_2^* - x_3$ illuminated from the point x_0 ($p-$ wave (2)) and received at the point x_3 with possible transformation $p - s - p$. The components of the displacement vector of $p-$ wave at the point x_3 are given by the following equation

$$u_k^{(p)}(x_3) = \iint_{S_2} \mathbf{T}_{y_2}\left[\mathbf{U}_p^{(k)}(y_2, x_3)\right] \cdot \mathbf{u}(y_2; s) dS_2 \tag{26}$$

Here $\mathbf{u}(y_2; s)$ is a full displacement vector at the point $y_2 \in S_2$ in the neighborhood of the specular reflection point y_2^*, which should be defined after the first $p - s$ transformation in the neighborhood S_1 of the point of specular reflection y_1^*.

For the purpose of the asymptotic estimation of the Kirchhoff integral in Eq. (26), the components of the full displacement field vector $\mathbf{u}_k(y_2; s),\ k = 1, 2, 3$ under the integration sign should be taken as a solution of the local diffraction problem for the scattering incident plane $s-$ wave, formed at $p - s$ transformation in the neighborhood S_1 of the point of specular reflection y_1^*.

$$\begin{aligned} u_m(y_2; s) &= \left(V_{ss}(y_2) - 1 - \mathrm{tg}\gamma_1 V_{sp}(y_2)\right) u_m^{(s)}(y_2), m = 1, 2 \\ u_3(y_2; s) &= \left(V_{ss}(y_2) + 1 + \frac{k_p}{k_s \sin\gamma_1}\sqrt{1 - \frac{k_s^2}{k_p^2}\sin^2\gamma_1 V_{sp}(y_2)}\right) u_3^{(s)}(y_2) \end{aligned} \tag{27}$$

where $V_{ss}(y_2)$ and $V_{sp}(y_2)$ are the coefficients of $s - s$ reflections and $s - p$ transformations [6, 8]. At the same time, the components of the displacement vector $u_k^{(s)}(y_2; s),\ k = 1, 2, 3$ are expressed by a similar equation

$$u_k^{(s)}(y_2) = \iint_{S_1} \mathbf{T}_{y_1}\left[\mathbf{U}_p^{(k)}(y_1, y_2)\right] \cdot \mathbf{u}(y_1; p) dS_1$$

where the vector of the full displacement field at the point $y_1 \in S_1$ of the y_1^* neighborhood is defined by Eqs. (11). After substitution of Eqs. (11), (27), (28) to Eq. (26), and then the transition to the spherical coordinate system r, θ, ψ at the point y_2^*, we can write out the displacement components at the point x_3.

$$u_r^{(p)}(x_3) = -\frac{Q_{\mathbf{q}}}{4\pi\mu}\frac{k_p}{k_s}\left(\frac{k_p}{2\pi}\right)^2\left(\frac{k_p}{2k_s}\right)^2\frac{\cos\gamma_1\cos\gamma_3}{L_0L_1L_2}V_{ps}(y_1^*)V_{sp}(y_2^*)\times \\ \times\iint_{S_2}\iint_{S_1} e^{ik_p\varphi_{psp}}dS_1dS_2 \tag{28}$$

$$u_\theta^{(p)}(x_3) = 0,\ u_\psi^{(p)}(x_3) = 0 \\ \varphi_{psp} = |x_0 - y_1| + k_sk_p^{-1}|y_1 - y_2| + |y_2 - x_3| \\ L_0 = |x_0 - y_1^*|, L_1 = |y_1^* - y_2^*|, L_2 = |y_2^* - x_3| \tag{29}$$

As in the case of single reflections, the neighborhoods S_1 and S_2 of the specular reflection points y_1^* and y_2^* are associated with the right Cartesian coordinate system $O_iX_1^{(i)}X_2^{(i)}X_3^{(i)}$, $i = 1, 2$ defined by normals $\mathbf{n}_1$(axe $O_1X_3^{(1)}$) and $\mathbf{n}_2$ (axis $O_2X_3^{(2)}$) and tangents to the curvature lines (axes $O_iX_1^{(i)}$, $O_iX_2^{(i)}$, $i = 1, 2$).

Lct the basis vectors in the local coordinate systems at the points y_1^* and y_2^* be $(\mathbf{i}_1, \mathbf{j}_1, \mathbf{k}_1)$ and $(\mathbf{i}_2, \mathbf{j}_2, \mathbf{k}_2)$. If the shapes of the reflecting surfaces are known, then the local coordinate systems and basis vectors can be recovered. Along the curvature lines we count the arc length $\Delta s_1^{(1)}$ and $\Delta s_2^{(1)}$ in the neighborhood S_1 of point y_1^* and $\Delta s_1^{(2)}$ and $\Delta s_2^{(2)}$ in the neighborhood S_2 of point y_2^*. The increments of the arcs $\Delta s_1^{(i)}, \Delta s_2^{(i)}$ $(i = 1, 2)$ are infinitely small quantities. The asymptotic representation of the stationary distances in the phase function (29) up to an infinitesimal of the second order inclusive has the form:

$$|x_0 - y_1| = L_0 - \Delta s_1^{(1)}\cos\alpha - \Delta s_2^{(1)}\cos\beta + 0.5\left(L_0^{-1}\sin^2\alpha + k_1^{(1)}\cos\gamma\right)\left(\Delta s_1^{(1)}\right)^2 - \\ - L_0^{-1}\cos\alpha\cos\beta\Delta s_1^{(1)}\Delta s_2^{(1)} + 0.5\left(L_0^{-1}\sin^2\beta + k_2^{(1)}\cos\gamma\right)\left(\Delta s_2^{(1)}\right)^2 \tag{30}$$

$$|y_2 - x_3| = L_2 + \Delta s_1^{(2)}\cos\alpha_3 + \Delta s_2^{(2)}\cos\beta_3 + 0.5\left(L_2^{-1}\sin^2\alpha_3 + k_1^{(2)}\cos\gamma_3\right)\left(\Delta s_1^{(2)}\right)^2 - \\ - L_2^{-1}\cos\alpha_3\cos\beta_3\Delta s_1^{(2)}\Delta s_2^{(2)} + 0.5\left(L_2^{-1}\sin^2\beta_3 + k_2^{(2)}\cos\gamma_3\right)\left(\Delta s_2^{(2)}\right)^2 \tag{31}$$

At $p - s - p$ transformation, $\{-\cos\alpha,\ -\cos\beta,\ -\cos\gamma\}$ *is* the vector which defines the incident direction of $p-$ wave (2), vector $\{-\cos\alpha_1,\ -\cos\beta_1,\ \cos\gamma_1\}$ defines the direction of the $p-$ wave reflected at the point y_1^*, vector

$\{-\cos\alpha_2, -\cos\beta_2, -\cos\gamma_2\}$ is the incident direction of $s-$ wave reflected at the point y_1^* relative to the coordinate system of the point y_2^*, $\{-\cos\alpha_3, -\cos\beta_3, \cos\gamma_3\}$ is the direction of $p-$ wave reflected at the point y_2^*.

Let us find the distance $|y_1 - y_2| = |\mathbf{y}_2\mathbf{y}_1|$ in the phase φ_{psp} (29). We find this distance in the right coordinate system $O_2X_1^{(2)}X_2^{(2)}X_3^{(2)}$ with the origin O_2 at the point y_2^*. In this coordinate system let us denote the coordinates of the points as $y_2(\xi_2, \eta_2, \zeta_2)$, $y_1(\xi_1, \eta_1, \zeta_1)$, $y_1^*(\xi_2^0, \eta_2^0, \zeta_2^0)$. With this in view the vectors $\mathbf{y}_2\mathbf{y}_1$ in this coordinate system are represented in the form:

$$\mathbf{y}_2\mathbf{y}_1 = \overset{*}{\mathbf{y}}_2\overset{*}{\mathbf{y}}_1 + \mathbf{A}\overset{*}{\mathbf{y}}_1\mathbf{y}_1 - \overset{*}{\mathbf{y}}_2\mathbf{y}_2$$

$$\mathbf{y}_2\mathbf{y}_1 = \{\xi_1 - \xi_2, \eta_1 - \eta_2, \varsigma_1 - \varsigma_2\}; \quad \overset{*}{\mathbf{y}}_2\overset{*}{\mathbf{y}}_1 = \{\xi_1^0, \eta_1^0, \varsigma_1^0\}$$

$$\overset{*}{\mathbf{y}}_i\mathbf{y}_i = \left\{\Delta s_1^{(i)}, \Delta s_2^{(i)}, -0.5\left(k_1^{(i)}\left(\Delta s_1^{(i)}\right)^2 + k_2^{(i)}\left(\Delta s_2^{(i)}\right)^2\right)\right\}, \quad \mathrm{i} = 1, 2$$

Here the coordinates of the vector $\overset{*}{\mathbf{y}}_i\mathbf{y}_i (i = 1, 2)$ are written out in the local coordinate systems at the points y_1^* and y_2^*. Matrix $\mathbf{A} = (a_{ij})$, $i, j = 1, 2, 3$ - orthogonal transition matrix from the basis of the rectangular coordinate system $O_2X_1^{(2)}X_2^{(2)}X_3^{(2)}$ at the point y_2^* to the basis of the Cartesian coordinates $O_1X_1^{(1)}X_2^{(1)}X_3^{(1)}$ at the point y_1^*. The asymptotic presentation of the distance (32) with the accuracy up to an infinitesimal of the second order inclusive is written out on the basis of its coordinate calculations.

$$\begin{aligned}
|y_1 - y_2| = L_1 &+ \\
&+ \cos\alpha_1 \Delta s_1^{(1)} + \cos\beta_1 \Delta s_2^{(1)} - 0.5\cos\gamma_1\left[k_1^{(1)}\left(\Delta s_1^{(1)}\right)^2 + k_2^{(1)}\left(\Delta s_2^{(1)}\right)^2\right] - \\
&- \cos\alpha_2 \Delta s_1^{(2)} - \cos\beta_2 \Delta s_2^{(2)} - 0.5\cos\gamma_2\left[k_1^{(2)}\left(\Delta s_1^{(2)}\right)^2 + k_2^{(2)}\left(\Delta s_2^{(2)}\right)^2\right] + \\
&+ L_1^{-1}\left[0.5\sin^2\alpha_1\left(\Delta s_1^{(1)}\right)^2 + 0.5\sin^2\beta_1\left(\Delta s_2^{(1)}\right)^2 - \cos\alpha_1\cos\beta_1\Delta s_1^{(1)}\Delta s_2^{(1)} + \right. \\
&+ 0.5\sin^2\alpha_2\left(\Delta s_1^{(2)}\right)^2 + 0.5\sin^2\beta_2\left(\Delta s_2^{(2)}\right)^2 - \cos\alpha_2\cos\beta_2\Delta s_1^{(2)}\Delta s_2^{(2)} + \\
&+ (\cos\alpha_1\cos\alpha_2 - a_{11})\Delta s_1^{(1)}\Delta s_1^{(2)} + (\cos\beta_1\cos\alpha_2 - a_{12})\Delta s_2^{(1)}\Delta s_1^{(2)} + \\
&\left. + (\cos\alpha_1\cos\beta_2 - a_{21})\Delta s_1^{(1)}\Delta s_2^{(2)} + (\cos\beta_1\cos\beta_2 - a_{22})\Delta s_2^{(1)}\Delta s_2^{(2)}\right]
\end{aligned} \tag{32}$$

where $a_{11} = (\mathbf{i}_1, \mathbf{j}_1)$, $a_{12} = (\mathbf{j}_1, \mathbf{i}_2)$, $a_{21} = (\mathbf{i}_1, \mathbf{j}_2)$, $a_{22} = (\mathbf{i}_2, \mathbf{j}_2)$. Then we use expressions for $|x_0 - y_1|$, $k_s k_p^{-1}|y_1 - y_2|$, $|y_2 - x_3|$, see (30)–(32), the Snell law of refraction $k_p \sin\gamma = k_s \sin\gamma_1$, $k_s \sin\gamma_2 = k_p \sin\gamma_3$, as well as relationships

$$\frac{\cos\alpha}{\sin\gamma}=\frac{\cos\alpha_1}{\sin\gamma_1},\quad \frac{\cos\beta}{\sin\gamma}=\frac{\cos\beta_1}{\sin\gamma_1},\quad \frac{\cos\alpha_1}{\sin\gamma_1}=\frac{\cos\alpha_2}{\sin\gamma_2},\quad \frac{\cos\beta_1}{\sin\gamma_1}=\frac{\cos\beta_2}{\sin\gamma_2},$$

It can be shown, that terms $\Delta s_1^{(i)}$ and $\Delta s_2^{(i)}$, $i=1,2$ are absent in the phase φ_{psp} (30). This means that points $y_1^*\in S_1$ and $y_2^*\in S_2$ of the direct ray reflection correspond to the stationary phase φ_{psp} value,

$$\begin{aligned}\varphi_{psp}&=|x_0-y_1|+k_sk_p^{-1}|y_1-y_2|+|y_2-x_3|=L_0+k_sk_p^{-1}L_1+L_2+\\&+0.5d_{11}\left(\Delta s_1^{(1)}\right)^2+d_{12}\Delta s_1^{(1)}\Delta s_2^{(1)}+d_{13}\Delta s_1^{(1)}\Delta s_1^{(2)}+d_{14}\Delta s_1^{(1)}\Delta s_2^{(2)}+\\&+0.5d_{22}\left(\Delta s_2^{(1)}\right)^2+d_{23}\Delta s_2^{(1)}\Delta s_1^{(2)}+d_{24}\Delta s_2^{(1)}\Delta s_2^{(2)}+\\&+0.5d_{33}\left(\Delta s_1^{(2)}\right)^2+d_{34}\Delta s_1^{(2)}\Delta s_2^{(2)}+0.5d_{44}\left(\Delta s_2^{(2)}\right)^2\end{aligned}\tag{33}$$

where

$$\left\{\begin{matrix}d_{11}\\d_{22}\end{matrix}\right\}=\frac{1}{L_0}\left\{\begin{matrix}\sin^2\alpha\\\sin^2\beta\end{matrix}\right\}+\frac{k_s}{k_pL_1}\left\{\begin{matrix}\sin^2\alpha_1\\\sin^2\beta_1\end{matrix}\right\}+2\left\{\begin{matrix}k_1^{(1)}\\k_2^{(1)}\end{matrix}\right\}\left(\cos\gamma+\frac{k_s}{k_p}\cos\gamma_1\right)$$

$$\left\{\begin{matrix}d_{33}\\d_{44}\end{matrix}\right\}=\frac{k_s}{k_pL_1}\left\{\begin{matrix}\sin^2\alpha_2\\\sin^2\beta_2\end{matrix}\right\}+\frac{1}{L_2}\left\{\begin{matrix}\sin^2\alpha_3\\\sin^2\beta_3\end{matrix}\right\}+2\left\{\begin{matrix}k_1^{(2)}\\k_2^{(2)}\end{matrix}\right\}\left(\frac{k_s}{k_p}\cos\gamma_2+\cos\gamma_3\right)$$

$$\begin{aligned}&d_{12}=-L_0^{-1}\cos\alpha\cos\beta-k_sk_p^{-1}L_1^{-1}\cos\alpha_1\cos\beta_1\\&d_{34}=-k_sk_p^{-1}L_1^{-1}\cos\alpha_2\cos\beta_2-L_2^{-1}\cos\alpha_3\cos\beta_3\\&d_{13}=k_sk_p^{-1}L_1^{-1}(\cos\alpha_1\cos\alpha_2-a_{11}),d_{14}=k_sk_p^{-1}L_1^{-1}(\cos\alpha_1\cos\beta_2-a_{21}),\\&d_{23}=k_sk_p^{-1}L_1^{-1}(\cos\beta_1\cos\alpha_2-a_{12})\\&d_{24}=k_sk_p^{-1}L_1^{-1}(\cos\beta_1\cos\beta_2-a_{22}).\end{aligned}$$

The final result is obtained from Eq. (28) by applying the four-dimensional stationary phase [7].

$$\begin{aligned}&u_r^{(p)}(x_3)=B_2\times\frac{\exp\left\{i\left[k_pL_0+k_sL_1+k_pL_2+\frac{\pi}{4}\left(\delta_4^{(psp)}-4\right)\right]\right\}}{L_0L_1L_2\sqrt{\left|\det\left(D_4^{(psp)}\right)\right|}}\\&B_2=\frac{Q_{\mathbf{q}}}{4\pi\mu}\frac{k_p}{k_s}V_{ps}\left(y_1^*\right)V_{sp}\left(y_2^*\right)\cos\gamma_1\cos\gamma_3\end{aligned}\tag{34}$$

where $D_4^{(psp)}=(d_{ij})$, $i,j=1,2,3,4$ is the Hessian matrix with its elements d_{ij}, $i\leq j$ provided in the Eq. (33), $\delta_4^{(psp)}=sign\,D_4^{(psp)}$. Applying the developed and presented above method, any other possible reflections and transformations of the ray $x_0 - y_1^* - y_2^* - x_3$ can be considered. It can also be used for ray reflections and transformations arbitrary finite (N times) number of times. As an example, for the repeated transformation $s - p - s$ the non-zero amplitude of the displacement $u_\theta^{(s)}(x_3)$ can be obtained from Eqs. (33)–(34) for $p - s - p$ transformation by swapping the types of waves p and s in all the indexes of the included characteristics and directions of the incidence and the reflection of waves at the specular reflection points.

7 Conclusion

The single and the double reflection of the high-frequency spherical wave from the surface of one cavity and the system with two cavities, located in the infinite elastic medium, is studied within the framework of the modified Kirchhoff approximation. The leading asymptotic term for the displacements in a doubly reflected wave with regard to different possible transformations is obtained in an explicit form, which is in good agreement with the GTD. The trajectory of the doubly reflected rays is represented by spatial polygonal lines. The amplitude of the re-reflected wave is expressed through the determinant of a banded matrix. The elements of the determinant are written out in an explicit form and defined by the geometrical and the mechanical parameters of the problem.

Acknowledgements The work is performed within the framework of the Project № 15-19-10008 of the Russian Science Foundation (RSCF).

References

1. Achenbach, J.D., Gautesen, A.K., McMaken, H.: Ray Methods for Waves in Elastic Solids With Applications to Scattering by Cracks. Pittman, New York (1982)
2. Babich, V.M., Buldyrev, V.S.: Asymptotic Methods in Short-Wavelength Diffraction Theory. Alpha Science, New York (2008)
3. Boyev, N.V., Sumbatyan, M.A.: A short-wave diffraction by bodies, bounded by arbitrary smooth surface. Russian Doklady. **392**, N **5**, (2003)
4. Bojarski, N.N.: A survey of the physical optics inverse scattering identity. IEEE Trans. Anten. Prop. AP **30**(5), 980–988 (1982)
5. Borovikov, V.A., Kinber, B.Y.: Geometrical Theory of Diffraction. The Institution of Electrical Engineers, London (1994)
6. Brekhovskikh, L.: Waves in Layered Media. Academic Press, New York (2012)
7. Fedorjuk, M.V.: Saddle point method. Nauka, Moscow (1977). (in Russian)
8. Grinchenko, V.T., Myaleshka, V.V: Harmonic oscillations and waves in elastic bodies. Naukova Dumka, Kiev (1981) (in Russian)
9. Kupradze, V.D.: Potential methods in the theory of elasticity. Israel Program Sci. Transl. Jerusalem (1965)

10. Liu, Z., Zhang, X., Mao, Y., et al.: Locally Resonant Sonic Materials. Science **289**, 1734–1736 (2000)
11. McNamara, D.A., Pistorius, C.W.I., Malherbe, I.A.G.: Introduction to the Uniform Geometrical Theory of Diffraction. Artech House, Norwood (1990)
12. Miles, J.W.: On Rayleigh scattering by a grating. Wave Motion **4**, 285–292 (1982)
13. Novatsky, V.: Theory of Elasticity. Mir, Moscow (1975). (in Russian)
14. Porter, R., Evans, D.V.: Wave scattering by periodic arrays of breakwaters. Wave Motion **23**, 95–120 (1996)
15. Scarpetta, E., Sumbatyan, M.A.: On wave propagation in elastic solids with a doubly periodic array of cracks. Wave Motion **25**, 61–72 (1997)
16. Scarpetta, E., Sumbatyan, M.A.: On the oblique wave penetration in elastic solids with a doubly periodic array of cracks. Q. Appl. Math. **58**(2), 239–250 (2001)
17. Scarpetta, E., Sumbatyan, M.A.: Wave propagation through elastic solids with a periodic array of arbitrarily shaped defects. Math. Comput. Model. **37**(1/2), 19–28 (2003)
18. Scarpetta, E., Tibullo, V.: Explicit results for scattering parameters in three-dimensional wave propagation through a doubly periodic system of arbitrary openings. Acta Mech. **185**, 1–9 (2006)
19. Scarpetta, E., Tibullo, V.: On the three-dimensional wave propagation through cascading screens having a periodic system of arbitrary openings. Int. J. Eng. Sci. **46**, 105–118 (2008)
20. Shenderov, E.L.: Wave tasks of hydroacoustics. Sudostroenie, Leningrad (1972). (in Russian)
21. Sumbatyan, M.A.: Low-frequency penetration of acoustic waves through a periodic arbitrary shaped grating: the three-dimensional problem. Wave Motion **22**, 133–144 (1995)
22. Sumbatyan, M.A., Boyev, N.V.: High-frequency diffraction by nonconvex obstacles. J. Acoust. Soc. Am. 95, N 5. (Part 1), 2347–2353 (1994)
23. Sumbatyan, M.A., Remizov, M.Y.: Low frequency penetration of elastic waves through a triple periodic array of cracks. Springer Proc. Phys. **175**, 459–474 (2016)

Finite Element Modeling and Computer Design of Anisotropic Elastic Porous Composites with Surface Stresses

A.V. Nasedkin and A.S. Kornievsky

Abstract The chapter presents mathematical modelling and computer design of effective properties of anisotropic porous elastic materials with a nanoscale random structure of porosity. This integrated approach includes the effective moduli method of composite mechanics, the simulation of representative volumes with stochastic porosity and the finite element method. In order to take into account nanoscale sizes of pores, the Gurtin-Murdoch model of surface stresses is used at the borders between material and pores. The general methodology for determination of effective mechanical properties of porous composites is produced for a two-phase bulk (mixture) composite with special conditions for stresses discontinuities at the phase interfaces. The mathematical statements of boundary value problems and the resulting formulas to determine the complete set of effective stiffness moduli of the two-phase composites with arbitrary anisotropy and with surface stresses are described; the generalized problem definitions are formulated and the finite element approximations are given. It is used, that the homogenization procedures for porous composites with surface effects can be considered as special cases of the corresponding procedures for the two-phase composites with interphase stresses if the moduli material of the second phase (nanoinclusions) are negligibly small. These approaches have been implemented in the finite element package ANSYS for a model of nanoporous silicon with cubic crystal system for various values of surface moduli, porosity and number of pores. Model of representative volume was built in the form of a cube, evenly divided into cubic solid finite elements, some of which had been declared as pores. Surface stresses on the boundaries between material and pores were modeled by shell finite elements with the options of membrane stresses. It has been noted that the magnitude of the area of the interphase boundaries has influence on the effective moduli of the porous materials with nanosized stochastic structure.

A.V. Nasedkin (✉) · A.S. Kornievsky
Institute of Mathematics, Mechanics and Computer Science,
Southern Federal University, Milchakova Street 8a, Rostov-on-Don 344090, Russia
e-mail: nasedkin@math.sfedu.ru

A.S. Kornievsky
e-mail: alexandr5koren@gmail.com

M.A. Sumbatyan (ed.), *Wave Dynamics and Composite Mechanics for Microstructured Materials and Metamaterials*,
Advanced Structured Materials 59, DOI 10.1007/978-981-10-3797-9_6

Keywords Porous elastic composite · Nanosized pore · Surface stress · Effective moduli · Representative volume · Stochastic porosity · Finite element method

1 Introduction

As it is known from many experiments, a scale effect can be observed for nanoscale bodies, which results in the change of effective stiffnesses and other material moduli compared to the corresponding macroscale bodies. Among various approaches that explain this phenomenon, the models of theory of elasticity with surface stresses are widely used now. The idea of surface stresses in solids has been formulated long time ago [30]. However, significant development of this idea was done later in [15, 28]. As it was shown further, the theory of surface stresses can be considered as a particular case of the models with imperfect interface boundaries.

At present the theory of surface stresses, commonly referred to as the model of Gurtin-Murdoch, has been become widely used for describing scale effects at nanolevel, which can be seen, for example, from overviews given in [8, 16, 32]. In a range of papers this theory was applied to model elastic nanoscale composites. For example, in [1–6, 9, 11, 20, 21, 26, 27, 33] and etc. the mechanical properties of composites with spherical nanoinclusions (nanopores), fiber and another nanocomposites were studied in the frames of the theory of elasticity with surface stresses. The methodology of finite element approximations for elastic materials with surface effects and the numerical examples was demonstrated in [10, 18, 19, 29, 31].

This chapter considers anisotropic elastic materials with randomly located nanopores. In order to take into account nanoscale level at the borders between material and pores, the Gurtin-Murdoch model of surface stresses is used. The paper is organized as follows. Section 2 presents the mathematical statement of a homogenization problem for two-phase elastic composites with conditions for stresses discontinuities at the phase interface boundaries. Both composite phases are assumed to be elastic materials of arbitrary classes of anisotropy. The boundary value problem statements, their generalized or weak formulations and the resulting formulas for determination of the full set of effective constants for a two-phase composite with arbitrary types of phase anisotropy and surface stresses are also described. We note that homogenization procedures for porous composites with surface stresses can be regarded as special cases of the corresponding procedures for two-phase composites with imperfect interface boundaries under negligibly small stiffnesses for nanoinclusions.

The finite element approximations of the considered homogenization problems are given in Sect. 3. We note that homogenization problems for the composites under investigation can be solved with the help of known finite element software, using shell finite elements with membrane stresses options in order to take into account interphase surface stresses.

Section 4 describes an implementation of the proposed approaches in the finite element software ANSYS with using its command language APDL. Model of representative volume was built in the form of a cube, evenly divided into cubic solid elastic finite elements, some of which had been declared as the elements with negligibly small stiffnesses, i.e. porous elements. We suggest an algorithm for automatic determination of interphase boundaries and location of shell elements on them, which will work for various sizes of representative volumes, built in a form of cubic lattice of hexahedral elastic finite elements. As an example, in Sect. 5 we consider the models of porous silicon material of cubic crystal system for various values of surface moduli, porosity and number of pores. As in [6, 8, 9], we note the influence of the magnitudes of the surface stiffness moduli and of area of interphase boundaries on the values of the effective moduli for porous material with nanoscale structure.

2 Homogenization of Two-Phase Elastic Nanocomposite by Effective Moduli Method

Let Ω be a representative volume of two-phase bulk (mixture) composite elastic body with nanodimensional inclusions; $\Omega = \Omega^{(1)} \cup \Omega^{(2)}$; $\Omega^{(1)}$ is the volume occupied by the primary material of the first phase (matrix); $\Omega^{(2)}$ is the set of the volumes occupied by the material of the second phase (inclusions); $\Gamma = \partial\Omega$ is the external boundary of the volume Ω; Γ^s is the set of frontier surfaces of materials with different phases ($\Gamma^s = \partial\Omega^{(1)} \cap \partial\Omega^{(2)}$); $\mathbf{n}$ is the external unit normal vector to the boundary, outward with respect to $\Omega^{(1)}$, i.e. to the region occupied by the material of the matrix; $\mathbf{x} = \{x_1, x_2, x_3\}$ is the vector of the Cartesian coordinates. We suppose that the volumes $\Omega^{(1)}$ and $\Omega^{(2)}$ are filled with different elastic materials of an arbitrary anisotropy classes. Then in the framework of linear static theory of elasticity we have the following system of differential equations

$$\sigma_{ij,j} = 0, \quad \sigma_{ij} = c_{ijkl}\varepsilon_{kl}, \quad \varepsilon_{ij} = (u_{i,j} + u_{j,i})/2, \tag{1}$$

where σ_{ij} are the components of the second rank stress tensor $\boldsymbol{\sigma}$; ε_{ij} are the components of the second rank strain tensor $\boldsymbol{\varepsilon}$; u_i are the components of the displacement vector $\mathbf{u}$; c_{ijkl} are the forth rank tensor of elastic stiffness moduli; $c_{ijkl} = c^{(m)}_{ijkl}$, $\sigma_{ij} = \sigma^{(m)}_{ij}$, $\varepsilon_{ij} = \varepsilon^{(m)}_{ij}$, $u_i = u^{(m)}_i$, $\mathbf{x} \in \Omega^{(m)}$.

The material moduli of the elastic medium have usual properties of symmetry ($c_{ijkl} = c_{jikl} = c_{klij}$) and positive definiteness

$$\exists\, c_\Omega > 0: \quad \forall\, \varepsilon_{ij} = \varepsilon_{ji}, \quad 2\Pi_\Omega(\boldsymbol{\varepsilon}) = c_{ijkl}\varepsilon_{ij}\varepsilon_{kl} \geq c_\Omega \varepsilon_{ij}\varepsilon_{ij}. \tag{2}$$

In vector-matrix symbols in R^3 the formulas (1) can be rewritten in the form

$$\mathbf{L}^*(\nabla) \cdot \mathbf{T} = 0, \quad \mathbf{T} = \mathbf{c} \cdot \mathbf{S}, \quad \mathbf{S} = \mathbf{L}(\nabla) \cdot \mathbf{u}, \tag{3}$$

where

$$\mathbf{L}^*(\nabla) = \begin{bmatrix} \partial_1 & 0 & 0 & 0 & \partial_3 & \partial_2 \\ 0 & \partial_2 & 0 & \partial_3 & 0 & \partial_1 \\ 0 & 0 & \partial_3 & \partial_2 & \partial_1 & 0 \end{bmatrix}, \quad \nabla = \begin{Bmatrix} \partial_1 \\ \partial_2 \\ \partial_3 \end{Bmatrix}, \tag{4}$$

$\mathbf{T} = \{\sigma_{11}, \sigma_{22}, \sigma_{33}, \sigma_{23}, \sigma_{13}, \sigma_{12}\}$ is the array of the stress components; $\mathbf{S} = \{\varepsilon_{11}, \varepsilon_{22}, \varepsilon_{33}, 2\varepsilon_{23}, 2\varepsilon_{13}, 2\varepsilon_{12}\}$ is the array of the strain components; $\mathbf{c}$ is the 6×6 matrix of elastic moduli; $c_{\alpha\beta} = c_{ijkl}$, $\alpha, \beta = 1, \ldots, 6$, $i, j, k, l = 1, 2, 3$ with correspondence law $\alpha \leftrightarrow (ij)$, $\beta \leftrightarrow (kl)$, $1 \leftrightarrow (11)$, $2 \leftrightarrow (22)$, $3 \leftrightarrow (33)$, $4 \leftrightarrow (23) \sim (32)$, $5 \leftrightarrow (13) \sim (31)$, $6 \leftrightarrow (12) \sim (21)$; $(\ldots)^*$ is the transpose operation; $(\ldots) \cdot (\ldots)$ is the scalar product operation between two vectors or matrix—vector multiplicator. The matrix $\mathbf{c}$ is symmetric and positive defined by virtue of the properties (2)

$$\exists\, c_\Omega > 0 : \quad \forall \mathbf{S} = \{S_1, S_2, S_3, S_4, S_5, S_6\}, \quad 2\Pi_\Omega(\boldsymbol{\varepsilon}) = \mathbf{S}^* \cdot \mathbf{c} \cdot \mathbf{S} \geq c_\Omega \mathbf{S}^* \cdot \mathbf{S}. \tag{5}$$

In accordance with Gurtin–Murdoch model for surface stresses we will assume that on nanosized interphase boundaries Γ^s the following equation is satisfied

$$\mathbf{L}^*(\mathbf{n}) \cdot [\mathbf{T}] = \mathbf{L}^*(\nabla^s) \cdot \mathbf{T}^s, \quad \mathbf{x} \in \Gamma^s, \tag{6}$$

where $[\mathbf{T}] = \mathbf{T}^{(1)} - \mathbf{T}^{(2)}$ is the stress jump; ∇^s is the surface gradient operator, associated with spatial nabla-operator by the formula $\nabla^s = \nabla - \mathbf{n}(\partial/\partial r)$, where r is the coordinate, measured by the normal Γ^s; $\mathbf{T}^s = \{\sigma^s_{11}, \sigma^s_{22}, \sigma^s_{33}, \sigma^s_{23}, \sigma^s_{13}, \sigma^s_{12}\}$ is the array of surface stresses σ^s_{ij}.

We adopt that the surface stresses $\mathbf{T}^s$ are related to the surface strains $\mathbf{S}^s$ by the surface Hooke's law

$$\mathbf{T}^s = \mathbf{c}^s \cdot \mathbf{S}^s, \quad \mathbf{S}^s = \mathbf{L}(\nabla^s) \cdot \mathbf{u}^s, \quad \mathbf{u}^s = \mathbf{A} \cdot \mathbf{u}, \quad \mathbf{A} = \mathbf{I} - \mathbf{n}\mathbf{n}^*, \tag{7}$$

where $\mathbf{c}^s$ is the 6×6 matrix of surface elastic moduli; $\mathbf{I}$ is the 3×3 identity matrix.

The properties of the matrix of surface elastic moduli $\mathbf{c}^s$ are analogous to the corresponding properties of the matrix $\mathbf{c}$, i.e. $c^s_{\alpha\beta} = c^s_{\beta\alpha}$, and in local coordinate system, attached with tangent orts $\boldsymbol{\tau}_1$, $\boldsymbol{\tau}_2$ and normal $\mathbf{n}$, the transformed matrix $\tilde{\mathbf{c}}^s$ has the following features

$$\exists\, c_\Gamma > 0 : \quad \forall \mathbf{S}^s = \{S^s_1, S^s_2, 0, 0, 0, S^s_6\}, \quad 2\Pi_\Gamma(\boldsymbol{\varepsilon}^s) = \mathbf{S}^{s*} \cdot \tilde{\mathbf{c}}^s \cdot \mathbf{S}^s \geq c_\Gamma \mathbf{S}^{s*} \cdot \mathbf{S}^s, \tag{8}$$

that follow from the condition of the positive definiteness of the surface energy density $\Pi_\Gamma(\boldsymbol{\varepsilon}^s)$ relative to surface strain $\mathbf{S}^s$.

Setting the appropriate boundary conditions at $\Gamma = \partial\Omega$, we can find the solutions of the problems (3)–(7) for heterogeneous medium in the representative volume Ω. Then the comparison of the averaged stress and strain fields with analogous values for homogeneous medium (the comparison medium) will permit to determine the effective moduli for the composite material. We note that for anisotropic media in order to determine the full set of the effective moduli it is necessary to solve several problems for different types of boundary conditions.

Here the main tasks consist in the choice of the representative volume and boundary problems for the heterogeneous medium and the comparison medium, as well as the technologies for solving the problems for heterogeneous media. According to the previously developed methods of modeling the composite materials of ordinary sizes [23, 24], we consider analogous approaches for the elastic problems with surface stresses.

For homogeneous elastic comparison medium we adopt that the same equations (1) or (3), (4) are satisfied with constant moduli $\mathbf{c}^{\mathrm{eff}}$, which are to be determined.

For determination of the moduli $\mathbf{c}^{\mathrm{eff}}$ let us assume that at the external boundary Γ the following relations take place

$$\mathbf{u} = \mathbf{L}^*(\mathbf{x}) \cdot \mathbf{S}_0, \quad \mathbf{x} \in \Gamma, \tag{9}$$

where $\mathbf{S}_0 = \{\varepsilon_{011}, \varepsilon_{022}, \varepsilon_{033}, 2\varepsilon_{023}, 2\varepsilon_{013}, 2\varepsilon_{012}\}$; ε_{0ij} are some constant values that do not depend on $\mathbf{x}$. Then $\mathbf{u} = \mathbf{L}^*(\mathbf{x}) \cdot \mathbf{S}_0$, $\mathbf{S} = \mathbf{S}_0$, $\mathbf{T} = \mathbf{T}_0 = \mathbf{c}^{\mathrm{eff}} \cdot \mathbf{S}_0$ will give the solution for the problem (3)–(7) in the volume Ω for the homogeneous comparison medium.

Let us solve now problem (3)–(9) for heterogeneous medium and assume that for this medium and for the comparison medium the averaged stresses are equal $\langle \mathbf{T} \rangle = \langle \mathbf{T}_0 \rangle$, where hereinafter the angle brackets $\langle (...) \rangle$ denote the averaged by the volume Ω and by the interface surfaces Γ^s values

$$\langle (...) \rangle = \frac{1}{|\Omega|} \left(\int_{\Omega} (...) \, d\Omega + \int_{\Gamma^s} (...)^s \, d\Gamma \right). \tag{10}$$

Therefore we obtain that for the effective moduli the equation $\mathbf{c}^{\mathrm{eff}} \cdot \mathbf{S}_0 = \langle \mathbf{T} \rangle$ is satisfied, where $\mathbf{S}_0$ are the given values from the boundary conditions (9). Hence, even in the assumption of the anisotropy of the general form for the comparison medium, all the stiffness moduli $c^{\mathrm{eff}}_{\alpha\zeta}$ can be computed. Indeed, setting in (6) $\mathbf{S}_0 = \varepsilon_0 \mathbf{h}_\zeta$, $\varepsilon_0 = \mathrm{const}$, where ζ is some fixed index mark ($\mathbf{h}_\zeta$ is the vector from six-dimensional basic set for the components for strain tensor basic set; $\mathbf{h}_j = \mathbf{e}_j \mathbf{e}_j$, $j = 1, 2, 3$; $\mathbf{h}_4 = (\mathbf{e}_2\mathbf{e}_3 + \mathbf{e}_3\mathbf{e}_2)/2$; $\mathbf{h}_5 = (\mathbf{e}_1\mathbf{e}_3 + \mathbf{e}_3\mathbf{e}_1)/2$; $\mathbf{h}_6 = (\mathbf{e}_1\mathbf{e}_2 + \mathbf{e}_2\mathbf{e}_1)/2$; $\mathbf{e}_j$ are the orts of the Cartesian coordinate system), we get the computation formulas for the elastic stiffness moduli: $c^{\mathrm{eff}}_{\alpha\zeta} = \langle T_\alpha \rangle / \varepsilon_0$, $\alpha = 1, \ldots, 6$.

Note that the boundary value problems (3)–(9) differ from the usual linear elastic problems by the presence of the interface boundary conditions (6), (7) which are typical for the Gurtin–Murdoch model of surface stresses for nanosized bodies.

For the numerical solution of the problem (3)–(9) we derive its weak or generalized statements. Previously we introduce the space of the vector functions $\mathbf{u}$, defined on Ω.

On the set of vector functions $\mathbf{u} \in C^1$ which satisfy the homogeneous boundary condition (9), i.e. $\mathbf{u} = 0$ on Γ, we introduce the scalar product $(\mathbf{v}, \mathbf{u})_{H^1_u} = \int_\Omega \mathbf{S}(\mathbf{v})^* \cdot$

$\mathbf{S}(\mathbf{u})\, d\Omega + \int_{\Gamma^s} \mathbf{S}^s(\mathbf{v})^* \cdot \mathbf{S}^s(\mathbf{u})\, d\Gamma$. The closure of this set of vector functions $\mathbf{u}$ in the norm generated by the indicated scalar product will be denoted by H^1_u.

In order to formulate the generalized or weak solution we scalar multiply the first equation (3) by arbitrary vector function $\mathbf{v} \in H^1_u$. After integration the obtained equation on Ω, and using the standard technique of the integration by parts with Eqs. (3)–(9), we obtain the following integral relations

$$c(\mathbf{v}, \mathbf{u}) = 0, \tag{11}$$

where

$$c(\mathbf{v}, \mathbf{u}) = c_\Omega(\mathbf{v}, \mathbf{u}) + c_{\Gamma^s}(\mathbf{v}, \mathbf{u}), \tag{12}$$

$$c_\Omega(\mathbf{v}, \mathbf{u}) = \int_\Omega \mathbf{S}^*(\mathbf{v}) \cdot \mathbf{c} \cdot \mathbf{S}(\mathbf{u})\, d\Omega, \quad c_{\Gamma^s}(\mathbf{v}, \mathbf{u}) = \int_{\Gamma^s} \mathbf{S}^{s*}(\mathbf{v}) \cdot \mathbf{c}^s \cdot \mathbf{S}^s(\mathbf{u})\, d\Gamma. \tag{13}$$

Further, we present the solution $\mathbf{u}$ of the problem (3)–(9) in the form

$$\mathbf{u} = \mathbf{u}_d + \mathbf{u}_b, \tag{14}$$

where $\mathbf{u}_d$ satisfies homogeneous boundary conditions and ad hoc fitted function $\mathbf{u}_b$ satisfies the inhomogeneous boundary conditions on Γ, i.e.

$$\mathbf{u}_d = 0, \quad \mathbf{u}_b = \mathbf{L}^*(\mathbf{x}) \cdot \mathbf{S}_0, \quad \mathbf{x} \in \Gamma, \tag{15}$$

and therefore, $\mathbf{u}_d \in H^1_u$.

By using (14) we can rewrite Eq. (11) in the form

$$c(\mathbf{v}, \mathbf{u}_d) = L(\mathbf{v}), \quad L(\mathbf{v}) = -c(\mathbf{v}, \mathbf{u}_b). \tag{16}$$

Now we can define the generalized or weak solution of the static elastic problem with surface stresses (3)–(9) using previously introduced functional space. Namely, the function $\mathbf{u}$, in the form (14), (15) is the weak solution of the problem (3)–(9), if Eq. (16) with (12), (13) is satisfied for $\forall\, \mathbf{v} \in H^1_u$.

So far we have been discussing the two-phase composites. However, we can note that the presented models also describe homogenization procedures for porous composites with surface effects, if we put the stiffness moduli negligible.

3 Finite Element Approximations

For solving problem (11) or (16) for elastic body with surface stresses in weak forms we will use classical finite element approximation techniques. Let Ω_h be a region of the corresponding finite element mesh composed of volume elements, $\Omega_h \approx \Omega, \Omega_h = \Omega_h^{(1)} \cup \Omega_h^{(2)}, \Omega_h^{(j)} \approx \Omega^{(j)}$, $\Omega_h = \cup_k \Omega_{ek}$, where Ω_{ek} is a separate volume finite element with number k. Let also Γ_h^s be a finite element mesh of surface elements conformable with the volume mesh Ω_h, $\Gamma_h^s = \partial\Omega_h^{(1)} \cap \partial\Omega_h^{(2)}$, $\Gamma_h^s \approx \Gamma^s$, $\Gamma_h^s = \cup_m \Gamma_{em}^s$, Γ_{em}^s is a separate surface finite element with number m, and the elements Γ_{em}^s are the faces of the suitable volume elements Ω_{ek} located on the interface boundaries.

We will use the classic Lagrangian or serendipity volume finite elements with nodal degrees of freedom of displacements. Note that due to the structure of surface mechanical fields (6), (7), for the elements Γ_{em}^s we can use standard shell or plate elements with elastic membrane stresses options, i.e. only with nodal degrees of freedom of displacements. For these elements we can take a fictitious unit thickness so that the surface moduli from (6) to (7) can be determined by the product of specially defined volume moduli and shell thickness.

On these finite element meshes we will find the approximation to the weak solutions $\{\mathbf{u}_h \approx \mathbf{u}\}$ for static elasic problem (11) in the form

$$\mathbf{u}_h(\mathbf{x}) = \mathbf{N}^*(\mathbf{x}) \cdot \mathbf{U}, \tag{17}$$

where $\mathbf{N}^*$ is the matrix of the shape functions for displacements, $\mathbf{U}$ is the vector of nodal displacements. Here, the surface shape functions are the reduction on the boundaries Γ_h^s of the volume shape functions.

According to conventional finite element technique, we approximate the continuous weak formulations of the elastic problems by the corresponding problems in finite-dimensional spaces. Substituting (17) and similar representations for projection functions into integral relations (11) for Ω_h, we obtain the following finite element system

$$\mathbf{K} \cdot \mathbf{U} = 0, \tag{18}$$

where

$$\mathbf{K} = \mathbf{K}_\Omega + \mathbf{K}_\Gamma, \tag{19}$$

$$\mathbf{K}_\Omega = \int_{\Omega_h} \mathbf{B}^* \cdot \mathbf{c} \cdot \mathbf{B}\, d\Omega, \quad \mathbf{K}_\Gamma = \int_{\Gamma_h^s} \mathbf{B}^{s*} \cdot \mathbf{c}^s \cdot \mathbf{B}^s\, d\Gamma, \tag{20}$$

$$\mathbf{B}^{(s)} = \mathbf{L}^*(\nabla^{(s)}) \cdot \mathbf{A} \cdot \mathbf{N}^*. \tag{21}$$

Also, we can represent the finite element solutions in the another variants considering the main boundary conditions similarly to the reforming (14)–(16): $\mathbf{u}_h = \mathbf{u}_{dh} + \mathbf{u}_{bh}$, $\mathbf{u}_{dh} \approx \mathbf{u}_d$, $\mathbf{u}_{dh} = \mathbf{N}_d^* \cdot \mathbf{U}_d$, $\mathbf{u}_{bh} \approx \mathbf{u}_b$, $\mathbf{u}_{bh} = \mathbf{N}_b^* \cdot \mathbf{U}_b$, $\mathbf{N} = \{\mathbf{N}_d, \mathbf{N}_b\}$,

$$\mathbf{K} = \begin{bmatrix} \mathbf{K}_{dd} & \mathbf{K}_{db} \\ \mathbf{K}_{bd} & \mathbf{K}_{bb} \end{bmatrix}, \quad \mathbf{U} = \begin{Bmatrix} \mathbf{U}_d \\ \mathbf{U}_b \end{Bmatrix}, \tag{22}$$

where $\mathbf{U}_b$ is the vector, known from the main boundary conditions.

So, after using these expressions we can obtain from Eq. (18) the reduced system with respect to only unknown vector $\mathbf{U}_d$:

$$\mathbf{K}_{dd} \cdot \mathbf{U}_d = \mathbf{F}_d, \quad \mathbf{F}_d = -\mathbf{K}_{db} \cdot \mathbf{U}_b. \tag{23}$$

Thus, the homogenizing problems for elastic composite with surface stresses can be solved by finite element approaches. The resulting finite element systems (23) with (19)–(22) differ from similar systems for the bodies of usual sizes by the matrix $\mathbf{K}_\Gamma$ in (20). This matrix arises due to the surface stresses.

4 Modeling of Representative Volumes

The presented approaches were implemented in the finite element software ANSYS for elastic composites of cubic anisotropy class. The representative volume Ω was taken in a shape of cube divided in $L \times L \times L$ geometrically equal small cubes which were 20-node hexahedral structured solid finite elements SOLID95. As a result in grand cube Ω there were $n \times n \times n$ finite elements of the size $l = L/n$, where n was an integer. In such two-phase composite the finite elements of the first phase had material properties of the original elastic material, and the pores had negligibly small elastic stiffness moduli by simple random method. Then the elements with the pore properties were selected according to the adopted microstructure of the composite. In the case of slightly porous material of irregular stochastic structure depending on the given porosity, some of the finite elements were announced as pores. It can be noted that such model is easy to build but it does not support the connectivity of the elements of the first phase and does not reflect the connectivity structure of the elements of the second phase (closed or open pores). Other methods that support the connectivity of the skeleton consisting of the elements of the first phase or supporting the cluster properties for the elements of the second phase are described [23].

For automatic coating of the inner boundaries of the pores in a cubic representative volume the following algorithm has been used. At first the finite elements with the pore material properties are selected. Then the external boundaries of the obtained array of elements are covered by target contact elements TARGE170 using the command TSHAP, QUA8. Hence, the edges of all finite elements with the pore properties lying on the external surfaces of the set of these elements get covered by eight node contact elements (TARGE170 of the shape QUA8). Then the contact elements, located on the external border of the whole representative volume, are deleted and the remaining finite elements are changed to the eight node shell elements SHELL281 with the membrane stress only option. As a result, all contact boundaries where elastic structural elements are in contact with pores have been coated by the membrane finite elements which simulate the effect of surface stresses.

At the next stage the problems (2)–(6) were solved for the representative volume by finite element method using the described above technology. Then in the ANSYS postprocessor the average stresses were calculated according to (7) both over solid

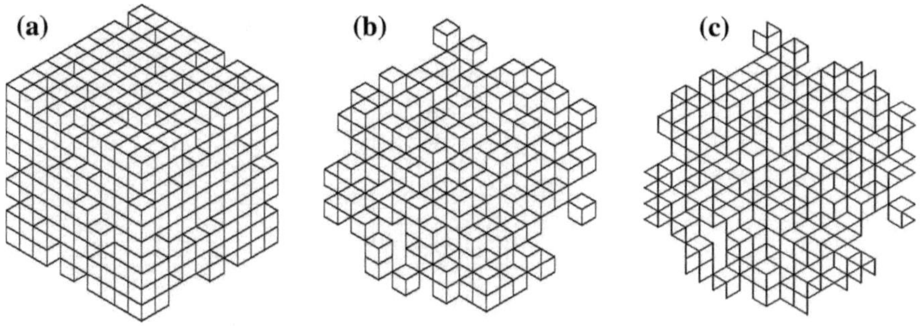

Fig. 1 One variant of cubic representative volume Ω for $p = 20\%, n = 10$: **a** matrix; **b** pores; **c** interphase boundaries

and surface finite elements. In the end, using the corresponding formulas for the averaged stresses, the effective stiffness modules of porous composite with surface effects were obtained.

We note that as the pores in the described above algorithm were chosen with the help of the random number generator, each new launch of the program will give different distribution of these elements. Therefore for different location of the pores the total area of the membranes Γ^s will change and consequently the total stiffness of the volume will change too. However, for small porosity p these effects will not have significant influence on the resulting values of the effective moduli for each new launch of the program.

Figure 1 shows the elements of the first phase (Fig. 1a), the elements of the second phase (Fig. 1b) and the interphase surface elements (Fig. 1c) for one run of the algorithm at $p = 20\%$ and $n = 10$, when the ratio $r(S_p) = |\Gamma^s|/|\Gamma|$ of the pores area to the total area of the volume is equal to 1.465.

It is interesting to note the fact that for the same size L of the volume Ω depending on the number of elements along the axes n, the size $l = L/n$ of the finite elements changes and therefore the size of the separate pores also changes. Hence, for the fixed percentage of porosity p when the value of n increases the size of the pores will decrease but the total amount of pores N_p will increase and as a result the total area of the boundary Γ^s will increase.

Figure 2 shows the surface elements for the porosity $p = 10\%$ for $n = 10$ (Fig. 2a), for $n = 15$ (Fig. 2b), and for $n = 20$ (Fig. 2c). In the shown runs of the algorithm the ratio of the pore area to the total area of the volume was equal to $r(S_p) = 0.803$ for $n = 10$, $r(S_p) = 1.121$ for $n = 15$, and $r(S_p) = 1.700$ for $n = 20$. As it can be seen, the last value obtained for $p = 10\%$, $n = 20$ is larger than the one obtained for Fig. 1 ($r(S_p) = 1.465$) for the larger porosity $p = 20\%$ and smaller number of elements $n = 10$.

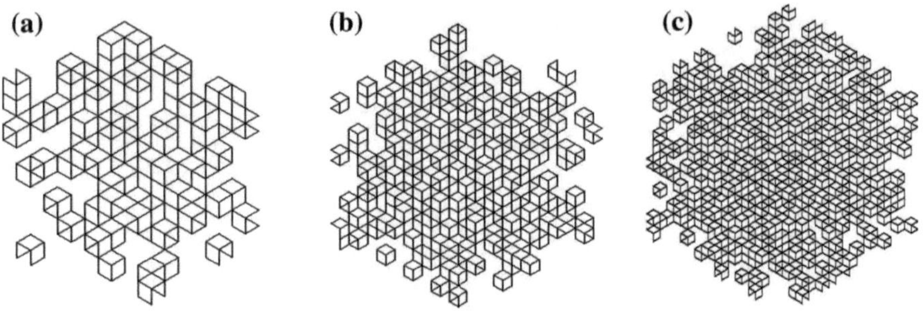

Fig. 2 Three variants of interphase boundaries for $p = 10\%$: **a** $n = 10$; **b** $n = 15$; **c** $n = 20$

5 Numerical Results and Discussion

As an example let us consider a nanoporous silicon. As it is known [17], the conventional silicon is an anisotropic material of cubic system, and therefore the structure of its material constants has the form

$$\mathbf{c} = \begin{bmatrix} c_{11} & c_{12} & c_{12} & 0 & 0 & 0 \\ c_{12} & c_{11} & c_{12} & 0 & 0 & 0 \\ c_{12} & c_{12} & c_{11} & 0 & 0 & 0 \\ 0 & 0 & 0 & c_{44} & 0 & 0 \\ 0 & 0 & 0 & 0 & c_{44} & 0 \\ 0 & 0 & 0 & 0 & 0 & c_{44} \end{bmatrix}. \tag{24}$$

For calculation we assume the following values of the material macro moduli of silicon with zero porosity ([17]): $c_{11} = 16.56 \times 10^{10}$ (N/m^2), $c_{12} = 6.39 \times 10^{10}$ (N/m^2), $c_{44} = 7.95 \times 10^{10}$ (N/m^2), where superscript $^{(1)}$ is absent ($c_{\alpha\beta} \sim c^{(1)}_{\alpha\beta}$, etc.) The material constants for the pores (marked by superscript $^{(2)}$) were taken equal to the following values: $c^{(2)}_{\alpha\beta} = \kappa c^{(1)}_{\alpha\beta}$, $\kappa = 1 \times 10^{-10}$.

As it can be seen, the considered elastic material of macrosize in the static problems is characterized by three material elastic stiffness moduli: c_{11}, c_{12}, c_{44}. For nanoporous silicon we can also introduce the surface moduli of the same cubic classes of anisotropy: c^s_{11}, c^s_{12} and c^s_{44}. However, currently there is not enough experimental data for the surface moduli values, and existing data is quite contradictory. In this connection we assume $c^s_{\alpha\beta} = h_e \kappa^s c_{\alpha\beta}$, where h_e is the formal multiplier, which we can accept as thicknesses for shell elements in ANSYS; κ^s is the multipliers, by which we will define the resulting values of the surface moduli.

Let us assume that the models of the representative volumes do not have explicit geometric anisotropy and hence the porous silicon also belongs to the class of anisotropic materials of cubic system. Then for the full set of effective stiffness moduli it is enough to solve two problems (3)–(9) with various boundary conditions in (9)

$$\text{I}\quad \mathbf{S}_0 = \varepsilon_0 \mathbf{h}_1 \quad \Rightarrow \quad c_{1j}^{\text{eff}} = \langle T_j \rangle / \varepsilon_0 = \langle \sigma_{jj} \rangle / \varepsilon_0, \quad j = 1,2,3, \tag{25}$$

$$\text{II}\quad \mathbf{S}_0 = \varepsilon_0 \mathbf{h}_4 \quad \Rightarrow \quad c_{44}^{\text{eff}} = \langle T_4 \rangle / \varepsilon_0 = \langle \sigma_{23} \rangle / \varepsilon_0, \tag{26}$$

where in (25) for the computations it should be $c_{12}^{\text{eff}} \approx c_{13}^{\text{eff}}$.

Thus, two boundary-value problems for the representative volumes, that differ by their boundary conditions, enable to calculate three main material elastic constants of nanoporous silicon: c_{11}^{eff}, c_{12}^{eff} and c_{44}^{eff}. From these values we can find other important constants: the elastic compliances $s_{11}^{\text{eff}} = (c_{11}^{\text{eff}} + c_{12}^{\text{eff}})/\Delta_c^{\text{eff}}$, $s_{12}^{\text{eff}} = -c_{12}^{\text{eff}}/\Delta_c^{\text{eff}}$, $\Delta_c^{\text{eff}} = (c_{11}^{\text{eff}})^2 + c_{11}^{\text{eff}} c_{12}^{\text{eff}} - 2(c_{12}^{\text{eff}})^2$; the Young's modulus $E^{\text{eff}} = 1/s_{11}^{\text{eff}}$; the Poisson's ratio $\nu^{\text{eff}} = -s_{12}^{\text{eff}}/s_{11}^{\text{eff}} = c_{12}^{\text{eff}}/(c_{11}^{\text{eff}} + c_{12}^{\text{eff}})$ and the bulk modulus $K^{\text{eff}} = E^{\text{eff}}/[3(1 - 2\nu^{\text{eff}})]$.

In the calculations, we have changed the values of multiplier κ^s from 10^{-4} to 1 with immobile factor $h_e = 1$ (m).

Figures 3, 4 and 5 show the dependencies of the relative effective moduli with respect to the factor k^s in the logarithmic scale ($\lg \equiv \log_{10}$): elastic stiffness modulus $r(c_{11}) = c_{11}^{\text{eff}}/c_{11}$ (Fig. 3a), Young's modulus $r(E) = E^{\text{eff}}/E$ (Fig. 3b), stiffness mod-

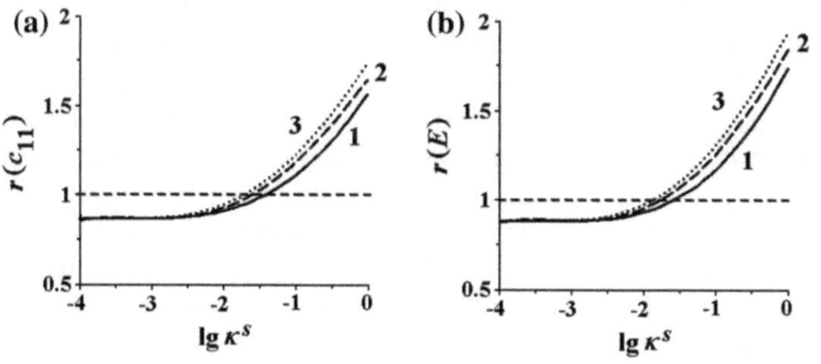

Fig. 3 Dependencies of the relative effective moduli c_{11}^{eff} (**a**) and E^{eff} (**b**) versus the multiplier k^s for porosity $p = 10\%$: *1*—$n = 10$; *2*—$n = 15$; *3*—$n = 20$

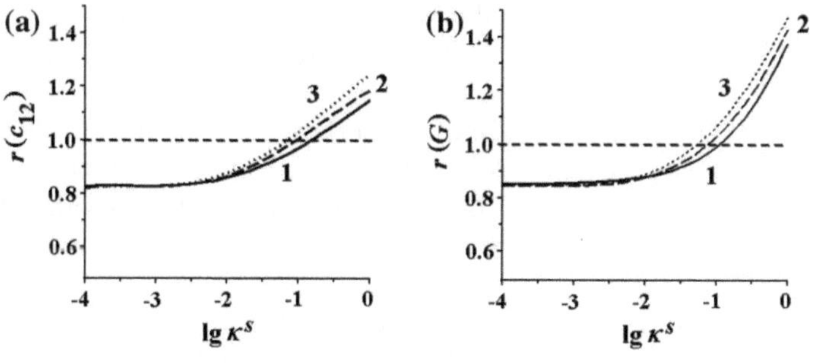

Fig. 4 Dependencies of the relative effective moduli c_{12}^{eff} (**a**) and G^{eff} (**b**) versus the multiplier k^s for porosity $p = 10\%$: *1*—$n = 10$; *2*—$n = 15$; *3*—$n = 20$

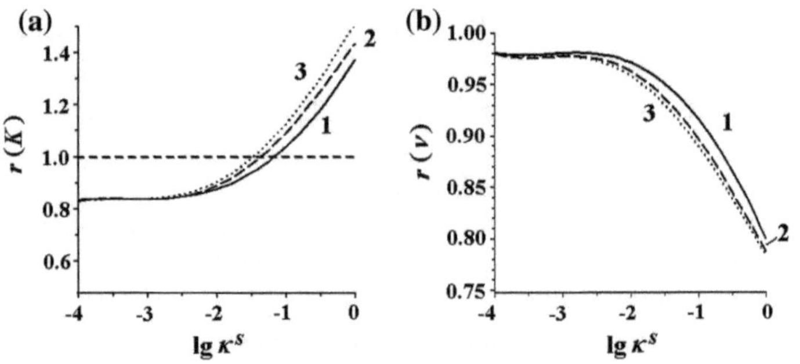

Fig. 5 Dependencies of the relative effective moduli K^{eff} (**a**) and ν^{eff} (**b**) versus the multiplier k^s for porosity $p = 10\%$: *1*—$n = 10$; *2*—$n = 15$; *3*—$n = 20$

ulus $r(c_{12}) = c_{12}^{\text{eff}}/c_{12}$ (Fig. 4a), shear modulus $r(G) = G^{\text{eff}}/G$, $G = c_{44}$ (Fig. 4b), bulk modulus $r(K) = K^{\text{eff}}/K$ (Fig. 5a), and Poisson's ratio $r(\nu) = \nu^{\text{eff}}/\nu$ (Fig. 5b). Here the porosity remains unchanged $p = 10\%$, and the curves 1, 2, 3 correspond to the values $n = 10$, $n = 15$ and $n = 20$, respectively. As we can see, the behaviors of effective moduli c_{11}^{eff}, c_{12}^{eff}, E^{eff}, G^{eff} and K^{eff}, depending on the surface moduli values, qualitative coincide, but moduli c_{11}^{eff} and E^{eff} increase a few rapidly when increasing k^s, and at that Young's modulus increases most rapidly. For $k^s \leq 10^{-2}$ the surface effects have only a slight influence on the values of stiffness moduli. Approximately from the values $k^s \approx 10^{-2}$, stiffness moduli begin to grow quite rapidly. Thus for fixed porosity $p = 10\%$ with increasing $n = 10, 15, 20$ the pore size decreased, but number of pores and its total area increased. As a result of the surface effects are manifested in a greater extent, and stiffness moduli (Figs. 3, 4 and 5a) increase (curves 1 are located lower than curves 2, and curves 2 are located lower than curves 3). It should be noted that for great values of surface stiffness moduli the overall effective stiffness modules of porous nanomaterials may exceed the relevant moduli of the continuous material. In Figs. 3, 4 and 5a, this effect is observed when the relative values of the moduli are more than 1.

Figure 5c shows that the behavior of the Poisson ratio with an increase of the surface module is the opposite to the behavior of stiffness moduli c_{11}^{eff}, c_{12}^{eff}, E^{eff}, G^{eff} and K^{eff}, that it can be explained to an increase of the overall stiffness of the material.

In order to analyze the influence of the surface effect on the effective modules with different porosity, we have calculated the relative effective stiffness moduli with a fixed number of elements $n = 10$, but with different porosity and with different, but not too large, values of the factor k^s. The results of these calculations are shown in Figs. 6, 7 and 8.

As is seen from these Figures, for the small values of the factor k^s (curves 3 and 4) the surface effects do not affect to the stiffness moduli. However, for any porosity the surface stresses increase the effective stiffnesses of the porous material. Moreover, as mentioned earlier, there are cases where the nano-porous material will have

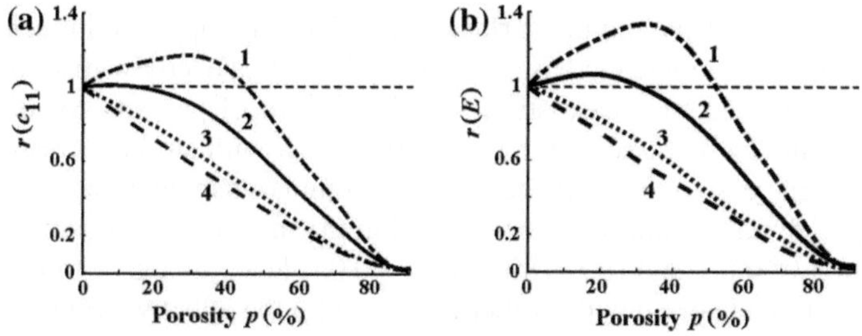

Fig. 6 Dependencies of the relative effective moduli c_{11}^{eff} (**a**) and E^{eff} (**b**) versus the porosity for $n = 10$ and for multiplier k^s: *1*—$k^s = 0.1$; *2*—$k^s = 0.05$; *3*—$k^s = 0.01$; *4*—$k^s = 0.001$

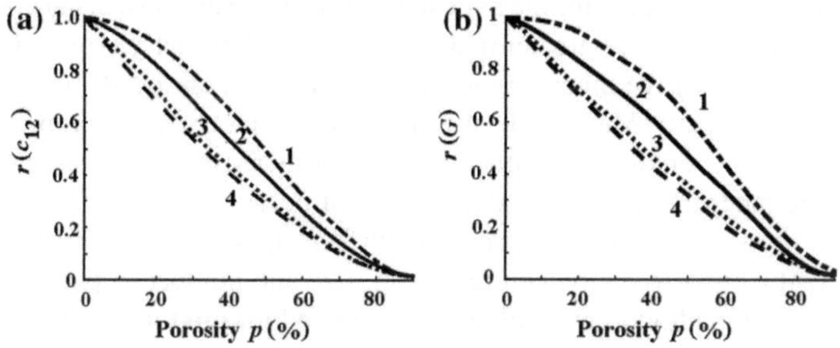

Fig. 7 Dependencies of the relative effective moduli c_{12}^{eff} (**a**) and G^{eff} (**b**) versus the porosity for $n = 10$ and for multiplier k^s: *1*—$k^s = 0.1$; *2*—$k^s = 0.05$; *3*—$k^s = 0.01$; *4*—$k^s = 0.001$

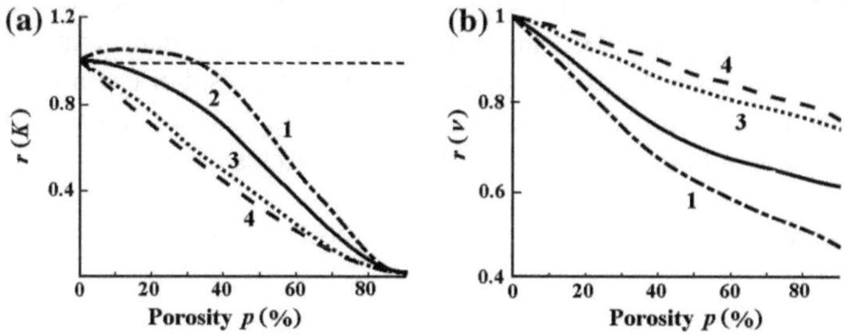

Fig. 8 Dependencies of the relative effective moduli K^{eff} (**a**) and ν^{eff} (**b**) versus the porosity for $n = 10$ and for multiplier k^s: *1*—$k^s = 0.1$; *2*—$k^s = 0.05$; *3*—$k^s = 0.01$; *4*—$k^s = 0.001$

greater stiffness than the same solid material. This situation takes place with $k^s = 0.1$ for stiffness modulus c_{11}^{eff} if $p \leq 45\%$, for Young's modulus E^{eff} if $p \leq 54\%$, and for shear modulus K^{eff} if $p \leq 33\%$, (see the dashed lines in the Figures, where we have the inequalities $r(\ldots) > 1$). When $k^s = 0.05$, that is more consistent with the experimental data for other materials, only the effective Young's modulus is much greater than the Young's modulus of solid material, but this effect is observed in a sufficiently wide range of porosity $p \leq 30\%$.

Note that the porosity and the surface effects have the opposite influence on the effective stiffness: a simple increase of porosity leads to a decrease in stiffness moduli, but surface effects increase stiffness. The growth of porosity for nanoporous materials entails an increase of the boundaries areas with surface stresses. Therefore, with the increase of porosity the stiffness moduli of nanomaterial may increase. For example, for large values of surface moduli the increase of porosity for small p ($p \leq 33\%$ for $k^s = 0.1$ and ($p \leq 18\%$ for $k^s = 0.05$) leads to a growth of the effective Young's modulus, and with further increase of p the effective Young's modulus begins to decrease.

Thus, as in [6, 8, 9], from the results of computational experiments the following trends have been observed. If we compare two similar bodies with ordinary dimensions and the nanoscale, then for the nanosized body at the expense of surface stresses the effective stiffness will be greater than for the body of ordinary size. Furthermore, for the porous body of the macroscopic size the effective elastic stiffness decreases with increase of porosity. Meanwhile, the effective stiffness of nanocomposite porous body with the same porosity may either decrease or increase depending upon the values of surface moduli, dimensions and number of pores. This effect is explained by the fact that the sizes of surface pore with surface stresses depend not only on the overall porosity, but also on the configuration, size and number of pores.

The described methodology could be also applied for mixture anisotropic nanostructured composites with other type of connectivity for different physic-mechanical fields, such as poroelastic, thermoelastic, piezoelectric, magnetoelectric (magneto-electroelastic) and other nanocomposites ([7, 12–14, 22, 25, 34], etc.) At the element level it allows to take into account local types of inhomogeneities, such as, for example, a rotation of the polarization vectors (element coordinate systems) in the vicinity of the pores for porous piezoceramic materials. For example, for porous piezoelectric nanosized composites the analogous approaches can be applied with taking into account both mechanical and electric surface effects.

Acknowledgements This work was supported by the Russian Science Foundation (grant number 15-19-10008).

References

1. Brisard, S., Dormieux, L., Kondo, D.: Hashin-Shtrikman bounds on the bulk modulus of a nanocomposite with spherical inclusions and interface effects. Comput. Mater. Sci. **48**, 589–596 (2010)
2. Brisard, S., Dormieux, L., Kondo, D.: Hashin-Shtrikman bounds on the shear modulus of a nanocomposite with spherical inclusions and interface effects. Comput. Mater. Sci. **50**, 403–410 (2010)
3. Chen, T., Dvorak, G.J., Yu, C.C.: Solids containing spherical nano-inclusions with interface stresses: effective properties and thermal-mechanical connections. Int. J. Solids Struct. **44**, 941–955 (2007)
4. Duan, H.L., Wang, J., Huang, Z.P., Karihaloo, B.L.: Eshelby formalism for nano-inhomogeneities. Proc. R. Soc. A. **461**, 3335–3353 (2005)

5. Duan, H.L., Wang, J., Huang, Z.P., Karihaloo, B.L.: Size-dependent effective elastic constants of solids containing nano-inhomogeneities with interface stress. J. Mech. Phys. Solids. **53**, 1574–1596 (2005)
6. Duan, H.L., Wang, J., Karihaloo, B.L., Huang, Z.P.: Nanoporous materials can be made stiffer than non-porous counterparts by surface modification. Acta Mater. **54**, 2983–2990 (2006)
7. Eremeev, V.A., Nasedkin, A.V.: Natural vibrations of nanodimensional piezoelectric bodies with contact-type boundary conditions. Mech. Solids. **50**(5), 495–507 (2015)
8. Eremeyev, V.A.: On effective properties of materials at the nano- and microscales considering surface effects. Acta Mech. **227**, 29–42 (2016)
9. Eremeyev, V., Morozov, N.: The effective stiffness of a nanoporous rod. Doklady Phys. **55**(6), 279–282 (2010)
10. Gao, W., Yu, S.W., Huang, G.Y.: Finite element characterization of the size-dependent mechanical behaviour in nanosystem. Nanotechnology **17**, 1118–1122 (2006)
11. Goudarzi, T., Avazmohammadi, R., Naghdabadi, R.: Surface energy effects on the yield strength of nanoporous materials containing nanoscale cylindrical voids. Mech. Mater. **42**, 852–862 (2010)
12. Gu, S.-T., Liu, J.-T., He, Q.-C.: Piezoelectric composites: imperfect interface models, weak formulations and benchmark problems. Comput. Mater. Sci. **94**, 182–190 (2014)
13. Gu, S.-T., Liu, J.-T., He, Q.-C.: The strong and weak forms of a general imperfect interface model for linear coupled multifield phenomena. Int. J. Eng. Sci. **85**, 31–46 (2014)
14. Gu, S.-T., Qin, L.: Variational principles and size-dependent bounds for piezoelectric inhomogeneous materials with piezoelectric coherent imperfect interfaces. Int. J. Eng. Sci. **78**, 89–102 (2014)
15. Gurtin, M.E., Murdoch, A.I.: A continuum theory of elastic material surfaces. Arch. Rat. Mech. Anal. **57**(4), 291–323 (1975)
16. Hamilton, J.C., Wolfer, W.G.: Theories of surface elasticity for nanoscale objects. Surface Sci. **603**, 1284–1291 (2009)
17. Hull, R. (ed.): Properties of Crystalline Silicon. INSPEC, London (1999)
18. Javili, A., Steinmann, P.: A finite element framework for continua with boundary energies. Part I: the two-dimensional case. Comput. Methods Appl. Mech. Eng. **198**, 2198–2208 (2009)
19. Javili, A., Steinmann, P.: A finite element framework for continua with boundary energies. Part II: the three-dimensional case. Comput. Methods Appl. Mech. Eng. **199**, 755–765 (2010)
20. Jeong, J., Cho, M., Choi, J.: Effective mechanical properties of micro/nano-scale porous materials considering surface effects. Interact. Multiscale Mech. **4**(2), 107–122 (2011)
21. Kushch, V.I., Mogilevskaya, S.G., Stolarski, H.K., Crouch, S.L.: Elastic fields and effective moduli of particulate nanocomposites with the Gurtin-Murdoch model of interfaces. Int. J. Solids Struct. **50**, 1141–1153 (2013)
22. Nasedkin, A.V., Eremeyev, V.A.: Some models for nanosized agnetoelectric bodies with surface effects. In: Parinov, I.A., Chang, S.-H., Topolov, V.Y. (eds.) Advanced Materials—Manufacturing, Physics, Mechanics and Applications. Springer Proceedings in Physics, vol. 175, pp. 373–391. Springer (2016)
23. Nasedkin, A.V., Nasedkina, A.A., Remizov, V.V.: Finite element modeling of porous thermoelastic composites with account for their microstructure. Vycisl. meh. splos. sred—Comput. Continuum Mech. **7**(1), 100–109 (2014)
24. Nasedkin, A.V., Nasedkina, A.A.: Finite element modeling and computer design of porous composites. In: Hellmich, C., Pichler, B., Adam, D. (eds.) Poromechanics V: Proceedings of the Fifth Biot Conference on Poromechanics, July 10–12, 2013, Vienna, Austria, pp. 608–617. ASCE (2013)
25. Nasedkin, A.V., Eremeyev, V.A.: Harmonic vibrations of nanosized piezoelectric bodies with surface effects. ZAMM **94**(10), 878–892 (2014)
26. Nazarenko, L., Bargmann, S., Stolarski, H.: Influence of interfaces on effective properties of nanomaterials with stochastically distributed spherical inclusions. Int. J. Solids Struct. **51**, 954–966 (2014)

27. Nazarenko, L., Bargmann, S., Stolarski, H.: Energy-equivalent inhomogeneity approach to analysis of effective properties of nanomaterials with stochastic structure. Int. J. Solids Struct. **59**, 183–197 (2015)
28. Povstenko, Y.Z.: Theoretical investigation of phenomena caused by heterogeneous surface-tension in solids. J. Mech. Phys. Solids. **41**, 1499–1514 (1993)
29. Riaz, U., Ashraf, S.M.: Application of finite element method for the design of nanocomposites. In: Musa, S.M. (ed.) Computational Finite Element Methods in Nanotechnology, pp. 241–290. CRC Press (2012)
30. Shuttleworth, R.: The surface tension of solid. Proc. Phys. Soc. A. **63**, 444–457 (1950)
31. Tian, L., Rajapakse, R.K.N.D.: Finite element modelling of nanoscale inhomogeneities in an elastic matrix. Comput. Mater. Sci. **41**, 44–53 (2007)
32. Wang, J., Huang, Z., Duan, H., Yu, S., Feng, X., Wang, G., Zhang, W., Wang, T.: Surface stress effect in mechanics of nanostructured materials. Acta Mech. Solida Sin. **24**(1), 52–82 (2011)
33. Wang, Z., Zhu, J., Jin, X.Y., Chen, W.Q., Zhang, C.: Effective moduli of ellipsoidal particle reinforced piezoelectric composites with imperfect interfaces. J. Mech. Phys. Solids. **65**, 138–156 (2014)
34. Xiao, J.H., Xu, Y.L., Zhang, F.C.: Size-dependent effective electroelastic moduli of piezoelectric nanocomposites with interface effect. Acta Mech. **222**(1–2), 59–67 (2011)

Acceleration Waves in Media with Microstructure

Victor A. Eremeyev

Abstract Within the unified approach to modelling of media with microstructure we discuss the propagation of acceleration waves. We describe a medium with microstructure as an elastic continuum with strain energy density which depends on deformations and additional internal variable and their first gradients. We use a Nth-order tensor as a kinematical descriptor of the microstructure. By acceleration wave we mean an isolated surface propagating in medium across which second derivatives of some fields undergo discontinuity jump. Here we formulate the conditions of existence of acceleration waves as algebraic inequality expressed using acoustic tensor.

Keywords Acceleration waves · Media with microstructure · Acoustic tensor · Micropolar medium

1 Introduction

Among many types of nonlinear waves observed in solids and fluids where the analytical solutions are rare, the acceleration waves are exceptional since their conditions of propagation can be reduced to algebraic equations. An acceleration wave called also wave of weak discontinuity of order two is a solution of motion equations with discontinuities in the second derivatives on some surfaces that are called singular. It means that the acceleration wave can be represented by an isolated traveling smooth enough surface which is a carrier of discontinuity jumps of the second derivatives with respect to the spacial coordinates and time whereas the solution and

V.A. Eremeyev (✉)
Institute of Mathematics, Mechanics and Computer Science,
Southern Federal University, Milchakova Street 8a, Rostov-on-Don 344090, Russia
e-mail: eremeyev.victor@gmail.com

V.A. Eremeyev
The Faculty of Mechanical Engineering, Rzeszów University of Technology,
al. Powstańców Warszawy 8, 35–959 Rzeszów, Poland

M.A. Sumbatyan (ed.), *Wave Dynamics and Composite Mechanics for Microstructured Materials and Metamaterials*,
Advanced Structured Materials 59, DOI 10.1007/978-981-10-3797-9_7

its first derivatives are continuous. Existence conditions of acceleration waves can be reduced to a spectral problem for an acoustic tensor and positivity of its eigenvalues. From the mathematical point of view the conditions of existence of acceleration waves coincide with the condition of strong ellipticity of the equilibrium equations. Ellipticity is a natural property of the equilibrium equations in the case of infinitesimal deformations. On the other hand the violation ellipticity for nonlinear media means that for certain deformations discontinuities may appear. Such discontinuous solutions may model such phenomena as shear-bands, phase transitions, slip surfaces, etc. Thus, analysis of conditions of propagation of acceleration waves plays an important role in the mechanics of materials.

Within the nonlinear elasticity including compressible, incompressible and media with constraints, acceleration waves are studied in many works, see, e.g., the original papers by [2, 6, 7, 19, 33, 34, 42–44], see also [21, 49, 50] where the generalization to thermoelasticity and viscoelasticity is also presented.

For the media with microstructure acceleration waves are considered in a number of papers. In particular, the propagation of acceleration waves is studied in porous media [3, 8–10, 22, 26, 46], in random materials [35, 36], in piezoelectric solids [29]. Acceleration waves are also studied in various types of fluids with complex constitutive equations, see [41, 45, 47] and the reference therein.

In nonlinear elastic micropolar media acceleration wave are studied in [23]. In [28] a generalization of these results in the case of is presented in elastic and viscoelastic micropolar media are given. Equivalence of existence of acceleration waves and the condition of strong ellipticity of the equilibrium equations is discussed in [12]. For micropolar thermoelasticity acceleration waves studied in [1, 14]. In order to describe the strain localizations in micropolar elastoplasticity the derivation of acoustic tensor and analysis of its properties is performed in [11].

The paper is organized ad follows. First we introduce the motion equations for a medium with microstructure. Then, using Maxwell's theorem we obtain the conditions of propagations of acceleration waves. As an example the acceleration waves in the micropolar medium is considered.

In what follows we use the direct tensor notations [24]. In particular, all vectors and tensors are denoted by semibold Roman font.

2 Basic Equations of a Hyperelastic Media with Internal Variables

Deformation of a non-linear elastic solid is described by the mapping from known state called initial configuration into another state called actual configuration. The mapping is given by

$$\mathbf{x} \equiv \mathbf{x}(\mathbf{X}, t) = \mathbf{X} + \mathbf{u}(\mathbf{X}, t). \tag{1}$$

Here vector $\mathbf{x}$ describes the position of a material point in the actual configuration at instant t, whereas $\mathbf{X}$ determines the position of the same material point in the reference configuration, and $\mathbf{u}$ is the displacement vector. In addition to deformation $\mathbf{x}$ for a medium with microstructure we introduce internal variable $\mathbf{W}$ which can be scalar, vector, second-order tensor, or even Nth-order tensor. Among examples of such internal variables are porosity in the theory of poroelasticity [4, 46], microrotation tensor in micropolar mechanics [13, 16], microdeformations in the micromorphic media [16–18], damage tensor [5, 25, 27]. As a result, the deformation of a solid with microstructure is determined by two fields

$$\mathbf{x} \equiv \mathbf{x}(\mathbf{X}, t), \quad \mathbf{W} = \mathbf{W}(\mathbf{X}, t). \tag{2}$$

Following the equipresence principle [50] for an hyperelastic medium we introduce the strain energy density

$$W = W(\nabla\mathbf{x}, \mathbf{W}, \nabla\mathbf{W}; \mathbf{X}). \tag{3}$$

Here ∇ is the gradient (nabla) operator in Lagrangian coordinates [24]. The specific form of functional dependence (3) depends on the nature of microstructural tensor $\mathbf{W}$. For brevity we omit here also discussion on the objectivity of $\mathbf{W}$ and form of W consistent with the principle material frame indifference. We assume W to be a twice continuously differentiable function. We use the following notations:

$$W_{,\nabla\mathbf{x}} = \frac{\partial W}{\partial \nabla\mathbf{x}}, \quad W_{,\mathbf{W}} = \frac{\partial W}{\partial \mathbf{W}}, \quad W_{,\nabla\mathbf{x}\nabla\mathbf{x}} = \frac{\partial^2 W}{\partial \nabla\mathbf{x} \partial \nabla\mathbf{x}}, \quad W_{,\nabla\mathbf{W}\nabla\mathbf{W}} = \frac{\partial^2 W}{\partial \nabla\mathbf{W} \partial \nabla\mathbf{W}}.$$

In addition to (3) we define the kinetic energy density as a positive quadratic form depending on velocity $\mathbf{v} \equiv \dot{\mathbf{x}}$ and $\dot{\mathbf{W}}$

$$K = \frac{1}{2}\rho\mathbf{v} \cdot \mathbf{v} + \mathbf{v} \cdot \mathbf{J}_1 : \dot{\mathbf{W}} + \dot{\mathbf{W}} : \mathbf{J}_2 : \dot{\mathbf{W}}. \tag{4}$$

Here the overdot stands for the derivative with respect to time t, $\cdot$ denotes scalar (inner) product, whereas $:$ denotes the full product in the space of tensors of arbitrary order. ρ is the density in the reference configuration, $\mathbf{J}_1$ and $\mathbf{J}_2$ are $N+1$th-order and $2N$th-order tensors of microinertia, respectively. The simplest form of K is

$$K = \frac{1}{2}\rho\mathbf{v} \cdot \mathbf{v} + j\dot{\mathbf{W}} : \dot{\mathbf{W}}, \tag{5}$$

where $j \geq 0$ is a scalar measure of microinertia.

Considering the principle of least action in the form

$$\delta\mathscr{H} = 0, \quad \mathscr{H}[\mathbf{x}, \mathbf{W}] = \int_{t_1}^{t_2} \int_V (K - W)\mathrm{d}V\mathrm{d}t, \tag{6}$$

where V is the volume occupied by the medium in the reference configuration, δ is the variation symbol, t_1 and t_2 are two instants, we derive motion equations for the medium with energies (3) and (5) in the following form

$$\nabla \cdot \mathbf{P} = \rho \dot{\mathbf{v}}, \quad \nabla \cdot \mathbf{G} - \frac{\partial W}{\partial \mathbf{W}} = j\ddot{\mathbf{W}}, \tag{7}$$

where we introduced the Lagrangian stress measures of Piola-Kirchhoff type

$$\mathbf{P} = \frac{\partial W}{\partial \nabla \mathbf{x}}, \quad \mathbf{G} = \frac{\partial W}{\partial \nabla \mathbf{W}}. \tag{8}$$

Let us note that here we neglect any volume loading.

3 Acceleration Waves

We consider such deformations of the medium when discontinuities of considered fields appear at a smooth surface $S(t)$ called singular, see Fig. 1. We assume existence of unilateral limit values at $S(t)$ for all considered quantities. We denote a jump of any quantity across $S(t)$ by the double squared brackets, for example, $[\![\mathbf{f}]\!] = \mathbf{f}_+ - \mathbf{f}_-$. Let us note that S is non-material surface propagating across material points.

From (7) it follows the following balance equations on S

$$\rho V [\![\mathbf{v}]\!] = -\mathbf{N} \cdot [\![\mathbf{P}]\!], \quad jV \left[\!\left[\dot{\mathbf{W}}\right]\!\right] = -\mathbf{N} \cdot [\![\mathbf{G}]\!]. \tag{9}$$

where $\mathbf{N}$ is the unit normal to S and V is the intrinsic speed of propagation of $S(t)$ in normal direction, see [49].

An acceleration wave (or *weak discontinuity wave*, or *singular surface of the second order*) is a traveling singular surface $S(t)$ at which the second spatial and time derivatives of the position vector $\mathbf{x}$ and of the microstructural tensor $\mathbf{W}$ have jumps, while $\mathbf{x}$ and $\mathbf{W}$ together with all first derivatives are continuous. So on $S(t)$ we have the following system of equations:

$$[\![\nabla \mathbf{x}]\!] = \mathbf{0}, \quad [\![\nabla \mathbf{W}]\!] = \mathbf{0}, \quad [\![\mathbf{v}]\!] = \mathbf{0}, \quad \left[\!\left[\dot{\mathbf{W}}\right]\!\right] = \mathbf{0}. \tag{10}$$

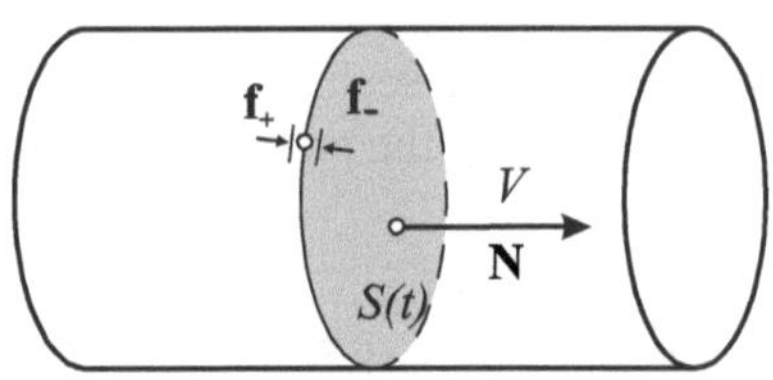

Fig. 1 Propagation of a singular surface

From (9) and (10) it follows that

$$\mathbf{N} \cdot [\![\mathbf{P}]\!] = \mathbf{0}, \quad \mathbf{N} \cdot [\![\mathbf{G}]\!] = \mathbf{0}.$$

Equation (10) imply continuity of **P** and **G** at $S(t)$:

$$[\![\mathbf{P}]\!] = \mathbf{0}, \quad [\![\mathbf{G}]\!] = \mathbf{0}.$$

Obviously, the balance equation (9) are also fulfilled.

In what follows we use *Maxwell's theorem* which states that, see [49, 50].

Theorem 1 (Maxwell) *For a continuously differentiable field* $\mathbf{Y}$ *such that* $[\![\mathbf{Y}]\!] = \mathbf{0}$ *the following relations hold*

$$\left[\!\left[\dot{\mathbf{Y}}\right]\!\right] = -V\mathbf{y}, \quad [\![\nabla\mathbf{Y}]\!] = \mathbf{N} \otimes \mathbf{y}, \tag{11}$$

where $\mathbf{y}$ *is the tensor amplitude of the jump of the first gradient of* $\mathbf{Y}$*; the tensor amplitude is a tensor of the order equal to the order of* $\mathbf{Y}$.

Here $\otimes$ is the tensor product.

Straightforward application of Maxwell's theorem to the continuous fields of $\mathbf{v}$, $\dot{\mathbf{W}}$, $\mathbf{P}$, and $\mathbf{G}$ results in the system of relations

$$\begin{gathered}
[\![\dot{\mathbf{v}}]\!] = -V\mathbf{a}, \quad [\![\nabla\mathbf{v}]\!] = \mathbf{N} \otimes \mathbf{a}, \\
\left[\!\left[\ddot{\mathbf{W}}\right]\!\right] = -V\mathbf{w}, \quad \left[\!\left[\nabla\dot{\mathbf{W}}\right]\!\right] = \mathbf{N} \otimes \mathbf{w}, \\
V\,[\![\nabla \cdot \mathbf{P}]\!] = -\mathbf{N} \cdot \left[\!\left[\dot{\mathbf{P}}\right]\!\right], \\
V\,[\![\nabla \cdot \mathbf{G}]\!] = -\mathbf{N} \cdot \left[\!\left[\dot{\mathbf{G}}\right]\!\right],
\end{gathered} \tag{12}$$

where $\mathbf{a}$ and $\mathbf{w}$ are the vectorial and tensorial amplitudes of the jumps. With these relations the motion equations transform into

$$\mathbf{N} \cdot \left[\!\left[\dot{\mathbf{P}}\right]\!\right] = \rho V^2\mathbf{a}, \quad \mathbf{N} \cdot \left[\!\left[\dot{\mathbf{G}}\right]\!\right] = jV^2\mathbf{w}. \tag{13}$$

Calculating $\dot{\mathbf{P}}$ and $\dot{\mathbf{G}}$ we obtain

$$\begin{aligned}
\mathbf{N} \cdot \left(W_{,\nabla\mathbf{x}\nabla\mathbf{x}} : (\mathbf{N} \otimes \mathbf{a}) + W_{,\nabla\mathbf{x}\nabla\mathbf{W}} : (\mathbf{N} \otimes \mathbf{w})\right) &= \rho V^2\mathbf{a}, \\
\mathbf{N} \cdot \left(W_{,\nabla\mathbf{W}\nabla\mathbf{x}} : (\mathbf{N} \otimes \mathbf{a}) + W_{,\nabla\mathbf{W}\nabla\mathbf{W}} : (\mathbf{N} \otimes \mathbf{w})\right) &= jV^2\mathbf{w}.
\end{aligned}$$

With matrix notation, we rewrite the latter equations in a more compact form:

$$\mathcal{Q}(\mathbf{N}) \cdot \xi = V^2 \mathcal{B} \cdot \xi, \tag{14}$$

where $\xi = (\mathbf{a}, \mathbf{w})$ and

$$\mathscr{Q}(\mathbf{N}) \equiv \begin{bmatrix} W_{,\nabla\mathbf{x}\nabla\mathbf{x}}\{\mathbf{N}\} & W_{,\nabla\mathbf{x}\nabla\mathbf{w}}\{\mathbf{N}\} \\ W_{,\nabla\mathbf{w}\nabla\mathbf{x}}\{\mathbf{N}\} & W_{,\nabla\mathbf{w}\nabla\mathbf{w}}\{\mathbf{N}\} \end{bmatrix}, \quad \mathscr{B} \equiv \begin{bmatrix} \rho\mathbf{I} & \mathbf{0} \\ \mathbf{0} & j\mathbf{I}^{(N)} \end{bmatrix}.$$

Here $\mathbf{I}$ and $\mathbf{I}^{(N)}$ are three-dimensional and N-dimensional unit tensors, respectively, and we introduced the following operation for arbitrary Mth-order tensor $\mathbf{H}$ and vector $\mathbf{N}$. For tensor $\mathbf{H}$ and vector $\mathbf{N}$ represented in a Cartesian basis $\mathbf{i}_k$ ($k = 1, 2, 3$), so that

$$\mathbf{H} = H_{i_1 i_2 \ldots i_M} \underbrace{\mathbf{i}_{i_1} \otimes \mathbf{i}_{i_2} \otimes \ldots \otimes \mathbf{i}_{i_M}}_{M \text{ times}}, \quad \mathbf{N} = N_k \mathbf{i}_k,$$

$\mathbf{H}\{\mathbf{N}\}$ denotes the following $(M-2)$th-order tensor:

$$\mathbf{H}\{\mathbf{N}\} \equiv H_{i_1 i_2 \ldots i_M} N_{i_1} N_{i_{M-N}} \underbrace{\mathbf{i}_{i_2} \otimes \ldots \otimes \mathbf{i}_M}_{M-2 \text{ times}}. \tag{15}$$

$\mathscr{Q}(\mathbf{N})$ is the *acoustic tensor* for the medium with microstructure. From existence of the strain energy W it follows that $\mathscr{Q}(\mathbf{N})$ is symmetric. This implies that the squared velocity of propagation of an acceleration wave is real-valued. For positivity of V we need additional requirement, that is $\mathscr{Q}(\mathbf{N})$ has to be positive definite

$$\xi \cdot \mathscr{Q}(\mathbf{N}) \cdot \xi > 0, \quad \forall\ \xi \neq \mathbf{0}, \quad \forall\ |\mathbf{N}| = 1. \tag{16}$$

Note that inequality (16) coincides with the condition of strong ellipticity of the equilibrium equations for considered elastic medium with microstructure.

Inequality (16) can be written in the form more convenient for calculations

$$\begin{aligned} &\mathbf{N} \cdot \left(W_{,\nabla\mathbf{x}\nabla\mathbf{x}} : (\mathbf{N} \otimes \mathbf{a}) + W_{,\nabla\mathbf{x}\nabla\mathbf{w}} : (\mathbf{N} \otimes \mathbf{w})\right) \cdot \mathbf{a} \\ &\quad + \mathbf{N} \cdot \left(W_{,\nabla\mathbf{w}\nabla\mathbf{x}} : (\mathbf{N} \otimes \mathbf{a}) + W_{,\nabla\mathbf{w}\nabla\mathbf{w}} : (\mathbf{N} \otimes \mathbf{w})\right) \cdot \mathbf{w} > 0, \\ &\quad \forall\ \mathbf{a} \neq \mathbf{0}, \quad \mathbf{w} \neq \mathbf{0}, \quad \forall\ |\mathbf{N}| = 1. \end{aligned}$$

The condition is also equivalent to

$$\begin{gathered} \frac{\mathrm{d}^2}{\mathrm{d}\varepsilon^2} W(\nabla\mathbf{x} + \varepsilon\mathbf{N} \otimes \mathbf{a}, \mathbf{W}, \nabla\mathbf{W} + \varepsilon\mathbf{N} \otimes \mathbf{W})\Big|_{\varepsilon=0} > 0, \\ \forall\ \mathbf{N} : |\mathbf{N}| = 1, \quad \mathbf{a} \neq \mathbf{0}, \quad \mathbf{W} \neq \mathbf{0}. \end{gathered} \tag{17}$$

The latter form relates with the convexity properties of W.

4 Example: Micropolar Media

Straightforward application of the presented above formulae depends on the form of constitutive equations and the nature of the microstructure tensor. As an example of presented approach let us consider a micropolar medium. For this medium the microrotation tensor $\mathbf{Q}$ plays a role of the microstructural tensor $\mathbf{W}$: $\mathbf{W} = \mathbf{Q}$, $\mathbf{Q}^{-1} = \mathbf{Q}^T$. Thus, the strain energy density of the micropolar medium is

$$W = W(\nabla\mathbf{x}, \mathbf{Q}, \nabla\mathbf{Q}).$$

Application of the principle of material frame indifference [50] leads to the following form [37]

$$W = W(\mathbf{E}, \mathbf{K}), \quad \mathbf{E} = \nabla\mathbf{x} \cdot \mathbf{Q}^T, \quad \mathbf{K} = -\frac{1}{2}\epsilon : (\nabla\mathbf{Q}) \cdot \mathbf{Q}^T,$$

where $\mathbf{E}$ and $\mathbf{K}$ are the strain tensor and wryness tensor, respectively, $\epsilon \equiv -\mathbf{I} \times \mathbf{I}$ is the permutation tensor. Detailed discussion on strain measures in the micropolar continuum is given in [37, 38].

For the physically linear isotropic material W is a quadratic function

$$W = W_1(\mathbf{E}) + W_2(\mathbf{K}), \tag{18}$$

$$\begin{aligned} 2W_1(\mathbf{E}) &= \alpha_1 \mathrm{tr}\left(\mathbf{E}\cdot\mathbf{E}^T\right) + \alpha_2 \mathrm{tr}\,\mathbf{E}^2 + \alpha_3 \mathrm{tr}^2\mathbf{E}, \\ 2W_2(\mathbf{K}) &= \beta_1 \mathrm{tr}\left(\mathbf{K}\cdot\mathbf{K}^T\right) + \beta_2 \mathrm{tr}\,\mathbf{K}^2 + \beta_3 \mathrm{tr}^2\mathbf{K}, \end{aligned} \tag{19}$$

where α_k and β_k $(k = 1, 2, 3)$ are elastic moduli.

Now acoustic tensor $\mathcal{Q}(\mathbf{N})$ is given by

$$\mathcal{Q}(\mathbf{N}) \equiv \begin{bmatrix} \mathcal{Q}_1(\mathbf{N}) & \mathbf{0} \\ \mathbf{0} & \mathcal{Q}_2(\mathbf{N}) \end{bmatrix}, \quad \mathcal{Q}_1(\mathbf{N}) = W_{1,\mathbf{EE}}\{\mathbf{N}\}, \quad \mathcal{Q}_2(\mathbf{N}) = W_{2,\mathbf{KK}}\{\mathbf{N}\}.$$

For (18), requirement (16) results in inequalities

$$\alpha_1 > 0, \quad \alpha_1 + \alpha_2 + \alpha_3 > 0, \quad \beta_1 > 0, \quad \beta_1 + \beta_2 + \beta_3 > 0. \tag{20}$$

Then the solutions of (14) are given in [12]

$$\begin{aligned} V_{1,2} &= \sqrt{\frac{\alpha_1}{\rho}}, \quad \xi_{1,2} = (\mathbf{e}_{1,2}, \mathbf{0}), \quad V_3 = \sqrt{\frac{\alpha_1+\alpha_2+\alpha_3}{\rho_0}}, \quad \xi_3 = (\mathbf{0}, \mathbf{N}), \\ V_{4,5} &= \sqrt{\frac{\beta_1}{j}}, \quad \xi_{4,5} = (\mathbf{e}_{4,5}, \mathbf{0}), \quad V_6 = \sqrt{\frac{\beta_1+\beta_2+\beta_3}{j}}, \quad \xi_6 = (\mathbf{0}, \mathbf{N}), \end{aligned} \tag{21}$$

where $\mathbf{e}_1$, $\mathbf{e}_2$, $\mathbf{e}_4$, $\mathbf{e}_5$ are arbitrary unit vectors in the tangential plane to $S(t)$ such that $\mathbf{e}_1 \cdot \mathbf{e}_2 = \mathbf{e}_1 \cdot \mathbf{N} = \mathbf{e}_2 \cdot \mathbf{N} = 0$, $\mathbf{e}_4 \cdot \mathbf{e}_5 = \mathbf{e}_4 \cdot \mathbf{N} = \mathbf{e}_5 \cdot \mathbf{N} = 0$.

Solutions $(21)_{1,2}$ describe transverse and longitudinal acceleration waves, respectively, whereas $(21)_{4,5}$ describe transverse and longitudinal acceleration waves of microrotation. V_k coincide with the limits of the phase velocities of plane harmonic waves (acoustic waves) in linear micropolar elasticity when the frequency of the waves tends to infinity, see [16, 32].

In a similar way the propagation of the acceleration waves can be analyzed in micromorphic media [16, 31] where the strain energy density depends on deformation gradient and microstrain $\mathbf{S}$ and its gradient: $W = W(\nabla\mathbf{x}, \mathbf{S}, \nabla\mathbf{S})$, where $\mathbf{S}$ is a second-order tensor. This analysis will be performed in forthcoming papers.

5 Conclusions

Within the unified approach based on the representation of the strain energy as a function of deformation gradient and the microstructural tensor and its gradient we derived the conditions of propagation of acceleration waves. The latter reduced to the analysis of generalized spectral problem for acoustic tensor. With presented approach one can formulate the conditions of existence of acceleration waves in more complex media. Let us note that violation of the conditions may results in discontinuous solutions with a priori unknown surfaces of discontinuity. For example, this may be important for analysis of damaged solids. In addition to detailed analysis of wave propagation near non-ideal interfaces, cracks, lattices, and other media with defects, see for example [15, 20, 30, 39, 40, 48] and the reference therein, these conditions may be used for nondestructive evaluation of damage.

Acknowledgements The author acknowledges the Russian Science Foundation (RSCF), for its support within the Project 15–19-10008.

References

1. Altenbach, H., Eremeyev, V.A., Lebedev, L.P., Rendón, L.A.: Acceleration waves and ellipticity in thermoelastic micropolar media. Arch. Appl. Mech. **80**(3), 217–227 (2010)
2. Bleach, G.P., Reddy, B.D.: The influence of constraints on the properties of acceleration waves in isotropic thermoelastic media. Arch. Ration. Mech. Anal. **98**, 31–64 (1987)
3. de Boer, R., Liu, Z.: Propagation of acceleration waves in incompressible saturated porous solids. Trans. Porous Media **21**, 163–173 (1995)
4. Capriz, G.: Continua with Microstructure. Springer, New York (1989)
5. Chaboche, J.L.: Continuum damage mechanics: Part II Damage growth, crack initiation, and crack growth. J. Appl. Mech. **55**(1), 65–72 (1988)
6. Chen, P.J.: Growth of acceleration waves in isotropic elastic materials. J. Acoust. Soc. Am. **43**, 982–987 (1968)
7. Chen, P.J.: One dimensional acceleration waves in inhomogeneous elastic non-conductors. Acta Mech. **17**, 17–24 (1973)

8. Ciarletta, M., Straughan, B.: Poroacoustic acceleration waves. Philos. Trans. R. Soc. Lond. A **462**, 3493–3499 (2006)
9. Ciarletta, M., Straughan, B.: Poroacoustic acceleration waves. Proc. R. Soc. Lond. A: Math. Phys. Eng. Sci. **462**(2075), 3493–3499 (2006)
10. Ciarletta, M., Straughan, B.: Thermo-poroacoustic acceleration waves in elastic materials with voids. J. Math. Anal. Appl. **333**(1), 142–150 (2007)
11. Dietsche, A., Steinmann, P., Willam, K.: Micropolar elastoplasticity and its role in localization. Int. J. Plast. **9**, 813–831 (1993)
12. Eremeyev, V.A.: Acceleration waves in micropolar elastic media. Dokl. Phys. **50**(4), 204–206 (2005)
13. Eremeyev, V.A., Lebedev, L.P., Altenbach, H.: Foundations of Micropolar Mechanics. Springer, Heidelberg (2013)
14. Eremeyev, V.A., Lebedev, L.P., Rendón, L.: On the propagation of acceleration waves in thermoelastic micropolar medias. Revista Colombiana de Matemáticas **41**(2), 397–406 (2007)
15. Eremeyev, V.A., Rosi, G., Naili, S.: Surface/interfacial anti-plane waves in solids with surface energy. Mech. Res. Commun. **74**, 8–13 (2016)
16. Eringen, A.C.: Microcontinuum Field Theory. I. Foundations and Solids. Springer, New York (1999)
17. Forest, S.: Micromorphic approach for gradient elasticity, viscoplasticity, and damage. J. Eng. Mech. **135**(3), 117–131 (2009)
18. Forest, S.: Micromorphic media. In: Altenbach, H., Eremeyev,V.A. (eds.) Generalized Continua from the Theory to Engineering Applications, CISM International Centre for Mechanical Sciences, vol. 541, pp. 249–300. Springer Vienna (2013)
19. Fu, Y.B., Scott, N.H.: The transistion from acceleration wave to shock wave. Int. J. Eng. Sci. **29**, 617–624 (1991)
20. Gorbushin, N., Mishuris, G.: Dynamic crack propagation along the interface with non-local interactions. J. Eur. Ceram. Soc. **36**(9), 2241–2244 (2016)
21. Hetnarski, R.B. (ed.): Encyclopedia of Thermal Stresses, vol. 1–11. Springer, Dorderchet (2014)
22. Jordan, P.M.: Growth and decay of acoustic acceleration waves in Darcy-type porous media. Philos. Trans. R. Soc. Lond. A **461**, 2749–2766 (2005)
23. Kafadar, C.B., Eringen, A.C.: Micropolar media - I. The classical theory. Int. J. Eng. Sci. **9**, 271–305 (1971)
24. Lebedev, L.P., Cloud, M.J., Eremeyev, V.A.: Tensor Analysis with Applications in Mechanics. World Scientific, New Jersey (2010)
25. Lemaitre, J.: A Course on Damage Mechanics. Springer, Berlin (1986)
26. Loret, B., Simões, F.M.F., Martins, J.A.C.: Growth and decay of acceleration waves in non-associative elastic-plastic fluid-saturated porous media. Int. J. Solids Struct. **34**, 1583–1608 (1997)
27. Lubarda, V., Krajcinovic, D.: Damage tensors and the crack density distribution. Int. J. Solids Struct. **30**(20), 2859–2877 (1993)
28. Maugin, G.A.: Acceleration waves in simple and linear viscoelastic micropolar materials. Int. J. Eng. Sci. **12**, 143–157 (1974)
29. McCarthy, M.F., Tiersten, H.F.: One-dimensional acceleration waves and acoustoelectric domains in piezoelectric semiconductors. J. Appl. Phys. **47**, 3389–3396 (1976)
30. Mishuris, G., Movchan, A., Slepyan, L.: Waves and fracture in an inhomogeneous lattice structure. Waves Random Complex Media **17**(4), 409–428 (2007)
31. Neff, P., Ghiba, I.D., Madeo, A., Placidi, L., Rosi, G.: A unifying perspective: the relaxed linear micromorphic continuum. Contin. Mech. Thermodynam. **26**(5), 639–681 (2014)
32. Nowacki, W.: Theory of Asymmetric Elasticity. Pergamon-Press, Oxford (1986)
33. Oden, J.T., Wellford, L.C.: J.: The finite element analysis of shocks and acceleration waves in nonlinearly elastic solids. The. Shock. Vib. Dig. **7**, 2–11 (1975)
34. Ogden, R.W.: Growth and decay of acceleration waves in incompresible elastic solids. Q. J. Mech. Appl. Math. **37**, 451–464 (1974)

35. Ostoja-Starzewski, M., Trębicki, J.: On the growth and decay of acceleration waves in random media. Philos. Trans. R. Soc. Lond. A **455**, 2577–2614 (1999)
36. Ostoja-Starzewski, M., Trębicki, J.: Stochastic dynamics of acceleration waves in random media. Mech. Mater. **38**, 840–848 (2006)
37. Pietraszkiewicz, W., Eremeyev, V.A.: On natural strain measures of the non-linear micropolar continuum. Int. J. Solids Struct. **46**(3–4), 774–787 (2009)
38. Pietraszkiewicz, W., Eremeyev, V.A.: On vectorially parameterized natural strain measures of the non-linear Cosserat continuum. Int. J. Solids Struct. **46**(11–12), 2477–2480 (2009)
39. Placidi, L., Rosi, G., Giorgio, I., Madeo, A.: Reflection and transmission of plane waves at surfaces carrying material properties and embedded in second-gradient materials. Math. Mech. Solids **19**(5), 555–578 (2014)
40. Quiligotti, S., Maugin, G.A., dell'Isola, F.: Wave motions in unbounded poroelastic solids infused with compressible fluids. ZAMP **53**(6), 1110–1138 (2002)
41. Saccomandi, G.: Acceleration waves in a thermo-microstretch fluid. Int. J. Non-linear Mech. **29**, 809–817 (1994)
42. Scott, N.H.: Acceleration waves in constrained elastic materials. Arch. Ration. Mech. Analys. **58**, 57–75 (1975)
43. Scott, N.H.: Acceleration waves in incompressible elastic solids. Q. J. Mech. Appl. Math. **29**, 295–310 (1976)
44. Scott, N.H., Hayes, M.: Constant amplitude acceleration waves in a prestrained incompressible isotropic elastic solid. Math. Mech. Solids **2**, 291–295 (1997)
45. Straughan, B.: Nonlinear waves in a general magnetic fluid. ZAMP **37**, 274–279 (1986)
46. Straughan, B.: Stability and Wave Motion in Porous Media, vol. 165. Springer, New York (2008)
47. Straughan, B.: Heat Waves. Springer, New York (2011)
48. Sumbatyan, M., Brigante, M.: Analysis of strength and wave velocity for micro-damaged elastic media. Eng. Fract. Mech. **145**, 43–53 (2015)
49. Truesdell, C.: Ration. Thermodyn., 2nd edn. Springer, New York (1984)
50. Truesdell, C., Noll, W.: The Non-linear Field Theories of Mechanics, 3rd edn. Springer, Berlin (2004)

Models of Active Bulk Composites and New Opportunities of the ACELAN Finite Element Package

N.V. Kurbatova, D.K. Nadolin, A.V. Nasedkin, A.A. Nasedkina, P.A. Oganesyan, A.S. Skaliukh and A.N. Soloviev

Abstract The chapter presents the mathematical models for thermopiezomagnetoelectric composite materials of an arbitrary anisotropy classes for computer engineering finite element package ACELAN. These homogenization models are based on the method of effective moduli with different boundary conditions, the approaches for generation of representative volumes with specified properties and the finite element method. Important features of ACELAN package also include the original models of irreversible processes of polarization and repolarization for polycrystalline ferroelectric materials. In this paper the software architecture of ACELAN package and its ability for creation of representative volumes with different structures are also described.

Keywords Coupled problems · Active material · Magnetoelectric composite · Thermopiezomagnetoelectricity · Inhomogeneous polarization · Effective moduli · Representative volume · Finite element software

---

N.V. Kurbatova · D.K. Nadolin · A.V. Nasedkin · A.A. Nasedkina ·
P.A. Oganesyan · A.S. Skaliukh
Institute of Mathematics, Mechanics and Computer Science, Southern Federal University,
Milchakova Street 8a, Rostov-on-Don 344090, Russia
e-mail: nvk-ru@yandex.ru

D.K. Nadolin
e-mail: nadolin@yandex.ru

A.V. Nasedkin
e-mail: nasedkin@math.sfedu.ru

A.A. Nasedkina
e-mail: nasedkina@math.sfedu.ru

P.A. Oganesyan
e-mail: wolwerine@yandex.ru

A.S. Skaliukh
e-mail: a.s.skaliukh@gmail.com

A.N. Soloviev (✉)
Don State Technical University, Department of Theoretical and Applied Mechanics,
Gagarin Sq., 1, Rostov-on-Don 344000, Russia
e-mail: solovievarc@gmail.com

M.A. Sumbatyan (ed.), *Wave Dynamics and Composite Mechanics for Microstructured Materials and Metamaterials*,
Advanced Structured Materials 59, DOI 10.1007/978-981-10-3797-9_8

1 Introduction

Finite element package ACELAN (ACoustoELectric ANalysis), developed about 15 years ago, was initially intended for computer design of piezoelectric devices for two-dimensional plane and axisymmetric compound areas, as well as three-dimensional regions of the generalized cylindrical shape. Basic methods and algorithms, underluimg in the first version ACELAN, were described in [1, 12]. Later releases of ACELAN package apart from the models of the elastic, piezoelectric and acoustic media, were also supplemented with more complicated problems with connectivity of mechanical, electrical, magnetic and thermal fields [20].

In the recent time an increased interest has been observed to the investigations of composite materials of complex structure that exhibit very effective properties for many practical applications. Two-phase magnetoelectric composites consisting of piezo- and magnetoactive phases demonstrate the ability to mutual transformation of magnetic and electric fields, whereas each single phase does not have such property. Modern magnetoelectric composites have high effectiveness of the magnetoelectric transformation, relatively high temperatures of phase transitions and considerable technological resource [11].

Recently magnetoelectric composites became of interest to many researchers which resulted in considerable increase in the number of works devoted to the modelling of the effective properties of these composites (see [7, 8, 11, 26, 29], etc.). Piezoelectric composites are studied much better and have numerous familiar practical applications [21, 27]. Meanwhile, the active composites with connectivity of mechanical, electrical, magnetic and thermal fields clearly have not been studied enough yet [6, 9, 10], and modern well-known finite element packages do not have the capabilities of their simulation.

Thermopiezomagnetoelectric composites as special cases allow us to investigate in ACELAN also elastic, thermoelastic, piezoelectric, piezomagnetic, pyroelectric and magnetoelectric composites of any type of anisotropy. In Sect. 2 similarly to [13–15, 19] we present different mathematical setting of the homogenization problems for active composite materials with connectivity between mechanical, electric, magnetic and heat fields. We formulate basic statements to validate the homogenization procedures. We derive the set of essential and natural boundary conditions for which the stress, strain, electric and magnetic fields, electric and magnetic flux densities, heat flow and heat gradient are constant for homogeneous congruence materials. At the end we formulate the set of boundary problems to define the full set of effective moduli for thermopiezomagnetoelectric materials of arbitrary anisotropic classes.

The finite element approaches to solve the stated homogenization problems in representative volumes are given in Sect. 3. We present the finite element systems of symmetrical structures with quasi-definite matrices (matrix structure for the problems with a saddle point), and we note that the set of effective algorithms for solving finite element systems with symmetric quasi-definite matrices are implemented in ACELAN package.

In Sect. 4 we describe the modelling procedure of inhomogeneous polarization processes which are more important for piezoelectric and piezomagnetic materials. We note that the effective original algorithms for modelling of irreversible processes of polarization and repolarization for composite ferroelectric materials are implemented in ACELAN. Sections 5 and 6 are devoted to software architecture of ACELAN package and software implementation for pre- and post- processors, for modelling the inhomogeneous polarization processes and for modelling the representative volumes with different basic structures of active multiphase composites.

2 Effective Moduli Method for Thermopiezomagnetoelectric Composites

Let Ω be a volume of thermopiezomagnetoelectric composite heterogeneous body; $\Gamma = \partial\Omega$ is the boundary of the region; $\mathbf{n}$ is the vector of the external unit normal to Γ; $\mathbf{u} = \mathbf{u}(\mathbf{x})$ is the displacement vector-function; $\varphi = \varphi(\mathbf{x})$ is the electric potential function; $\phi = \phi(\mathbf{x})$ is the magnetic potential function; $\theta = \theta(\mathbf{x})$ is the increment of the temperature $T(\mathbf{x})$ from the temperature T_0 of the natural state, i.e. $\theta = T - T_0$.

We denote by $\boldsymbol{\varepsilon}$ the second-order strain tensor; $\boldsymbol{\sigma}$ is the second-order stress tensor; $\mathbf{E}$ is the electric field intensity vector or the electric field vector; $\mathbf{D}$ is the electric flux density vector or the electric displacement vector; $\mathbf{H}$ is the magnetic field intensity vector or magnetic field vector; $\mathbf{B}$ is the magnetic flux density vector, $\mathbf{G}$ is the thermal gradient vector, $\mathbf{q}$ is the heat flow. The fields $\boldsymbol{\varepsilon}$, $\mathbf{E}$, $\mathbf{H}$ and $\mathbf{G}$ are determined from the functions $\mathbf{u}$, φ, ϕ and θ by the following form

$$\boldsymbol{\varepsilon} = (\nabla\mathbf{u} + (\nabla\mathbf{u})^*)/2, \quad \mathbf{E} = -\nabla\varphi, \quad \mathbf{H} = -\nabla\phi, \quad \mathbf{G} = \nabla\theta\,. \tag{1}$$

where superscript "*" denotes the transposition operation.

On the boundary Γ we will consider the mechanical stress vector $\mathbf{p}$, surface densities of electric charges σ_e, magnetic (pseudo-) charges σ_m and heat flow q_s:

$$\mathbf{p} = \mathbf{n}\cdot\boldsymbol{\sigma}, \quad \sigma_e = -\mathbf{n}\cdot\mathbf{D}, \quad \sigma_m = -\mathbf{n}\cdot\mathbf{B}, \quad q_s = -\mathbf{n}\cdot\mathbf{q}\,. \tag{2}$$

As usual, we will denote the volume-averaged quantities in the broken brackets as:

$$\langle(...)\rangle = \frac{1}{|\Omega|}\int_\Omega (...)\,d\Omega\,. \tag{3}$$

We will formulate auxiliary lemmas for thermopiezomagnetoelectric body following the proof of the effective moduli method for more simple medium. These affirmations are substantiated under similar techniques that are used for poroelastic, thermoelastic [18], piezoelectric [15, 19] and magnetoelectric bodies [13–15].

Lemma 1 *These representations take place for the field characteristics (1) averaged in the volume Ω by means of the appropriate values on the boundary Γ:*

$$
\begin{aligned}
&(a)\ \ \forall\, \boldsymbol{\varepsilon}=\boldsymbol{\varepsilon}^*: \ \langle\boldsymbol{\varepsilon}\rangle = \tfrac{1}{2|\Omega|}\int_\Gamma(\mathbf{un}^*+\mathbf{nu}^*)\,d\Gamma;\\
&(b)\ \ \forall\, \mathbf{E},\mathbf{H},\mathbf{G}: \langle\mathbf{E}\rangle = -\tfrac{1}{|\Omega|}\int_\Gamma \mathbf{n}\varphi\,d\Gamma,\quad \langle\mathbf{H}\rangle = -\tfrac{1}{|\Omega|}\int_\Gamma \mathbf{n}\phi\,d\Gamma,\\
&\qquad\qquad\qquad\ \langle\mathbf{G}\rangle = \tfrac{1}{|\Omega|}\int_\Gamma \mathbf{n}\theta\,d\Gamma;\\
&(c)\ \ \forall\ \boldsymbol{\sigma}=\boldsymbol{\sigma}^*: \nabla\cdot\boldsymbol{\sigma}=0,\quad \langle\boldsymbol{\sigma}\rangle = \tfrac{1}{2|\Omega|}\int_\Gamma(\mathbf{pn}^*+\mathbf{np}^*)\,d\Gamma;\\
&(d)\ \ \forall\ \mathbf{D},\mathbf{B},\mathbf{q}: \nabla\cdot\mathbf{D}=0,\quad \langle\mathbf{D}\rangle = -\tfrac{1}{|\Omega|}\int_\Gamma \mathbf{x}\sigma_e\,d\Gamma,\\
&\qquad\qquad\qquad\ \nabla\cdot\mathbf{B}=0,\quad \langle\mathbf{B}\rangle = -\tfrac{1}{|\Omega|}\int_\Gamma \mathbf{x}\sigma_m\,d\Gamma,\\
&\qquad\qquad\qquad\ \nabla\cdot\mathbf{q}=0,\quad \langle\mathbf{q}\rangle = -\tfrac{1}{|\Omega|}\int_\Gamma \mathbf{x}q_s\,d\Gamma
\end{aligned}
$$

Here, for (a), (b) *the formulae (1) are required; while* $\mathbf{p}$, σ_e, σ_m *and* q_s *in* (c), (d) *are connected with* $\boldsymbol{\sigma}$, $\mathbf{D}$, $\mathbf{B}$ *and* $\mathbf{q}$ *by representations (2).*

Lemma 2 *For* $\mathbf{x}\in\Gamma$ *we have the following relations*:
(a) *if* $\mathbf{u}=\mathbf{x}\cdot\boldsymbol{\varepsilon}_0$, *where* $\boldsymbol{\varepsilon}_0=\boldsymbol{\varepsilon}_0^*=\text{const}$, *i.e.* $\boldsymbol{\varepsilon}_0$ *is the arbitrary symmetric independent on* $\mathbf{x}$ *second rank tensor, then* $\langle\boldsymbol{\varepsilon}\rangle=\boldsymbol{\varepsilon}_0$;
(b) *if* $\varphi=-\mathbf{x}\cdot\mathbf{E}_0$, *where* $\mathbf{E}_0=\text{const}$, *i.e.* $\mathbf{E}_0$ *is the arbitrary independent on* $\mathbf{x}$ *vector, then* $\langle\mathbf{E}\rangle=\mathbf{E}_0$;
(c) *if* $\phi=-\mathbf{x}\cdot\mathbf{H}_0$, *where* $\mathbf{H}_0=\text{const}$, *i.e.* $\mathbf{H}_0$ *is the arbitrary independent on* $\mathbf{x}$ *vector, then* $\langle\mathbf{H}\rangle=\mathbf{H}_0$;
(d) *if* $\theta=\mathbf{x}\cdot\mathbf{G}_0$, *where* $\mathbf{G}_0=\text{const}$, *i.e.* $\mathbf{G}_0$ *is the arbitrary independent on* $\mathbf{x}$ *vector, then* $\langle\mathbf{G}\rangle=\mathbf{G}_0$;
(e) *if* $\mathbf{p}=\mathbf{n}\cdot\boldsymbol{\sigma}_0$, *where* $\mathbf{p}$ *is the stress vector from (2),* $\boldsymbol{\sigma}_0=\boldsymbol{\sigma}_0^*=\text{const}$, *then* $\langle\boldsymbol{\sigma}\rangle=\boldsymbol{\sigma}_0$;
(f) *if* $\sigma_e=-\mathbf{n}\cdot\mathbf{D}_0$, *where* σ_e *is the surface density of electric charges from (2),* $\mathbf{D}_0=\text{const}$, *then* $\langle\mathbf{D}\rangle=\mathbf{D}_0$;
(g) *if* $\sigma_m=-\mathbf{n}\cdot\mathbf{B}_0$, *where* σ_m *is the surface density of magnetic charges from (2),* $\mathbf{B}_0=\text{const}$, *then* $\langle\mathbf{B}\rangle=\mathbf{B}_0$;
(h) *if* $q_s=-\mathbf{n}\cdot\mathbf{q}_0$, *where* q_s *is the heat flow from (2),* $\mathbf{q}_0=\text{const}$, *then* $\langle\mathbf{q}\rangle=\mathbf{q}_0$.

Lemma 3 *If for* $\mathbf{x}\in\Gamma$:
(a) $\mathbf{u}=\mathbf{x}\cdot\boldsymbol{\varepsilon}_0$, $\boldsymbol{\varepsilon}_0=\boldsymbol{\varepsilon}_0^*=\text{const}$, *and the equilibrium equation* $\nabla\cdot\boldsymbol{\sigma}=0$ *takes place for any given symmetric second rank stress tensor* $\boldsymbol{\sigma}$, *then we have* $\langle\boldsymbol{\sigma}:\boldsymbol{\varepsilon}\rangle=\langle\boldsymbol{\sigma}\rangle:\langle\boldsymbol{\varepsilon}\rangle$, *where the symbol": "denotes the double scalar product operation between two tensors* ($\boldsymbol{\sigma}:\boldsymbol{\varepsilon}=\sigma_{ij}\varepsilon_{ji}$);
(b) $\varphi=-\mathbf{x}\cdot\mathbf{E}_0$, $\mathbf{E}_0=\text{const}$, *and the equation of electrostatics* $\nabla\cdot\mathbf{D}=0$ *takes place for any given electric flux density vector* $\mathbf{D}$, *then we have* $\langle\mathbf{D}\cdot\mathbf{E}\rangle=\langle\mathbf{D}\rangle\cdot\langle\mathbf{E}\rangle$;
(c) $\phi=-\mathbf{x}\cdot\mathbf{H}_0$, $\mathbf{H}_0=\text{const}$, *and the equation of magnetostatics* $\nabla\cdot\mathbf{B}=0$ *takes place for any given magnetic flux density vector* $\mathbf{B}$, *then we have* $\langle\mathbf{B}\cdot\mathbf{H}\rangle=\langle\mathbf{B}\rangle\cdot\langle\mathbf{H}\rangle$;
(d) $\theta=\mathbf{x}\cdot\mathbf{G}_0$, $\mathbf{G}_0=\text{const}$, *and the equation of stationary thermal conductivity* $\nabla\cdot\mathbf{q}=0$ *takes place for any given heat flux vector* $\mathbf{q}$, *then we have* $\langle\mathbf{q}\cdot\mathbf{G}\rangle=\langle\mathbf{q}\rangle\cdot\langle\mathbf{G}\rangle$;
(e) $\mathbf{p}=\mathbf{n}\cdot\boldsymbol{\sigma}_0$, $\boldsymbol{\sigma}_0=\boldsymbol{\sigma}_0^*=\text{const}$, *where* $\mathbf{p}$ *is the stress vector from (2), and the equilibrium equation* $\nabla\cdot\boldsymbol{\sigma}=0$ *takes place, then we have* $\langle\boldsymbol{\sigma}:\boldsymbol{\varepsilon}\rangle=\langle\boldsymbol{\sigma}\rangle:\langle\boldsymbol{\varepsilon}\rangle$;
(f) $\sigma_e=-\mathbf{n}\cdot\mathbf{D}_0$, $\mathbf{D}_0=\text{const}$, *where* σ_e *is the surface density of electric charges from (2), and the equation of electrostatics* $\nabla\cdot\mathbf{D}=0$ *takes place, then we have* $\langle\mathbf{D}\cdot\mathbf{E}\rangle=\langle\mathbf{D}\rangle\cdot\langle\mathbf{E}\rangle$;

(g) $\sigma_m = -\mathbf{n} \cdot \mathbf{B}_0$, $\mathbf{B}_0 = \text{const}$, *where* σ_m *is the surface density of magnetic charges from (2), and the equation of magnetostatics* $\nabla \cdot \mathbf{B} = 0$ *takes place, then we have* $\langle \mathbf{B} \cdot \mathbf{H} \rangle = \langle \mathbf{B} \rangle \cdot \langle \mathbf{H} \rangle$;
(h) $q_s = -\mathbf{n} \cdot \mathbf{q}_0$, $\mathbf{q}_0 = \text{const}$, *where* q_s *is the surface heat flow from (2), and the equation of stationary thermal conductivity* $\nabla \cdot \mathbf{q} = 0$ *takes place, then we have* $\langle \mathbf{q} \cdot \mathbf{G} \rangle = \langle \mathbf{q} \rangle \cdot \langle \mathbf{G} \rangle$.

In accordance with eight equivalent fundamental forms of constitutive relations between mechanical, electric and magnetic field variables with thermal effects we will introduce the moduli of thermopiezomagnetoelectric medium:
—$\varepsilon EH\theta$—form ($\boldsymbol{\varepsilon}$, $\mathbf{E}$, $\mathbf{H}$, and θ are considered as independent variables)

$$\boldsymbol{\sigma} = \mathbf{c}^{E,H,\theta} : \boldsymbol{\varepsilon} - \mathbf{e}^{H,\theta *} \cdot \mathbf{E} - \mathbf{h}^{E,\theta *} \cdot \mathbf{H} - \boldsymbol{\gamma}^{E,H} \theta\,, \tag{4}$$

$$\mathbf{D} = \mathbf{e}^{H,\theta} : \boldsymbol{\varepsilon} + \boldsymbol{\kappa}^{\varepsilon,H,\theta} \cdot \mathbf{E} + \boldsymbol{\beta}^{\varepsilon,\theta} \cdot \mathbf{H} + \mathbf{g}^{\varepsilon,H} \theta\,, \tag{5}$$

$$\mathbf{B} = \mathbf{h}^{E,\theta} : \boldsymbol{\varepsilon} + \boldsymbol{\beta}^{\varepsilon,\theta *} \cdot \mathbf{E} + \boldsymbol{\mu}^{\varepsilon,E,\theta} \cdot \mathbf{H} + \mathbf{m}^{\varepsilon,E} \theta\,, \tag{6}$$

—$\sigma EH\theta$—form ($\boldsymbol{\sigma}$, $\mathbf{E}$, $\mathbf{H}$, and θ are considered as independent variables)

$$\boldsymbol{\varepsilon} = \mathbf{s}^{E,H,\theta} : \boldsymbol{\sigma} + \mathbf{d}^{H,\theta *} \cdot \mathbf{E} + \mathbf{b}^{E,\theta *} \cdot \mathbf{H} + \boldsymbol{\alpha}^{E,H} \theta\,, \tag{7}$$

$$\mathbf{D} = \mathbf{d}^{H,\theta} : \boldsymbol{\sigma} + \boldsymbol{\kappa}^{\sigma,H,\theta} \cdot \mathbf{E} + \boldsymbol{\gamma}^{\sigma,\theta} \cdot \mathbf{H} + \mathbf{p}^{\sigma,H} \theta\,, \tag{8}$$

$$\mathbf{B} = \mathbf{b}^{E,\theta} : \boldsymbol{\sigma} + \boldsymbol{\gamma}^{\sigma,\theta *} \cdot \mathbf{E} + \boldsymbol{\mu}^{\sigma,E,\theta} \cdot \mathbf{H} + \mathbf{n}^{\sigma,E} \theta\,, \tag{9}$$

and similarly for $\varepsilon DH\theta$, $\sigma DH\theta$, $\varepsilon EB\theta$, $\sigma EB\theta$, $\varepsilon DB\theta$, $\sigma DB\theta$—forms with corresponding independent variables.

Here in Eqs. (4)–(6) $\mathbf{c}^{E,H,\theta}$ is the fourth rank tensor of elastic stiffness moduli, $\mathbf{e}^{H,\theta}$ is the third rank tensor of piezoelectric moduli, $\mathbf{h}^{E,\theta}$ is the third rank tensor of magnetostriction moduli, $\boldsymbol{\gamma}^{E,H}$ is the second-order tensor of thermal stresses, $\boldsymbol{\kappa}^{\varepsilon,H,\theta}$ is the second rank tensor of dielectric permittivity moduli, $\boldsymbol{\beta}^{\varepsilon,\theta}$ is the second rank tensor of magnetoelectric coupling coefficients, $\mathbf{g}^{\varepsilon,H}$ is the vector of pyroelectric constants, $\boldsymbol{\mu}^{\varepsilon,E,\theta}$ is the second rank tensor of magnetic permittivity moduli, $\mathbf{m}^{\varepsilon,E}$ is the vector of pyromagnetic constants, and the upper indexes indicate for which constant fields calculate these moduli.

For indicated above constitutive relations we have to add the Fourier's law

$$\mathbf{q} = -\mathbf{k} \cdot \mathbf{G}\,, \tag{10}$$

where $\mathbf{k}$ is the second rank tensor of thermal conductivities.

Let now Ω be the representative volume of heterogeneous thermopiezomagnetoelectric composite material. We will determine the effective moduli $\tilde{\mathbf{c}}^{E,H,\theta}$, $\tilde{\mathbf{e}}^{H,\theta}$, $\tilde{\mathbf{h}}^{E,\theta}$, $\tilde{\boldsymbol{\gamma}}^{E,H}$, $\tilde{\boldsymbol{\kappa}}^{\varepsilon,H,\theta}$, $\tilde{\boldsymbol{\beta}}^{\varepsilon,\theta}$, $\tilde{\mathbf{g}}^{\varepsilon,H}$, $\tilde{\boldsymbol{\mu}}^{\varepsilon,E,\theta}$, $\tilde{\mathbf{m}}^{\varepsilon,E}$, $\tilde{\mathbf{k}}$ by the following technique, similar to the well-known procedures for poroelastic, thermoelastic, piezoelectric and magnetoelectric composites [13–19].

At first we consider the static thermopiezomagnetoelectric problem for representative volume Ω with constant thermal field ($\theta = \theta_0$):

$$\nabla \cdot \boldsymbol{\sigma} = 0, \quad \nabla \cdot \mathbf{D} = 0, \quad \nabla \cdot \mathbf{B} = 0, \quad \nabla \cdot \mathbf{q} = 0, \quad \mathbf{x} \in \Omega, \tag{11}$$

$$\mathbf{u} = \mathbf{x} \cdot \boldsymbol{\varepsilon}_0, \quad \varphi = -\mathbf{x} \cdot \mathbf{E}_0, \quad \phi = -\mathbf{x} \cdot \mathbf{H}_0, \quad \theta = \theta_0, \quad \mathbf{x} \in \Gamma. \tag{12}$$

We call the problem (11), (12) with (1), (4)–(6), (10) as the problem I, and denote the solution of this problem by $\mathbf{u}^I$, φ^I, ϕ^I, θ^I ($\theta^I = \theta_0$). After solving this problem and by using (1), (4)–(6) we find $\boldsymbol{\varepsilon}^I$, $\mathbf{E}^I$, $\mathbf{H}^I$, $\boldsymbol{\sigma}^I$, $\mathbf{D}^I$ and $\mathbf{B}^I$, where $\boldsymbol{\sigma}^I = \boldsymbol{\sigma}(\mathbf{u}^I, \varphi^I, \phi^I, \theta^I)$, etc. We note from Lemma 2, that for the problem I: $\langle \boldsymbol{\varepsilon}^I \rangle = \boldsymbol{\varepsilon}_0$, $\langle \mathbf{E}^I \rangle = \mathbf{E}_0$ and $\langle \mathbf{H}^I \rangle = \mathbf{H}_0$.

We supply in conformity to the initial heterogeneous medium some "equivalent" homogeneous medium with effective moduli $\tilde{\mathbf{c}}^{E,H,\theta}$, $\tilde{\mathbf{e}}^{H,\theta}$, $\tilde{\mathbf{h}}^{E,\theta}$, $\tilde{\boldsymbol{\gamma}}^{E,H}$, $\tilde{\boldsymbol{\kappa}}^{\varepsilon,H,\theta}$, $\tilde{\boldsymbol{\beta}}^{\varepsilon,\theta}$, $\tilde{\mathbf{g}}^{\varepsilon,H}$, $\tilde{\boldsymbol{\mu}}^{\varepsilon,E,\theta}$, $\tilde{\mathbf{m}}^{\varepsilon,E}$, $\tilde{\mathbf{k}}$. The constitutive equations for "equivalent" medium, similar to (4)–(6), are in the forms:

$$\boldsymbol{\sigma}_0 = \tilde{\mathbf{c}}^{E,H,\theta} : \boldsymbol{\varepsilon}_0 - \tilde{\mathbf{e}}^{H,\theta*} \cdot \mathbf{E}_0 - \tilde{\mathbf{h}}^{E,\theta*} \cdot \tilde{\mathbf{H}}_0 - \tilde{\boldsymbol{\gamma}}^{E,H} \theta_0, \tag{13}$$

$$\mathbf{D}_0 = \tilde{\mathbf{e}}^{H,\theta} : \boldsymbol{\varepsilon}_0 + \tilde{\boldsymbol{\kappa}}^{\varepsilon,H,\theta} \cdot \mathbf{E}_0 + \tilde{\boldsymbol{\beta}}^{\varepsilon,\theta} \cdot \mathbf{H}_0 + \tilde{\mathbf{g}}^{\varepsilon,H} \theta_0, \tag{14}$$

$$\mathbf{B}_0 = \tilde{\mathbf{h}}^{E,\theta} : \boldsymbol{\varepsilon}_0 + \tilde{\boldsymbol{\beta}}^{\varepsilon,\theta*} \cdot \mathbf{E}_0 + \tilde{\boldsymbol{\mu}}^{\varepsilon,E,\theta} \cdot \mathbf{H}_0 + \tilde{\mathbf{m}}^{\varepsilon,E} \theta_0. \tag{15}$$

For the problem I we accept the following equations such as relations for definition of effective moduli from (13) to (15)

$$\langle \boldsymbol{\sigma}^I \rangle = \boldsymbol{\sigma}_0, \quad \langle \mathbf{D}^I \rangle = \mathbf{D}_0, \quad \langle \mathbf{B}^I \rangle = \mathbf{B}_0. \tag{16}$$

The moduli, found from these conditions, are marked with superscripts "I". Note, that due to Lemma 3 the average energies are equal for heterogeneous and for "equivalent" homogeneous thermopiezomagnetoelectric media:

$$\langle \boldsymbol{\sigma}^I : \boldsymbol{\varepsilon}^I + \mathbf{D}^I \cdot \mathbf{E}^I + \mathbf{B}^I \cdot \mathbf{H}^I \rangle / 2 = (\boldsymbol{\sigma}_0 : \boldsymbol{\varepsilon}_0 + \mathbf{D}_0 \cdot \mathbf{E}_0 + \mathbf{B}_0 \cdot \mathbf{H}_0)/2. \tag{17}$$

Now, by using Eqs. (13)–(15), we can select such boundary conditions, at which the obvious expressions for effective moduli are obtained. For example, we consider the problem I (11), (12) with

$$\boldsymbol{\varepsilon}_0 = \varepsilon_0 (\mathbf{e}_k \mathbf{e}_m + \mathbf{e}_m \mathbf{e}_k)/2, \quad \mathbf{E}_0 = 0, \quad \mathbf{H}_0 = 0, \quad \theta_0 = 0, \tag{18}$$

where k, m are some fixed numbers ($k, m = 1, 2, 3$); $\mathbf{e}_k$ are the unit vectors of Cartesian basis. Then, from (13)–(15), (18) we obtain:

$$\tilde{c}_{ijkm}^{E,H,\theta I} = \langle \sigma_{ij}^I \rangle / \varepsilon_0, \quad \tilde{e}_{jkm}^{H,\theta I} = \langle D_j^I \rangle / \varepsilon_0, \quad \tilde{h}_{jkm}^{E,\theta I} = \langle B_j^I \rangle / \varepsilon_0 . \tag{19}$$

If in I (11), (12) we accept

$$\varepsilon_0 = 0, \quad \mathbf{E}_0 = E_0 \mathbf{e}_k, \quad \mathbf{H}_0 = 0, \quad \theta_0 = 0 , \tag{20}$$

then from (13) to (15), (20) we find

$$\tilde{e}_{kij}^{H,\theta I} = -\langle \sigma_{ij}^I \rangle / E_0, \quad \tilde{\kappa}_{kj}^{\varepsilon,H,\theta I} = \langle D_j^I \rangle / E_0, \quad \tilde{\beta}_{kj}^{\varepsilon,\theta I} = \langle B_j^I \rangle / E_0 . \tag{21}$$

Similarly, if in the problem I (11), (12) we accept

$$\varepsilon_0 = 0, \quad \mathbf{E}_0 = 0, \quad \mathbf{H}_0 = H_0 \mathbf{e}_k, \quad \theta_0 = 0 , \tag{22}$$

then from (13) to (15), (22) we find

$$\tilde{h}_{kij}^{E,\theta I} = -\langle \sigma_{ij}^I \rangle / H_0, \quad \tilde{\beta}_{jk}^{\varepsilon,\theta I} = \langle D_j^I \rangle / H_0, \quad \tilde{\mu}_{kj}^{\varepsilon,E,\theta I} = \langle B_j^I \rangle / H_0 . \tag{23}$$

Finally, if in I (11), (12) we suppose

$$\varepsilon_0 = 0, \quad \mathbf{E}_0 = 0, \quad \mathbf{H}_0 = 0, \quad \theta_0 \neq 0 , \tag{24}$$

then from (13) to (15), (24) we obtain

$$\tilde{\gamma}_{ij}^{E,H I} = -\langle \sigma_{ij}^I \rangle / \theta_0, \quad \tilde{g}_j^{\varepsilon,H I} = \langle D_j^I \rangle / \theta_0, \quad \tilde{m}_j^{\varepsilon,E,I} = \langle B_j^I \rangle / \theta_0 . \tag{25}$$

Note, that the quantities σ_{ij}^I, D_j^I and B_j^I in (19), (21), (23) and (25) are different, since they are calculated from the solutions of the problem I with different boundary conditions (12): (18), (20), (22) and (24).

In order to determine the effective coefficients of the tensor $\tilde{\mathbf{k}}$ we solve the uncoupled thermal problem tI

$$\nabla \cdot \mathbf{q} = 0, \quad \mathbf{q} = -\mathbf{k} \cdot \mathbf{G}, \quad \mathbf{G} = \nabla \theta, \quad \mathbf{x} \in \Omega; \quad \theta = \mathbf{x} \cdot \mathbf{G}_0, \quad \mathbf{x} \in \Gamma , \tag{26}$$

where $\mathbf{G}_0$ is some constant vector that does not depend on $\mathbf{x}$.

It is obvious that $\theta = \mathbf{x} \cdot \mathbf{G}_0$, $\mathbf{G} = \nabla \theta = \mathbf{G}_0$, $\mathbf{q} = \mathbf{q}_0 = -\tilde{\mathbf{k}}^{tI} \cdot \mathbf{G}_0$ will give the solution of the problem (26) in the volume Ω for the homogeneous comparison medium. Then having solved the problem (26) for heterogeneous medium, we can set that for this medium and for the comparison medium the averaged heat fluxes are equal $\langle \mathbf{q} \rangle = \mathbf{q}_0$. As a result we get the equation for the effective thermal conductivity moduli of the composite: $\tilde{\mathbf{k}}^{tI} \cdot \mathbf{G}_0 = -\langle \mathbf{q} \rangle$. Then for the comparison medium with anisotropy of general form it is not difficult to obtain computation formulas for thermal conductivity moduli $\tilde{k}_{il}^{tI}$. Indeed, setting in (26),

$$\mathbf{G}_0 = G_0\mathbf{e}_l\,, \tag{27}$$

where $G_0 = \text{const}$, $l = 1, 2, 3$ is some fixed index, we get computation formulas for the moduli $\tilde{k}^{tI}_{il}$:

$$\tilde{k}^{tI}_{il} = -\langle q_i\rangle / G_0\,. \tag{28}$$

We note from Lemmas 2, 3, (*d*), that for the problem *tI* between the heat flux and temperature gradient fields for the heterogeneous medium and the comparison medium the following equalities take place $\langle\mathbf{G}\rangle = \mathbf{G}_0$, $\langle\mathbf{q}\cdot\mathbf{G}\rangle = \mathbf{q}_0\cdot\mathbf{G}_0$, where the last relation has the energy sense.

In addition to described technique, by analogy with poroelastic, thermoelastic, piezoelectric and magnetoelectric composite media [13–19], for thermopiezomagnetoelectric media it is possible to suggest the other ways of introducing effective moduli, by considering the problems with other mechanical, electric and magnetic boundary conditions from Lemma 2. Indeed, we can consider the following problems:

—the problem *II* with boundary conditions for mechanical stress vector **p**, electric potential φ, magnetic potential ϕ and temperature θ

$$\mathbf{p} = \mathbf{n}\cdot\boldsymbol{\sigma}_0, \quad \varphi = -\mathbf{x}\cdot\mathbf{E}_0, \quad \phi = -\mathbf{x}\cdot\mathbf{H}_0, \quad \theta = \theta_0, \quad \mathbf{x}\in\Gamma\,; \tag{29}$$

—the problem *III* with boundary conditions for displacement **u**, surface density of electric charges σ_e, magnetic potential ϕ and temperature θ

$$\mathbf{u} = \mathbf{x}\cdot\boldsymbol{\varepsilon}_0, \quad \sigma_e = -\mathbf{n}\cdot\mathbf{D}_0, \quad \phi = -\mathbf{x}\cdot\mathbf{H}_0, \quad \theta = \theta_0, \quad \mathbf{x}\in\Gamma\,; \tag{30}$$

—the problem *IV* with boundary conditions for mechanical stress vector **p**, surface density of electric charges σ_e, magnetic potential ϕ and temperature θ

$$\mathbf{p} = \mathbf{n}\cdot\boldsymbol{\sigma}_0, \quad \sigma_e = -\mathbf{n}\cdot\mathbf{D}_0, \quad \phi = -\mathbf{x}\cdot\mathbf{H}_0, \quad \theta = \theta_0, \quad \mathbf{x}\in\Gamma\,; \tag{31}$$

—the problems *V*–*VIII*, where as distinct from the problems *I*–*IV* the boundary conditions for magnetic potential $\phi = -\mathbf{x}\cdot\mathbf{H}_0$ are replaced by the boundary conditions for surface density of magnetic charges $\sigma_m = -\mathbf{n}\cdot\mathbf{B}_0$.

In the all these problems the field equations of equilibrium, electrostatic, magnetostatic and steady-state heat (11) are considered. Here for the problem *II* the constitutive equations (7)–(9) are used and originally the effective moduli $\tilde{\mathbf{s}}^{E,H,\theta\,II}$, $\tilde{\mathbf{d}}^{H,\theta\,II}$, $\tilde{\mathbf{b}}^{E,\theta\,II}$, $\tilde{\boldsymbol{\alpha}}^{E,H\,II}$, $\tilde{\boldsymbol{\kappa}}^{\sigma,H,\theta\,II}$, $\tilde{\boldsymbol{\gamma}}^{\sigma,\theta\,II}$, $\tilde{\mathbf{p}}^{\sigma,H\,II}$, $\tilde{\boldsymbol{\mu}}^{\sigma,E,\theta\,II}$ and $\tilde{\mathbf{n}}^{\sigma,E\,II}$ are defined.

Indeed, we can solve the problems *II* (11), (29) with (1), (7)–(9), (10) for different boundary conditions in (29), and in the result we obtain the full set of effective moduli from (7)–(9):

— problems *II*, *a*

$$\boldsymbol{\sigma}_0 = \sigma_0(\mathbf{e}_k\mathbf{e}_m + \mathbf{e}_m\mathbf{e}_k)/2, \quad \mathbf{E}_0 = 0, \quad \mathbf{H}_0 = 0, \quad \theta_0 = 0\,, \tag{32}$$

$$\tilde{s}_{ijkm}^{E,H,\theta\,II} = \langle \varepsilon_{ij}^{II} \rangle / \sigma_0, \quad \tilde{d}_{jkm}^{H,\theta\,II} = \langle D_j^{II} \rangle / \sigma_0, \quad \tilde{b}_{jkm}^{E,\theta\,II} = \langle B_j^{II} \rangle / \sigma_0 \,, \tag{33}$$

— problems II, b

$$\sigma_0 = 0, \quad \mathbf{E}_0 = E_0 \mathbf{e}_k, \quad \mathbf{H}_0 = 0, \quad \theta_0 = 0 \,, \tag{34}$$

$$\tilde{d}_{kij}^{H,\theta\,II} = \langle \varepsilon_{ij}^{II} \rangle / E_0, \quad \tilde{\kappa}_{kj}^{\sigma,H,\theta\,II} = \langle D_j^{II} \rangle / E_0, \quad \tilde{\gamma}_{kj}^{\sigma,\theta\,II} = \langle B_j^{II} \rangle / E_0 \,, \tag{35}$$

— problems II, c

$$\sigma_0 = 0, \quad \mathbf{E}_0 = 0, \quad \mathbf{H}_0 = H_0 \mathbf{e}_k, \quad \theta_0 = 0 \,, \tag{36}$$

$$\tilde{b}_{kij}^{H,\theta\,II} = \langle \varepsilon_{ij}^{II} \rangle / H_0, \quad \tilde{\gamma}_{kj}^{\sigma,\theta\,II} = \langle D_j^{II} \rangle / H_0, \quad \tilde{\mu}_{kj}^{\sigma,E,\theta\,II} = \langle B_j^{II} \rangle / H_0 \,, \tag{37}$$

— problem II, d

$$\sigma_0 = 0, \quad \mathbf{E}_0 = 0, \quad \mathbf{H}_0 = 0, \quad \theta_0 \neq 0 \,, \tag{38}$$

$$\tilde{\alpha}_{ij}^{E,H\,II} = \langle \varepsilon_{ij}^{II} \rangle / \theta_0, \quad \tilde{p}_{j}^{\sigma,H\,II} = \langle D_j^{II} \rangle / \theta_0, \quad \tilde{n}_{j}^{\sigma,E\,II} = \langle B_j^{II} \rangle / \theta_0 \,. \tag{39}$$

Respectively, for the problems *III–VIII* the constitutive equations for $\varepsilon DH\theta$, $\sigma DH\theta$, $\varepsilon EB\theta$, $\sigma EB\theta$, $\varepsilon DB\theta$, $\sigma DB\theta$—forms are used and corresponding effective moduli are defined.

Note, that from Lemma 3 for all these problems, similarly to the problem *I*, the average energy is conserved, i.e. the relation (17) is satisfied with replace of the superscript "*I*" by "*II*"–"*VIII*". This fact is fundamental for the effective moduli method.

In any of these problems from the obtained effective moduli we can find other moduli from constitutive equations for $\varepsilon EH\theta$, $\sigma EH\theta$, $\varepsilon DH\theta$, $\sigma DH\theta$, $\varepsilon EB\theta$, $\sigma EB\theta$, $\varepsilon DB\theta$, $\sigma DB\theta$—forms for "equivalent" homogeneous medium. We note, that the effective moduli, found from different problems, will differ, i.e., $\mathbf{c}^{E,H,\theta\,I} \neq \mathbf{c}^{E,H,\theta\,II} \neq \ldots \neq \mathbf{c}^{E,H,\theta\,VIII}$, etc.

Similarly, instead of determination of effective thermal conductivities coefficients from (26) we can consider thermal problem *tII*

$$\nabla \cdot \mathbf{q} = 0, \quad \mathbf{q} = -\mathbf{k} \cdot \mathbf{G}, \quad \mathbf{G} = \nabla \theta, \quad \mathbf{x} \in \Omega; \quad q_s = -\mathbf{n} \cdot \mathbf{q} = -\mathbf{n} \cdot \mathbf{q}_0, \quad \mathbf{x} \in \Gamma \,, \tag{40}$$

where $\mathbf{q}_0$ is some constant vector that does not depend on $\mathbf{x}$. Then $\mathbf{q} = \mathbf{q}_0$, $\theta = \mathbf{x} \cdot \mathbf{G}_0$, $\mathbf{G} = \nabla\theta$, $\mathbf{G} = \mathbf{G}_0$, $\mathbf{G}_0 = -\tilde{\mathbf{r}}^{\,tII} \cdot \mathbf{q}_0$ give the solution of the problem (40) for homogeneous comparison medium, moreover, by Lemma 2, (*h*) for any solution of the problem (40) with heterogeneous coefficients $\mathbf{k}(\mathbf{x}) = \mathbf{r}^{-1}(\mathbf{x})$ the following equality holds $\langle \mathbf{q} \rangle = \langle \mathbf{q}_0 \rangle$. Therefore for the determination of the effective inverse thermal conductivities coefficients $\tilde{r}_{il}^{\,tII}$ we can set: $\langle \mathbf{G} \rangle = \langle \mathbf{G}_0 \rangle$, where $\mathbf{G}$ is the gradient of the temperature field, calculated for heterogeneous medium. This condition leads to the relation $\tilde{\mathbf{r}}^{\,tII} \cdot \mathbf{q}_0 = -\langle \mathbf{G} \rangle$, from which, assuming in (40)

$$\mathbf{q}_0 = q_0 \mathbf{e}_l \,, \tag{41}$$

where $q_0 = \text{const}$, $l = 1, 2, 3$ is some fixed index, we get the computation formulas for the effective thermal resistance (inverse thermal conductivities) coefficients:

$$\tilde{r}_{il}^{tII} = -\langle G_i \rangle / q_0 \,. \tag{42}$$

As for the problem *tI*, by Lemma 3, (*h*) the choice of the boundary condition $\mathbf{n} \cdot \mathbf{q} = \mathbf{n} \cdot \mathbf{q}_0$ in (40) can be justified by the energy equality for the heterogeneous medium and the comparison medium: $\langle \mathbf{q} \cdot \mathbf{G} \rangle = \langle \mathbf{q}_0 \cdot \mathbf{G}_0 \rangle$.

Note that use of different constitutive equations of problems *II*–*VIII* and *tI*, *tII* can be useful to determine effective moduli of the inhomogeneous structures by dealing with mainly one- or two-dimensional movements, for example, for thermopiezomagnetoelectric rods, plates and disks, etc.

In addition to the described above models, in the finite element package ACELAN the model of active composite materials with interface boundary conditions we also plan to implement. Such models with the interface and surface effects can be used to simulate nanoscale multi–field composites.

3 Finite Element Technologies

For solving thermopiezomagnetoelectric problems (11) with corresponding boundary conditions for heterogeneous two-phase composite material in the representative volume Ω we can pass to their weak settings and use classical technique of finite element approximations. Let Ω_h be the region filled by the finite element mesh, $\Omega_h \subset \Omega$, $\Omega_h = \cup_m \Omega^{em}$, where Ω^{em} is a separate finite element with the number m. For static problems we can find the approximate solution $\{\mathbf{u}_h \approx \mathbf{u},\ \varphi_h \approx \varphi,\ \phi_h \approx \phi,\ \theta_h \approx \theta\}$ at the finite element mesh Ω_h in the form

$$\mathbf{u}_h(\mathbf{x}) = \mathbf{N}_u^*(\mathbf{x}) \cdot \mathbf{U}, \quad \varphi_h(\mathbf{x}) = \mathbf{N}_\varphi^*(\mathbf{x}) \cdot \boldsymbol{\Phi}, \quad \phi_h(\mathbf{x}) = \mathbf{N}_\phi^*(\mathbf{x}) \cdot \mathbf{A}, \quad \theta_h(\mathbf{x}) = \mathbf{N}_\theta^*(\mathbf{x}) \cdot \mathbf{T} \,, \tag{43}$$

where $\mathbf{N}_u^*$ is the matrix of the shape functions for the displacements, $\mathbf{N}_\varphi^*$ is the row vector of the shape functions for the electric potential, $\mathbf{N}_\phi^*$ is the row vector of the shape functions for the magnetic potential, $\mathbf{N}_\theta^*$ is the row vector of the shape functions for the temperature, $\mathbf{U}$, $\boldsymbol{\Phi}$, $\mathbf{A}$, $\mathbf{T}$ are the global vectors of the nodal displacements, the electric potential, the magnetic potential and the temperature, respectively.

According to usual technique we approximate the continuum weak formulation of the thermopiezomagnetoelectric problem in finite-dimensional spaces related to the basis functions $\mathbf{N}_u^*$, $\mathbf{N}_\varphi^*$, $\mathbf{N}_\phi^*$, $\mathbf{N}_\theta^*$. Substituting (43) and analogous representations for projecting functions into the weak setting of the thermopiezomagnetoelectric problem for Ω_h, we get the finite element system

$$\mathbf{K}_{uu} \cdot \mathbf{U} + \mathbf{K}_{u\varphi} \cdot \boldsymbol{\Phi} + \mathbf{K}_{u\phi} \cdot \mathbf{A} - \mathbf{K}_{u\theta} \cdot \mathbf{T} = \mathbf{F}_u \,, \tag{44}$$

$$-\mathbf{K}^{*}_{u\varphi}\cdot\mathbf{U}+\mathbf{K}_{\varphi\varphi}\cdot\boldsymbol{\Phi}+\mathbf{K}_{\varphi\phi}\cdot\mathbf{A}-\mathbf{K}_{\varphi\theta}\cdot\mathbf{T}=\mathbf{F}_{\varphi}, \tag{45}$$

$$-\mathbf{K}^{*}_{u\phi}\cdot\mathbf{U}+\mathbf{K}^{*}_{\varphi\phi}\cdot\boldsymbol{\Phi}+\mathbf{K}_{\phi\phi}\cdot\mathbf{A}-\mathbf{K}_{\phi\theta}\cdot\mathbf{T}=\mathbf{F}_{\phi}, \tag{46}$$

$$\mathbf{K}_{\theta\theta}\cdot\mathbf{T}=\mathbf{F}_{\theta}. \tag{47}$$

Here, $\mathbf{K}_{uu}=\sum_m^a\mathbf{K}^{em}_{uu}$, $\mathbf{K}_{u\varphi}=\sum_m^a\mathbf{K}^{em}_{u\varphi}$, $\mathbf{K}_{u\phi}=\sum_m^a\mathbf{K}^{em}_{u\phi}$ etc. are the global matrices, obtained from the corresponding element matrices by the ensemble operation ($\sum_m^a$), and the element matrices are in the forms

$$\mathbf{K}^{em}_{uu}=\int_{\Omega^{em}}\mathbf{B}^{em*}_{u}\cdot\mathbf{c}\cdot\mathbf{B}^{em}_{u}\,d\Omega,\quad \mathbf{K}^{em}_{u\varphi}=\int_{\Omega^{em}}\mathbf{B}^{em*}_{u}\cdot\mathbf{e}^{*}\cdot\mathbf{B}^{em}_{\varphi}\,d\Omega, \tag{48}$$

$$\mathbf{K}^{em}_{u\phi}=\int_{\Omega^{em}}\mathbf{B}^{em*}_{u}\cdot\mathbf{h}^{*}\cdot\mathbf{B}^{em}_{\phi}\,d\Omega,\quad \mathbf{K}^{em}_{u\theta}=\int_{\Omega^{em}}\mathbf{B}^{em*}_{u}\cdot\boldsymbol{\gamma}\,\mathbf{N}^{em*}_{\theta}\,d\Omega, \tag{49}$$

$$\mathbf{K}^{em}_{\varphi\varphi}=\int_{\Omega^{em}}\mathbf{B}^{em*}_{\varphi}\cdot\boldsymbol{\kappa}\cdot\mathbf{B}^{em}_{\varphi}\,d\Omega,\quad \mathbf{K}^{em}_{\varphi\phi}=\int_{\Omega^{em}}\mathbf{B}^{em*}_{\varphi}\cdot\boldsymbol{\beta}\cdot\mathbf{B}^{em}_{\phi}\,d\Omega, \tag{50}$$

$$\mathbf{K}^{em}_{\varphi\theta}=\int_{\Omega^{em}}\mathbf{B}^{em*}_{\varphi}\cdot\mathbf{g}\,\mathbf{N}^{em*}_{\theta}\,d\Omega,\quad \mathbf{K}^{em}_{\phi\phi}=\int_{\Omega^{em}}\mathbf{B}^{em*}_{\phi}\cdot\boldsymbol{\mu}\cdot\mathbf{B}^{em}_{\phi}\,d\Omega, \tag{51}$$

$$\mathbf{K}^{em}_{\phi\theta}=\int_{\Omega^{em}}\mathbf{B}^{em*}_{\phi}\cdot\mathbf{m}\,\mathbf{N}^{em*}_{\theta}\,d\Omega,\quad \mathbf{K}^{em}_{\theta\theta}=\int_{\Omega^{em}}\mathbf{B}^{em*}_{\theta}\cdot\mathbf{k}\cdot\mathbf{B}^{em}_{\theta}\,d\Omega, \tag{52}$$

$$\mathbf{B}^{em}_{u}=\mathbf{L}(\nabla)\cdot\mathbf{N}^{em*}_{u},\quad \mathbf{B}^{em}_{\varphi}=\nabla\mathbf{N}^{em*}_{\varphi},\quad \mathbf{B}^{em}_{\phi}=\nabla\mathbf{N}^{em*}_{\phi},\quad \mathbf{B}^{em}_{\theta}=\nabla\mathbf{N}^{em*}_{\theta}, \tag{53}$$

$$\mathbf{L}^{*}(\nabla)=\begin{bmatrix}\partial_1 & 0 & 0 & 0 & \partial_3 & \partial_2\\ 0 & \partial_2 & 0 & \partial_3 & 0 & \partial_1\\ 0 & 0 & \partial_3 & \partial_2 & \partial_1 & 0\end{bmatrix}, \tag{54}$$

where $\mathbf{N}^{em*}_{u}$, $\mathbf{N}^{em*}_{\varphi}$, $\mathbf{N}^{em*}_{\phi}$, $\mathbf{N}^{em*}_{\theta}$ are the matrices and the row vectors of approximate shape functions for $\mathbf{u}_h$, φ_h, ϕ_h, θ_h, respectively, defined on separate finite elements with number m. The vectors $\mathbf{F}_u$, $\mathbf{F}_{\varphi}$, $\mathbf{F}_{\phi}$, $\mathbf{F}_{\theta}$ in (44)–(47) are obtained from the boundary conditions, the corresponding right parts of the weak statements, and the finite element approximations.

In (48)–(54) we use vector-matrix forms for the moduli [5]: $\mathbf{c}$ is the 6x6 matrix of elastic moduli, $c_{\alpha\beta}=c^{E,H,\theta}_{ijkl}$; $\alpha,\beta=1,\ldots,6$; $i,j,k,l=1,2,3$ with the correspondence law $\alpha\leftrightarrow(ij)$, $\beta\leftrightarrow(kl)$, $1\leftrightarrow(11)$, $2\leftrightarrow(22)$, $3\leftrightarrow(33)$, $4\leftrightarrow(23)=(32)$, $5\leftrightarrow(13)=(31)$, $6\leftrightarrow(12)=(21)$; $\mathbf{e}$ is the 3x6 matrix of piezoelectric moduli ($e_{i\beta}=e^{H,\theta}_{ikl}$); $\mathbf{h}$ is the 3x6 matrix of piezomagnetic moduli ($h_{i\beta}=h^{E,\theta}_{ikl}$); $\boldsymbol{\gamma}=\{\gamma^{E,H}_{11},\gamma^{E,H}_{22},\gamma^{E,H}_{33},\gamma^{E,H}_{23},\gamma^{E,H}_{13},\gamma^{E,H}_{12}\}$.

Note that for problem *I–VIII* the vector of nodal temperature is known $\mathbf{T}=\mathbf{T}_0$, because $\theta=\theta_0$, and from (44)–(46) we have the finite element system relative to unknown vector $\mathbf{a} = \{\mathbf{U}, \boldsymbol{\Phi}, \mathbf{A}\}$:

$$\mathbf{K} \cdot \mathbf{a} = \mathbf{F}, \tag{55}$$

where

$$\mathbf{K} = \begin{bmatrix} \mathbf{K}_{uu} & \mathbf{K}_{u\varphi} & \mathbf{K}_{u\phi} \\ \mathbf{K}^*_{u\varphi} & -\mathbf{K}_{\varphi\varphi} & -\mathbf{K}_{\varphi\phi} \\ \mathbf{K}^*_{u\phi} & -\mathbf{K}^*_{\varphi\phi} & -\mathbf{K}_{\phi\phi} \end{bmatrix}, \quad \mathbf{F} = \begin{Bmatrix} \mathbf{F}_u + \mathbf{K}_{u\theta} \cdot \mathbf{T}_0 \\ -\mathbf{F}_\varphi - \mathbf{K}_{\varphi\theta} \cdot \mathbf{T}_0 \\ -\mathbf{F}_\phi - \mathbf{K}_{\phi\theta} \cdot \mathbf{T}_0 \end{Bmatrix}. \tag{56}$$

Then for the problems *I–VIII* we can solve system (55) with symmetric quasi-definite matrix (matrix structure for the problems with a saddle point [3, 4, 28]) **K** from (56). For the problems *tI* and *tII* we have the system (47) with symmetric positive definite matrix.

Thus, as has been described in [1, 12, 15], in ACELAN we can use effective algorithms with symmetric positive definite and quasi-definite matrices for solving finite element Eqs. (47) and (55), (56). Additionally, all the procedures, that we need in finite element manipulations (the degree of freedom rotations, mechanical and electric boundary condition settings, etc.), can be also provided in a symmetric form.

4 Modelling of Inhomogeneous Polarization

In the simulation of the effective properties of active composites original inhomogeneous structure plays important role. For example, the moduli of piezoelectric and piezomagnetic phases may be determined by the polarization vectors, which depend on the spatial coordinates. For the calculation of the initial heterogeneity associated with polarization, ACELAN package has capabilities for solving nonlinear problems of irreversible processes of polarization of polycrystalline ferroelectric materials. The pre-polarized ferroelectric ceramics are examples of such materials.

This section presents the basic mathematical models of nonlinear problems of irreversible processes of polarization that are in the form of software moduli, implanted in the ACELAN package. They are described in detail in [2, 23, 24]. Image graphics software package and the results of the respective modules will be presented in the following sections. We consider a body of polycrystalline ferroelectric material the boundary of which is affected by slow time-varying external electrical and mechanical stresses of high intensity, as shown in Fig. 1. The polycrystalline ferroelectrics have very complex structure, and their structure is being changed in the process of polarization. Therefore, it is obvious that for the modeling of irreversible processes it is necessary to take into account that the changes occur in the body.

Figure 2 shows the ferroelectric perovskite structure at the macro-, micro- and nano- levels. Nevertheless, the developed model may be successfully applied for other types of ferroelectrics.

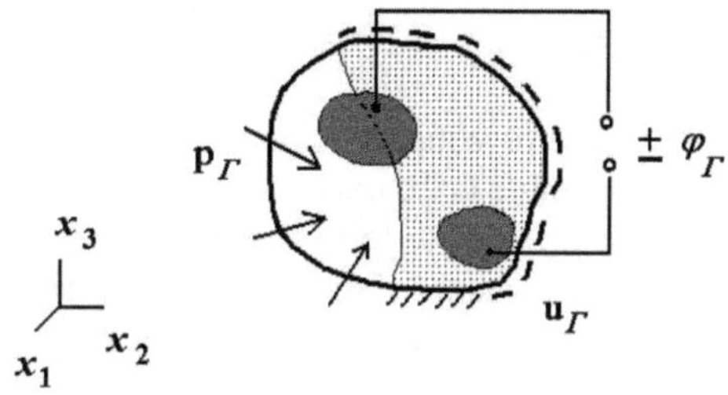

Fig. 1 General view and boundary conditions

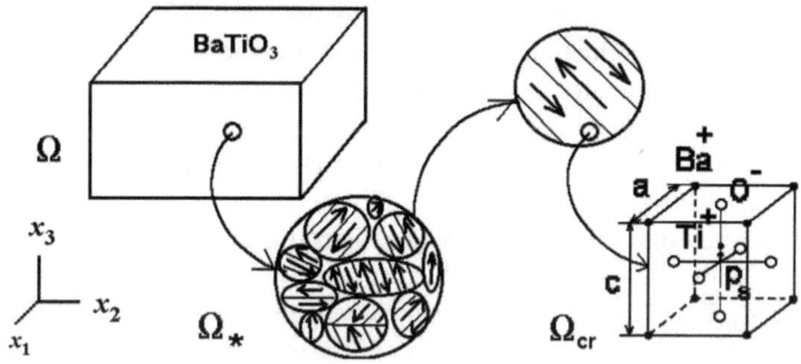

Fig. 2 Structure at the macro-, micro- and nano- levels

As in Sect. 2, let $\mathbf{u}$ be the macroscopic displacement field, φ is the macroscopic scalar potential, $\boldsymbol{\sigma}$ is the macroscopic mechanical stress and $\mathbf{D}$ is the electrical displacement. In order to formulate the electromechanical boundary value problem, the surface Γ of the considered body is divided into mechanical and electrical parts $\Gamma = \Gamma_\sigma \cup \Gamma_u$ and $\Gamma = \Gamma_D \cup \Gamma_\varphi$, respectively. The essential boundary conditions on Γ_u and Γ_φ as well as the natural boundary conditions on $\Gamma = \Gamma_\sigma$ and $\Gamma = \Gamma_D$ are $\mathbf{u} = \mathbf{u}_\Gamma$ on Γ_u, $\varphi = \varphi_\Gamma$ on Γ_φ, $\mathbf{n} \cdot \boldsymbol{\sigma} = \mathbf{p}_\Gamma$ on Γ_σ, $\mathbf{n} \cdot \mathbf{D} = 0$ on $\Gamma = \Gamma_D$.

We need to determine the displacements $\mathbf{u}$, stresses $\boldsymbol{\sigma}$, electric field $\mathbf{E}$ in the area of body, and arising field of the residual polarization $\mathbf{P}_0$ and residual strain $\boldsymbol{\varepsilon}_0$ for any values of boundary functions, i.e. of the displacement vector $\mathbf{u}_\Gamma$, the vector mechanical stresses $\mathbf{p}_\Gamma$ and the electric potential φ_Γ. The basic equations are obtained from the relations of balance of impulse, momentum, Gauss's law and Faraday's law. In differential form we obtain the field equations:

$$\nabla \cdot \boldsymbol{\sigma} + \rho \mathbf{f} = 0, \quad \nabla \cdot \mathbf{D} = 0,$$

to which we add two first geometric relations between displacement vector $\mathbf{u}$ and electric potential φ with strain tensor $\boldsymbol{\varepsilon}$ and electric field vector $\mathbf{E}$ from Eq. (1).

The polarization process, as well as plasticity is an irreversible process. Therefore along with the parameters of state we have to input the parameters of process. For this purpose, the strain tensor and the polarization vector is presented in the form of reversible (elastic) and irreversible (remaining) parts

$$\boldsymbol{\varepsilon} = \boldsymbol{\varepsilon}_e + \boldsymbol{\varepsilon}_0, \qquad \mathbf{D} = \varepsilon_0 \mathbf{E} + \mathbf{P}_e + \mathbf{P}_0. \tag{57}$$

To obtain a closed system of equations, one must add the defining relations. Firstly, this is a ratio for reversible components of the polarization vector and strain tensor. And, secondly, it is necessary to formulate the law to determinate the rate of change for the irreversible components of the polarization and deformation. We also note that the vector $\mathbf{P}_0$ of ceramics polarization together with the tensor $\boldsymbol{\varepsilon}$ describes the changes of the structure material.

Constitutive relations for reversible components are obtained by using a common approach of thermodynamics of irreversible processes, as described in detail in [24]. Such relationships are the linear algebraic equations relating the reversible part of the strain and electric induction on the one hand and the stress tensor and vector of the electric field on the other hand. They have the form:

$$\boldsymbol{\varepsilon} = \boldsymbol{\varepsilon} - \boldsymbol{\varepsilon}_0 = \mathbf{s} : \boldsymbol{\sigma} + \mathbf{d}^* \cdot \mathbf{E}, \quad \mathbf{D}_e = \mathbf{D} - \mathbf{P}_0 = \mathbf{d} : \boldsymbol{\sigma} + \boldsymbol{\kappa} \cdot \mathbf{E}, \tag{58}$$

where the elastic compliance tensor $\mathbf{s}$, piezoelectric coefficients $\mathbf{d}$ and the dielectric constants $\boldsymbol{\kappa}$ depend on the parts of irreversible polarization and strain $\mathbf{S}(\boldsymbol{\varepsilon}_0, \mathbf{P}_0)$, $\mathbf{d}(\boldsymbol{\varepsilon}_0, \mathbf{P}_0)$, $\boldsymbol{\kappa}(\boldsymbol{\varepsilon}_0, \mathbf{P}_0)$. These components are changed with the change of irreversible components. It is known that thermally depolarized ceramic is an isotropic material in terms of mechanical and electrical properties, and does not have piezoelectric properties. Therefore, we adopted the following law for these tensors

$$\mathbf{s} = \mathbf{s}_0 + \frac{|\mathbf{P}_0|}{p_{sat}} \mathbf{s}_1, \quad \mathbf{d} = \frac{|\mathbf{P}_0|}{p_{sat}} \mathbf{d}_1, \quad \boldsymbol{\kappa} = \boldsymbol{\kappa}_0 + \frac{|\mathbf{P}_0|}{p_{sat}} \boldsymbol{\kappa}_1 .$$

In the unpolarized state these tensors are

$$\mathbf{s} = \mathbf{s}_0, \quad \mathbf{d} = 0, \quad \boldsymbol{\kappa} = \boldsymbol{\kappa}_0 ,$$

and in a state of saturation polarization

$$\mathbf{s} = \mathbf{s}_{sat}, \quad \mathbf{d} = \mathbf{d}_{sat}, \quad \boldsymbol{\kappa} = \boldsymbol{\kappa}_{sat}.$$

Then the correction tensor

$$\mathbf{s}_1 = \mathbf{s}_{sat} - \mathbf{s}_0, \quad \mathbf{d}_1 = \mathbf{d}_{sat}, \quad \boldsymbol{\kappa}_1 = \boldsymbol{\kappa}_{sat} - \boldsymbol{\kappa}_0 .$$

The appearance of the residual polarization changes the structure of the material, resulting in a phase transition “body – the body”, in which the class of an anisotropy of the material is changed. If we polarize an initially unpolarized ceramic, then the isotropic material becomes anisotropic. The appearance of the residual polarization vector in the material induces polar electrical and mechanical properties. Therefore, it is natural to assume that even a partially polarized ceramics belongs to a class of transversely isotropic body. If the electric fields, causing polarization in the ceramic are heterogeneous, it is customary to speak of locally transversely isotropic body in which the anisotropy axis coincides with the direction of the residual polarization. If mechanical stresses with electric field in the depolarization of ceramics are present, then they fundamentally alter residual strain tensor. And if none of the principal axes of the tensor residual deformation coincides with the direction of residual polarization, then in this case the class of anisotropy is not defined. Such situation requires additional research to identify a class of anisotropy and piezoelectric properties of the material.

The most complex and least developed part of this research is the construction of constitutive relations for irreversible components of strain and polarization. Currently proposed several models, among which we mention the model associated with the methods of the theory of plasticity, the effective medium model, the models of orientational and energetical switchings, models of locked walls and some others. As previously noted, the irreversible parameters are the process parameters and therefore the constitutive relations do not include the parameters themselves but their velocities. In the case of quasi-static processes instead of velocities the increments of these parameters are examined. In our studies the dynamics of dipole switching is not taken into account, so the quasi-static polarization process is presented as a sequence of equilibrium states. For each state we determine the residual polarization and the residual strain. The increments of irreversible parameters are determined in the transition from one equilibrium state to another. On the other hand, all models are built for the representative volume. Therefore, at a finite element implementation each single finite element is considered as the representative volume.

For plane and axisymmetric problems we use one-dimensional model of the locked wall, which was generalized to the two-dimensional case. This model is a further development of Jile-Atherton model [2]. This model assumes that the mechanical stresses are small and do not affect the polarization process that is carried out only by electric field. In this case, the residual strain can be easily found from the condition of incompressibility of the material, and the residual strain tensor does not affect elastic, piezoelectric and dielectric properties of partially polarized material.

For three-dimensional models we also use a model of the locked wall, but with the addition of the energy criterion. This approach allowes us to find the values of the residual polarization and the residual strain for any concurrent acting stress and electric field in the current state.

The fundamental principle of our study is the assumption that the whole process of polarization can be represented as a sequence of equilibrium states $C^{(0)}$, $C^{(1)}$, ..., $C^{(N)}$, where the $C^{(0)}$ is initial state, and $C^{(N)}$ is the final state. This permits to shorten the process of integration over time and replace it by finding of the finite increments. To determine the conditions of equilibrium for the current state $C^{(i)}$ we use the principle of virtual displacements, which also allows us to use the finite element method. This approach, on the one hand, the execute integration by coordinates, and, on the other hand, enables to build the non-uniform field of residual polarization and strain,

$$\int_{\Omega} (\triangle \boldsymbol{\sigma}^{(i)} : \delta \triangle \boldsymbol{\varepsilon}^{(i)} - \triangle \mathbf{D}^{(i)} \cdot \delta \triangle \mathbf{E}^{(i)} - \rho \triangle \mathbf{f}^{(i)} \cdot \delta \triangle \mathbf{u}^{(i)})\, d\Omega -$$

$$\int_{\Gamma_\sigma} \triangle \mathbf{p}_{\Gamma}^{(i)} \cdot \delta \triangle \mathbf{u}^{(i)} d\Gamma = 0,$$

where the symbol "$\triangle$" denotes increment, "δ" is the variation. The parameters are subjected to variation only during the transition from one equilibrium state to another. Applying the principle of virtual displacements should be supplemented by constitutive relations for increments of known and unknown parameters. For this purpose we use the relations (58) in two neighboring equilibrium states. Neglecting of magnitudes of smaller order magnitude, such relations in increments can be written as

$$\triangle \boldsymbol{\varepsilon}_e{}^{(i)} = \mathbf{s}(P_0^{(i)}) : \triangle \boldsymbol{\sigma}^{(i)} + \mathbf{d}^*(P_0^{(i)}) \cdot \triangle \mathbf{E}^{(i)},$$

$$\triangle \mathbf{D}_e{}^{(i)} = \mathbf{d}(P_0^{(i)}) : \triangle \boldsymbol{\sigma}^{(i)} + \boldsymbol{\kappa}(P_0^{(i)}) \cdot \triangle \mathbf{E}^{(i)}.$$

These relations can be also written in the form of

$$\triangle \boldsymbol{\sigma}^{(i)} = \mathbf{c}(P_0^{(i)}) : \triangle \boldsymbol{\varepsilon}_e{}^{(i)} - \mathbf{e}^*(P_0^{(i)}) \cdot \triangle \mathbf{E}^{(i)},$$

$$\triangle \mathbf{D}_e{}^{(i)} = \mathbf{e}(P_0^{(i)}) : \triangle \boldsymbol{\varepsilon}_e{}^{(i)} + \mathbf{g}(P_0^{(i)}) \cdot \triangle \mathbf{E}^{(i)},$$

where

$$\mathbf{c}(P_0^{(i)}) = \mathbf{s}^{-1}(P_0^{(i)}), \quad \mathbf{e}(P_0^{(i)}) = \mathbf{d}(P_0^{(i)}) : \mathbf{s}^{-1}(P_0^{(i)}),$$

$$\mathbf{g}(P_0^{(i)}) = \boldsymbol{\kappa}(P_0^{(i)}) - \mathbf{d}(P_0^{(i)}) : \mathbf{s}^{-1}(P_0^{(i)}) : \mathbf{d}^*(P_0^{(i)}).$$

It is easy to see that the constitutive equations contain the increments of reversible strains and electrical displacements, so they may be represented by the increments of the residual parts of (58), but now we should take the value of the increments of residuals parts from the previous state. Then we get

$$\triangle \boldsymbol{\sigma}^{(i)} = \mathbf{c}(P_0^{(i)}) : \triangle \boldsymbol{\varepsilon}^{(i)} - \mathbf{e}^*(P_0^{(i)}) \cdot \triangle \mathbf{E}^{(i)} + \mathbf{c}(P_0^{(i)}) : \triangle \boldsymbol{\varepsilon}_0{}^{(i)},$$

$$\triangle \mathbf{D}^{(i)} = \mathbf{e}(P_0^{(i)}) : \triangle \boldsymbol{\varepsilon}_e{}^{(i)} + \mathbf{g}(P_0^{(i)}) \cdot \triangle \mathbf{E}^{(i)} - \triangle \mathbf{D}^{(i-1)} - \mathbf{e}(P_0^{(i)}) : \triangle \boldsymbol{\varepsilon}_0{}^{(i-1)}.$$

To determine the increments of irreversible parts in our study we use a locked wall model, which also is called a Jile-Atherton model. This model was generalized to the three-dimensional case (for plane and axisymmetric problems on the two-dimensional case). In addition we introduced an amendment concerning the definition of the effective field. The main positions of this model can be found in

[2, 23]. We would like to note that in order to determine the increment of the residual polarization it is necessary to solve a system of equations in differentials

$$-\mathbf{P}_0 = -\mathbf{P}_\infty + k\frac{d\mathbf{P}_0}{|d\mathbf{E}^{(\mathrm{eff})}|}.$$

The solution of this system of equations in two-dimensional and axially symmetric case can be formally written as

$$P_{O\zeta}^{(i)} = P_{O\zeta}^{(i-1)} + \int_{(E_\zeta^{(i-1)},P_{O\zeta}^{(i-1)})}^{\triangle E_\zeta^{(i-1)}} f(E_\zeta^{(i-1)}, P_{O\zeta}^{(i-1)}, \mathrm{sign}(\triangle E_\zeta^{(i-1)}))d\zeta,$$

$$P_{O\eta}^{(i)} = P_{O\eta}^{(i-1)} + \int_{(E_\eta^{(i-1)},P_{O\eta}^{(i-1)})}^{\triangle E_\eta^{(i-1)}} f(E_\eta^{(i-1)}, P_{O\eta}^{(i-1)}, \mathrm{sign}(\triangle E_\eta^{(i-1)}))d\eta,$$

$$P_{O\xi}^{(i)} = 0.$$

In that case, when sufficiently small mechanical stresses are considered, the strain in local axes are determined from a condition of incompressibility material. This means that in the direction of remanent polarization vector a particle undergoes elongation deformation, and the strain in the transverse direction is the strain of compression, i.e.

$$\boldsymbol{\varepsilon}_0^{(i)} = \frac{\varepsilon_{sat}\mathbf{P}_0^{(i)}}{p_{sat}}\left(\mathbf{e}_1\mathbf{e}_1 - \frac{1}{2}\mathbf{e}_2\mathbf{e}_2 - \frac{1}{2}\mathbf{e}_3\mathbf{e}_3\right).$$

Then, using the standard scheme of the finite element method, we obtain a system of algebraic equations for determining all of the desired characteristics.

$$\mathbf{K}_{uu} \cdot \mathbf{U} + \mathbf{K}_{u\varphi} \cdot \boldsymbol{\Phi} = \mathbf{f}_1 + \mathbf{f}_\varepsilon,$$

$$\mathbf{K}_{u\varphi}^* \cdot \mathbf{U} - \mathbf{K}_{\varphi\varphi} \cdot \boldsymbol{\Phi} = \mathbf{f}_2 + \mathbf{f}_\varphi,$$

where $\mathbf{K}_{uu}$, $\mathbf{K}_{u\varphi}$, $\mathbf{K}_{\varphi\varphi}$ are the same matrices, as in Sect. 3,

$$\mathbf{f}_1 = \int_{\Gamma_\sigma} \mathbf{N}_u \cdot \triangle\mathbf{p}^{(i)}\, d\Gamma + \int_{\Omega_m} \mathbf{N}_u \cdot \triangle\mathbf{f}^{(i)} d\Omega, \quad \mathbf{f}_\varepsilon = \int_{\Omega_m} \mathbf{B}_u^* \cdot \mathbf{c} \cdot \triangle\varepsilon_0^{(i-1)}\, d\Omega,$$

$$\mathbf{f}_2 = \int_{\Gamma_D} \mathbf{N}_\varphi \cdot \triangle\mathbf{D}_n^{(i)}\, d\Gamma, \quad \mathbf{f}_\varphi = \int_{\Omega_m} \mathbf{B}_\varphi^* \cdot \mathbf{e} \cdot \triangle\varepsilon_0^{(i-1)}\, d\Omega - \int_{\Omega_m} \mathbf{B}_\varphi^* \cdot \triangle\mathbf{P}_0^{(i-1)} d\Omega.$$

The right parts of this system include not only the increment of boundary conditions and the increment of the residual parameters. The determinant of the system is changed with changing of values of residual parameters, but it satisfies all the conditions of solvability, as is the case of the model of electroelasticity with polarization to saturation.

All these algorithms were implemented in ACELAN software package. A problems definition of residual fields polarization and strain are solved in two stages.

At the first step we have to create a geometric image of the object with appropriate boundary conditions. Here we should determine the elastic and the ferroelectric regions, and introduce physical characteristics of materials in these regions in a non-polarized state and in a state of polarization for saturation. For unpolarized state we define elastic and dielectric constants. For the state of polarization for saturation we introduce a complete set of elastic, piezoelectric and dielectric constants, as described in [2]. At the end of the first stage the input data for all selected areas is saved.

At the second step we need to specify the variation of intensity of the electrical potential across the electrodes in the form of a function that varies over time (the package allows users to select some elementary functions). Further actions are standard, and they are associated with the usual actions for any problem solving process. In the pop-up menu we need to select action "polarization" and after that the program calculates the field of residual polarization at each finite element, saving information about the residual polarization vector and the residual strain in the principal axes. For example, for the plane stress or plane strain problem a table with dimension $5 \times N$ is saved, where N it the number of finite elements for ferroelectric area. It should be noted that the change of the field of the residual polarization is indicated on the second screen.

The final step of the second stage is to save all data associated with finite element cells for all considered areas and residual polarization and strain. Some examples and the fields of residual polarization and strain will be provided in subsequent sections of this paper.

5 New Opportunities of ACELAN

Computer-aided engineering (CAE) programs are usually based on distributed moduli architecture that allows to simplify development of individual units and make reusable agile code base. Basic functionality of ACELAN package was achieved by developing the following parts: problem storage format, pre-processing graphical user interface (GUI), script language for imperative model description, material library, finite element library, boundary condition library, mesh generator, linear algebra package, post-processing model, common GUI shell and internal and external program interfaces. Starting from the first versions of the ACELAN package we have been updating its programming languages and tools. Following, programming languages were used in different versions of the package: two-dimensional modelling tools: Fortran, C+11 (ISO/IEC 14882:2011), three-dimensional modelling tools—C# and .Net libraries. For advanced web-based client visualization application JavaScript was used.

In order to discuss specific features of the described package, let us focus on inhomogeneous materials models, custom storage formats and three-dimensional material distribution models for some classes of mixed composites.

Inhomogeneous properties can be taken into account during the process of building the local stiffness matrices. Each inhomogeneous property can be described as one or several scalar fields, common form for such distribution presented below: moduli $\mathbf{c} = \mathbf{c}^{E,H,\theta}$, $\mathbf{e} = \mathbf{e}^{H,\theta}$, $\boldsymbol{\kappa} = \boldsymbol{\kappa}^{\varepsilon,H,\theta}$, others material constants from (4)–(6) and density ρ will be evaluated as functions depending on geometrical position of point:

$$\rho = \rho(\mathbf{x}), \quad \mathbf{c} = \mathbf{c}(\mathbf{x}), \quad \mathbf{e} = \mathbf{e}(\mathbf{x}), \quad \boldsymbol{\kappa} = \boldsymbol{\kappa}(\mathbf{x}).$$

In the case of piezoelectric materials the mechanical and electrical properties can depend on inhomogeneous polarization:

$$\mathit{for}\ \ \mathbf{c}\ \ \mathit{and}\ \ \boldsymbol{\kappa}: \quad \mathbf{g} = \mathbf{g}^{(i)} + |P|\,(\mathbf{g}^{(a)} - \mathbf{g}^{(i)}),$$

$$\mathit{for}\ \ \mathbf{e}: \quad \mathbf{g} = |P|\,\mathbf{g}^{(a)},$$

where $\mathbf{g}$ stands for the corresponding tensors, superscript (i) describes the isotropic non-polarized state of the material, superscript (a) describes the anisotropic properties, factor $|P|$ affects the properties which will be manifested to a greater extent, and will be zeroes for isotropic materials. There can be several rules describing different properties in single body and the whole model can consist of several bodies with independent properties. In the case of vector field (e.g. polarization) it can be separated into scalar fields. In two-dimensional models we use polar coordinates to describe polarization as modulus and angle in every node of the mesh.

Inhomogeneous polarization problems are mostly two-dimensional due to limitations of applied mathematical model. There are two visualization tools: one for setting inhomogeneous polarization for problems, and one for post-processing analysis in case of determining polarization for predefined electrode configuration.

Important part of modelling continuous functions as material properties in finite element method is a discretization technique. We used two approaches: the first one was based on an assumption that functions are constant inside finite elements. In the second approach the nodes of numerical integration were used to achieve better accuracy. Depending on element type, the number of additional operations to handle the second approach can grow, but CPU and memory consumption remains relatively small compared to the procedures of global matrix assembling and solving.

GUI for describing inhomogeneous properties was developed on .NET platform with Windows Forms tools. Program has built-in visualization tool for two-dimensional case. Screen shots are presented in Figs. 3 and 4 with interface in Russian.

In some cases, analytical representation of the function is available for specific purposes. Therefore, there are some classes of functions that can be used in

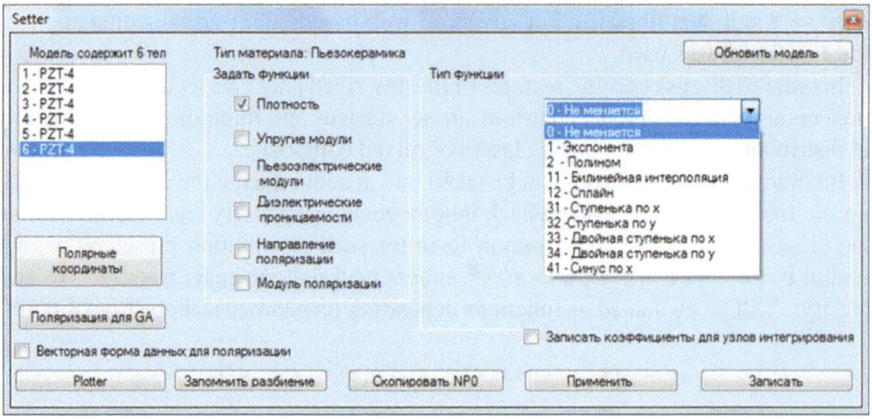

Fig. 3 GUI for setting inhomogeneous properties

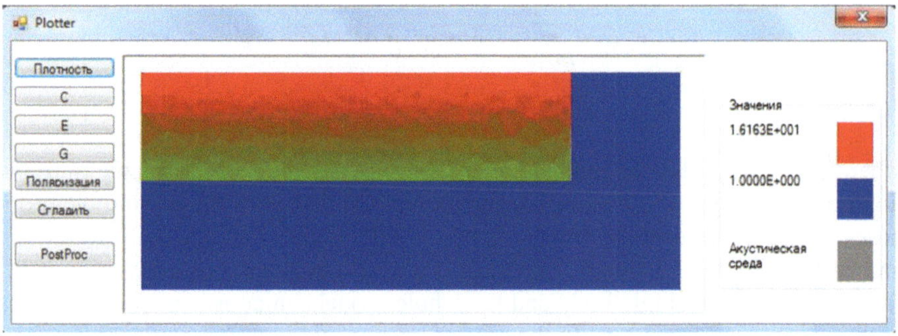

Fig. 4 GUI for viewing inhomogeneous properties

ACELAN by specifying coefficients: exponential function, polynomial function, Heaviside function, trigonometrical functions. In other cases, we have to deal with tabular functions or sets of values in some points. Transferring such data to different meshes can be achieved with splines.

Program interface and library of classes for importing discrete geometry representations was developed. Imported geometry can be converted to internal in-memory ACELAN model, serialized binary file based on .NET library. Geometry data can be supplemented with material properties and boundary conditions. Several widely used in industry CAD formats were considered for importing geometry. Currently Gmsh format is used for importing external geometries. The model can also be represented as relational database. Local databases created with SQLite are used for such purpose and also provide material editor information storage. Elastic, electrical, magnetic, thermal and acoustic properties can be presented in materials in different combinations. Simplified model database is shown in Fig. 5.

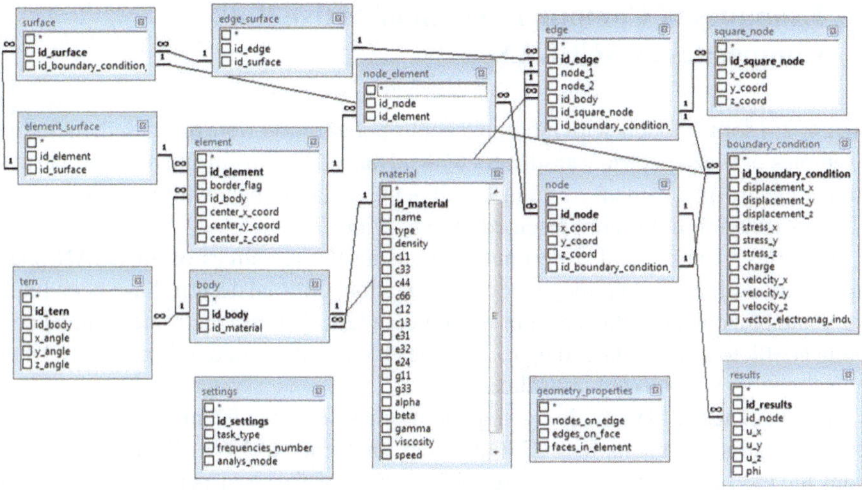

Fig. 5 Simplified relational model for mesh, materials, boundary conditions and results

Idea of using relational database as storage format instead of plain text or binary format is based on such characteristics of finite element models, as large sizes, separated multiple results sets, multiple sets of initial and boundary conditions. The possibility to load sets of data separately is essential for improving performance of post-processing tools and for back-end part of client-server application. Variety of available storage formats allows to perform complicated post-processing analysis with reduced resource requirements.

Linear algebra operations, including sparse matrices creation and editing, were performed by custom algorithms in two-dimensional versions. In three-dimensional version several solutions were used: in C++ version Intel MKL and CSparse libraries and custom triplet-based storage scheme were tested, in C# version AlgLib package was used with additional specific functions for operation with large sparse matrices. Time efficiency is achieved by in-built multithreading of linear algebra packages. Application interface allows to switch between linear algebra packages with the same storage formats, including mixing managed and unmanaged libraries.

Mixed composite materials were simulated using cubic elements (octants) in the representative volume. The absence of intersections between adjacent octants leads to the requirement of distinct elements at the atomic or molecular level. Initial volume is divided into octants using Octree algorithm, size and number of elements may vary depending on physical properties of material or needed accuracy. Octants can easily be used as finite elements on regular mesh. Structural elements of the same material are united into the composite component. Each component has a communication property: any element belonging to the component, is reachable from any other component element of the transition between the adjacent elements. This approach allows to model specific materials created with mixtures.

6 Examples of Computer Design of Active Bulk Composites in ACELAN

In [24, 25] some models with inhomogeneous polarization were discussed, including examples of piezoelectric transducers with advanced electro-mechanical properties: higher electro-mechanical coupling coefficient and larger bandwidth. Some of modeled polarization fields are presented in Figs. 6 and 7.

In series of numerical experiments, by using the resources of ACELAN package for modelling the inhomogeneous polarization, optimal designs have been found for some piezoelectric devices. For example, in [24, 25] the electromechanical coupling coefficient was enlarged up to 3.5 times for specific modes of one-layered nonhomogeneously polarized transducer.

The package allows building regular meshes for specific material with up to 262144 distinct elements with 274625 nodes, that leads to 1373125 degrees of freedom for the most complicated case of 5 degrees of freedom per node for electro-magneto-elastic materials. Large number of degrees of freedom is one of the reasons for using the client-server architecture available in the newest versions of the package.

Different types of available topologies are presented in Figs. 8, 9 and 10.

Thus, the current version of ACELAN package has capability of modelling active materials with inhomogeneous properties, including inhomogeneous polarization. This feature allows optimizing output properties of piezoelectric transducers, for example, storage devices for green energy. Module for modelling multi-component composite materials with different connectivity types can be used for two purposes: direct problems solving with such materials and building representative volumes to define physical properties of the materials.

Fig. 6 Predefined polarization described by analytical function

Fig. 7 Results of solving the polarization direct problem

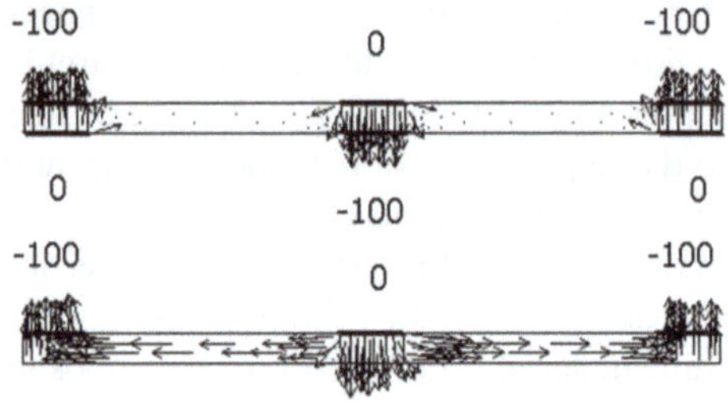

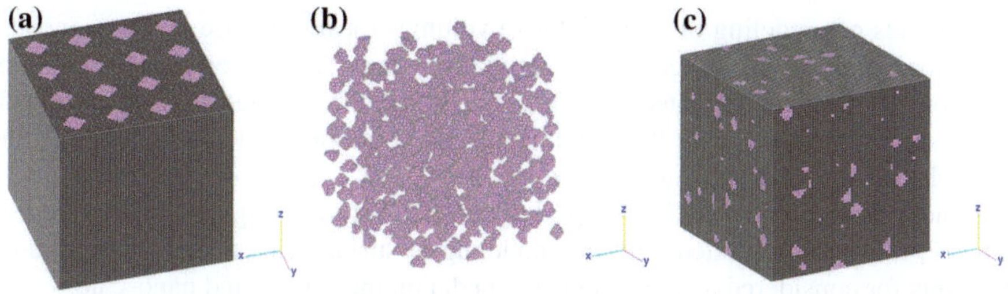

Fig. 8 Regular mesh for composite with pillar topology (**a**) and material with inclusions: inclusions only (**b**) and whole composite (**c**)

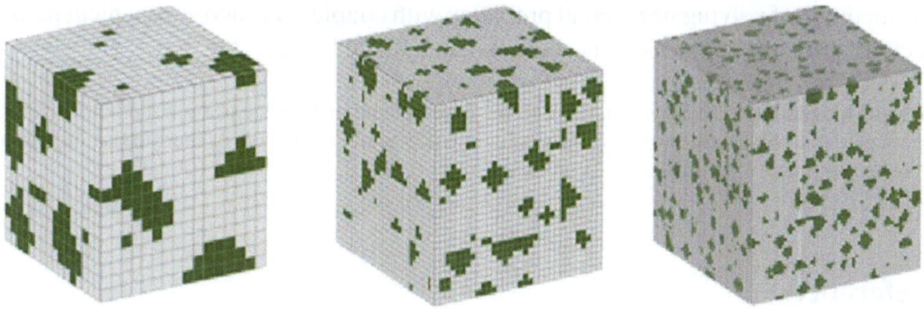

Fig. 9 Inclusions topology depending on number of elements

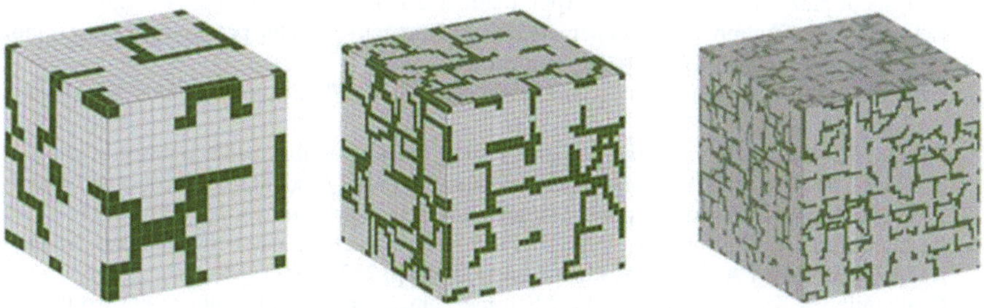

Fig. 10 Two components of material, each has connectivity property, depending on number of elements

7 Conclusion

Thus, in this paper, we have described the models of active composites and the possibilities of their simulation in the finite element software ACELAN. Other features of ACELAN package for solving common problems for active devices with coupled different physical and mechanical fields have been presented previously [12, 15, 18, 20]. Here we have restricted ourself only to simulation of the homogenization problems for active composites. The approach allows us to find the effective moduli on the basis of the main statements of composite mechanics and based on advanced possibilities of modeling of representative volumes for multiphase inhomogeneous anisotropic media with different connectivity of physical and mechanical fields. In addition, we have implemented in ACELAN package the calculation procedures and the basis for further use of the inhomogeneous polarization fields [2, 23–25] for piezoelectric and piezomagnetic phases.

Other possibilities of ACELAN package, which is now being actively developed [16, 17, 20], are associated with the modeling of surface effects and surface finite elements for considered active composite media on the micro- and nanoscale.

Analysis of the well-known finite element software shows that the models and technologies adopted in new version of ACELAN package significantly increase the facilities for analyzing complex active multiphase composite materials and provide the methods of solving new actual problems with coupled physico-mechanical fields, including the problems for nanoscale and composite bodies.

Acknowledgements The investigations on the development of homogenization models for active composite materials have been carried by A. Nasedkin with the support of the Russian Scientific Foundation (RSCF) by Project 15-19-10008.

References

1. Belokon, A.V., Nasedkin, A.V., Soloviev, A.N.: New schemes for the finite-element dynamic analysis of piezoelectric devices. J. Appl. Math. Mech. (PMM) **66**(3), 481–490 (2002)
2. Belokon, A.V., Skaliuh, A.S.: Mathematical modeling of irreversible processes of polarization. M., FIZMATLIT, 1–328 (2010) (Russian edition)
3. Benzi, M., Golub, G.H., Liesen, J.: Numerical solution of saddle point problems. Acta Numerica **14**, 1–137 (2005)
4. Benzi, M., Wathen, A.J.: Some preconditioning techniques for saddle point problems. In: Schilders, W.H.A., van der Vorst, H.A., Rommes, J. (eds.) Model Order Reduction: Theory, Research Aspects and Applications. Mathematics in Industry, vol. 13, pp. 195–211
5. Berlincourt, D.A., Curran, D.R., Jaffe, H.: Piezoelectric and piezomagnetic materials. In: Physical Acoustics, Part A, vol. 1. Academic Press, NY, pp. 233–256 (1964)
6. Challagulla, K.S., Georgiades, A.V.: Micromechanical analysis of magneto-electro-thermo-elastic composite materials with applications to multilayered structures. Int. J. Eng. Sci. **49**, 85–104 (2011)
7. Lee, J., Boyd, J.G., Lagoudas, D.C.: Effective properties of three-phase electro-magneto-elastic composites. Int. J. Eng. Sci. **43**, 790–825 (2005)
8. Li, J.Y.: Magnetoelectroelastic multi-inclusion and inhomogeneity problems and their applications in composite materials. Int. J. Eng. Sci. **38**, 1993–2011 (2000)
9. Hadjiloizi, A.D., Georgiades, A.V., Kalamkarov, A.V., Jothi, S.: Micromechanical modeling of piezo-magneto-thermo-elastic composite structures: part I-Theory. Eur. J. Mech. A-Solids **39**, 298–312 (2013)
10. Hadjiloizi, D.A., Georgiades, A.V., Kalamkarov, A.L., Jothi, S.: Micromechanical modeling of piezo-magneto-thermo-elastic composite structures: Part II–Applications. Eur. J. Mech. A-Solids **39**, 313–327 (2013)
11. Nan, C.-W., Bichurin, M.I., Dong, S., Viehland, D., Srinivasan, G.: Multiferroic magnetoelectric composites: historical perspective, status, and future directions. J. Appl. Phys. **103**, 031101–1–35 (2008)

12. Nasedkin, A.V.: Some finite element methods and algorithms for solving acousto-piezoelectric problems. In: Parinov, I.A. (ed.) Piezoceramic Materials and Devices, pp. 177–218. Nova Science Publishers, New York (2010)
13. Nasedkin, A.V.: Modeling of magnetoelectric composites by effective moduli and finite element methods. Theoretical approaches. Ferroelectrics. **461**(1), 106–112 (2014)
14. Nasedkin, A.V.: Multiscale computer design of piezomagnetoelectric mixture composite structures. AIP Conf. Proc. **1627**, 64–69 (2014)
15. Nasedkin, A.V.: Finite element design of piezoelectric and magnetoelectric composites by using symmetric saddle algorithms. In: Parinov, I.A., Chang, S.-H., Theerakulpisut, S. (eds.) Advanced Materials-Studies and Applications, pp. 109–124. Nova Science Publishers, New York (2015)
16. Nasedkin, A.V., Eremeyev, V.A.: Modeling of nanosized piezoelectric and magnetoelectric bodies with surface effects. AIP Conf. Proc. **1627**, 70–75 (2014)
17. Nasedkin, A.V., Eremeyev, V.A.: Some models for nanosized magnetoelectric bodies with surface effects. In: Parinov, I.A., Chang, S.-H., Topolov, V.Y. (eds.) Advanced Materials-Manufacturing, Physics, Mechanics and Applications, Series "Springer Proceedings in Physics, vol. 175, pp. 373–391. Springer, Heidelberg, New York, Dordrecht, London (2016)
18. Nasedkin, A.V., Nasedkina, A.A.: Finite element modeling and computer design of porous composites. In: Hellmich, C., Pichler, B., Adam, D. (eds.) Poromechanics V. Proceedings of the Fifth Biot Conference on Poromechanics, pp. 608–617, 10–12 July 2013, Vienna, Austria. Publ. ASCE (2013)
19. Nasedkin, A.V., Shevtsova, M.S.: Improved finite element approaches for modeling of porous piezocomposite materials with different connectivity. In: Parinov, I.A. (ed.) Ferroelectrics and Superconductors: Properties and Applications, pp. 231–254. Nova Science Publishers, New York (2011)
20. Nasedkin, A., Skaliukh, A., Soloviev, A.: New models of coupled active materials for finite element package ACELAN. AIP Conf. Proc. **1637**, 714–723 (2014)
21. Rybyanets, A.N.: Ceramic piezocomposites: modeling, technology, characterization. In: Parinov, I.A. (ed.) Piezoceramic Materials and Devices, pp. 115–174. Nova Science Publishers, New York (2010)
22. Rybyanets, A.N.: Porous piezoeramics: theory, technology, and properties. IEEE Trans. UFFC. **58**, 1492–1507 (2011)
23. Skaliukh, A.S.: Modeling of polarisation of the polycrystilline ferroelectrics. In: Parinov, I.A. (ed.) Piezoelectric materials and devices, pp. 50–102. Nova Science Publishers, New York (2012)
24. Skaliukh, A.S., Soloviev, A.N., Oganesyan, P.A.: Modeling of piezoelectric elements with inhomogeneous polarization in ACELAN. Ferroelectrics **483**(1), 95–101 (2015)
25. Soloviev, A.N., Oganesyan, P.A., Skaliukh, A.S.: Modeling of piezoelectric elements with inhomogeneous polarization by using ACELAN. In: Parinov, I.A., Chang, S.-H., Theerakulpisut, S. (eds.) Advanced Materials-Studies and Applications, pp. 169–192. Nova Science Publishers, New York (2015)
26. Tang, T., Yu, W.: Variational asymptotic homogenization of heterogeneous electromagnetoelastic materials. Int. J. Eng. Sci. **46**, 741–757 (2008)
27. Topolov, V.Yu., Bowen, C.R.: Electromechanical Properties in Composites Based on Ferroelectrics, pp. 1–202. Springer, London (2009)
28. Vanderbei, R.J.: Symmetric quasidefinite matrices. SIAM J. Optim. **5**, 100–113 (1995)
29. Zhang, Z.K., Soh, A.K.: Micromechanics predictions of the effective moduli of magnetoelectroelastic composite materials. Eur. J. Mech. A-Solids **24**, 1054–1067 (2005)

On the Models of Three-Layered Plates and Shells with Thin Soft Core

Victor A. Eremeyev and Konstantin Naumenko

Abstract We discuss here the mechanics of thin three-layered plates and shallow shells with thin soft core. Recently such thin-walled structures are widely used in engineering, among examples are laminated glasses and photovoltaic panels. We briefly consider layer-wise and first-order shear deformable plates and shells theories in order to model these structures.

Keywords First order shear deformable plate · Layer-wise theory · Three-layered plate · Soft core · Effective stiffness

1 Introduction

Recently the interest grows to application of mechanics of plates and shells to such thin-walled structures as laminated glasses and photovoltaic panels [1–6]. Among them are three-layered structures with soft thin internal layer (core) with high contrast in the mechanical properties of faces and core. For example, the ratio of the shear moduli G_c/G_f for materials used in photovoltaic panels is in the range between 10^{-5} and 10^{-2}, where G_f and G_c are the shear moduli of the glass faces and the polymeric core, respectively, see [7–9].

Among models of layered plates and shells there various theories based on different approximations of displacement fields along the thickness. One of the often used models is the first order shear deformation theory (FSDT) of plates [10–17]. Within this model one uses the hypothesis of straight normal. In other words, the

V.A. Eremeyev (✉)
Institute of Mathematics, Mechanics and Computer Science, Southern Federal University, Milchakova Street 8a, Rostov-on-Don 344090, Russia
e-mail: eremeyev.victor@gmail.com

V.A. Eremeyev
Rzeszów University of Technology, al. Powstańców Warszawy 8, 35959 Rzeszów, Poland

K. Naumenko
Institut für Mechanik, Otto von Guericke University Magdeburg, Universitätplatz 2, 39106 Magdeburg, Germany
e-mail: konstantin.naumenko@ovgu.de

M.A. Sumbatyan (ed.), *Wave Dynamics and Composite Mechanics for Microstructured Materials and Metamaterials*,
Advanced Structured Materials 59, DOI 10.1007/978-981-10-3797-9_9

normal fiber to the middle surface of a plate or shell behaves like rigid rod. In this theory the basic kinematic variables are three translations of the middle surface and two rotations. The so-called drilling moment is not taken into account here.

Another approach uses zig-zag and/or layer-wise approximations of displacements along the thickness. Within this approach the displacements are approximated by piecewise functions with respect to the thickness coordinate such that the compatibility in displacements between the layers is fulfilled. Then the governing equations of the three-dimensional elasticity are reduced to the two-dimensional plate equations by means of variational methods or asymptotic techniques [1, 3, 5, 8, 9, 18–25]. Analysis of differences between theories can be found for example in [22, 26, 27].

The aim of the paper is to briefly discuss both approaches. The paper is organized as follows. In Sects. 2 and 3 we consider layer-wise model of plates and shells, respectively. Within this approach for each layer thin-walled structure we introduce independently translation and rotation fields. As a result we derived the system of equations for few basic variables describing mean deflection and shearing between layers. The influence of boundary conditions is also discussed. In Sect. 4 we consider the Mindlin-Reissner plates. Here the model contains five degrees of freedom, that are three translations and two rotations. We present formulas for effective tangential, bending and transverse stiffness parameters. In particular, the dependence of the transverse shear stiffness on the elastic moduli relations is discussed. Through the paper we use the coordinate-free tensor notations [28].

2 Layer-Wise Models of Plates

Following [22] we present the basic equations of the layer-wise theory of plates. For the derivation of the two-dimensional equations, the through-the-thickness integration of the three-dimensional equilibrium equations was used.

2.1 Equilibrium Conditions

The stress resultants can be obtained by the through-the-thickness integration of the stress tensor $\boldsymbol{\sigma}$ as follows, e.g. [28]

$$\mathbf{T}_k = \langle \mathbf{P}\cdot\sigma\rangle_k = \mathbf{N}_k + \mathbf{Q}_k \otimes \mathbf{n}, \quad \mathbf{M}_k = -\langle z\mathbf{P}\cdot\sigma\times\mathbf{n}\rangle_k, \tag{1}$$

$$N_{k_{\alpha\beta}} = \langle\sigma_{\alpha\beta}\rangle_k, \quad Q_{k_\alpha} = \langle\sigma_{\alpha 3}\rangle_k, \quad M_{k_{\alpha\beta}} = \langle z\sigma_{\alpha\beta}\rangle_k,$$

$$\mathbf{P} = \mathbf{i}_\alpha \otimes \mathbf{i}_\alpha, \quad \langle\ldots\rangle_k = \int_{-h_k/2}^{h_k/2} (\ldots)\mathrm{d}z, \qquad \mathrm{k} = \mathrm{T}, \mathrm{C}, \mathrm{B},$$

where $\mathbf{P}$ is the projector, $\mathbf{i}_1, \mathbf{i}_2, \mathbf{n}$ are the Cartesian base vectors and the corresponding coordinates x_1, x_2 and z are used. h_T, h_C, and h_B denote the thicknesses of the top, core and bottom layers, respectively. Hereinafter all quantities related to the top, core and bottom layers, will be denoted by subscripts T, C, and B, respectively. The origin for z-coordinate is placed in the midplane of the core layer, so $-h_B - h_C/2 \leq z \leq h_C/2 + h_T$.

The equilibrium equations for each layer take the form

$$\nabla \cdot \mathbf{T}_T + (q_T + q)\mathbf{n} + \mathbf{s}_T + \mathbf{s} = \mathbf{0}, \quad \nabla \cdot \mathbf{M}_T + \mathbf{T}_{T\times} + \mathbf{m}_T = \mathbf{0}, \tag{2}$$

$$\nabla \cdot \mathbf{T}_C + (q_B - q_T)\mathbf{n} + \mathbf{s}_B - \mathbf{s}_T = \mathbf{0}, \quad \nabla \cdot \mathbf{M}_C + \mathbf{T}_{C\times} + \mathbf{m}_C = \mathbf{0}, \tag{3}$$

$$\nabla \cdot \mathbf{T}_B - q_B\mathbf{n} - \mathbf{s}_B = \mathbf{0}, \quad \nabla \cdot \mathbf{M}_B + \mathbf{T}_{B\times} + \mathbf{m}_B = \mathbf{0}, \tag{4}$$

where $\nabla = \mathbf{i}_\alpha \dfrac{\partial}{\partial x_\alpha}$ is the plane nabla operator, $\mathbf{T}_\times$ denotes the vectorial invariant of the stress resultant tensor $\mathbf{T}$, see [28], the components of interaction force vectors between the layers include the tangentially distributed forces $\pm\mathbf{s}_T$ (interactions between the layers T and C) and $\pm\mathbf{s}_B$ (interaction between the layers C and B) as well as the corresponding normally distributed forces $\pm q_T\mathbf{n}$ and $\pm q_B\mathbf{n}$. We assume that the top face of the plate is subjected to the normal $\mathbf{q} = q\mathbf{n}$ and tangential $\mathbf{s}$ loads, whereas the bottom face is free. In addition we introduced moments

$$\mathbf{m}_T = \frac{h_T}{2}\mathbf{n} \times (\mathbf{s} + \mathbf{s}_T), \quad \mathbf{m}_C = \frac{h_C}{2}\mathbf{n} \times (\mathbf{s}_T + \mathbf{s}_B), \quad \mathbf{m}_B = \frac{h_B}{2}\mathbf{n} \times \mathbf{s}_B \tag{5}$$

The equilibrium equations can be transformed into

$$\nabla \cdot \mathbf{N}_T + \mathbf{s}_T + \mathbf{s} = \mathbf{0}, \quad \nabla \cdot \mathbf{Q}_T + q_T + q = 0, \tag{6}$$

$$\nabla \cdot \mathbf{N}_C + \mathbf{s}_B - \mathbf{s}_T = \mathbf{0}, \quad \nabla \cdot \mathbf{Q}_C + q_B - q_T = 0, \tag{7}$$

$$\nabla \cdot \mathbf{N}_B - \mathbf{s}_B = \mathbf{0}, \quad \nabla \cdot \mathbf{Q}_B - q_B = 0 \tag{8}$$

$$\nabla \cdot \mathbf{L}_T - \mathbf{Q}_T + \frac{h_T}{2}(\mathbf{s}_T + \mathbf{s}) = \mathbf{0}, \quad \nabla \cdot \mathbf{L}_C - \mathbf{Q}_C + \frac{h_C}{2}(\mathbf{s}_B + \mathbf{s}_T) = \mathbf{0}, \tag{9}$$

$$\nabla \cdot \mathbf{L}_B - \mathbf{Q}_B + \frac{h_B}{2}\mathbf{s}_B = \mathbf{0}, \tag{10}$$

where instead of $\mathbf{M}_k$ we introduce $\mathbf{L}_k$ as follows

$$\mathbf{L}_k = \mathbf{M}_k \times \mathbf{n} = M_{k_{\alpha\beta}} \mathbf{i}_\alpha \otimes \mathbf{i}_\beta.$$

2.2 Constitutive Equations

For the linear-elastic materials the constitutive equations for the stress resultants of thin plate layers can be formulated as follows

$$\mathbf{N}_k = \mathbf{C}_k \cdot\cdot \boldsymbol{\varepsilon}_k, \quad \mathbf{Q}_k = \boldsymbol{\Gamma}_k \cdot \boldsymbol{\gamma}_k, \quad \mathbf{L}_k = \mathbf{D}_k \cdot\cdot \boldsymbol{\chi}_k, \tag{11}$$

where the strain measures are defined as follows

$$\boldsymbol{\varepsilon}_k \equiv \frac{1}{2}\left(\nabla \mathbf{u}_k + (\nabla \mathbf{u}_k)^T\right) = \varepsilon_{k_{\alpha\beta}} \mathbf{i}_\alpha \otimes \mathbf{i}_\beta, \quad \boldsymbol{\gamma}_k \equiv \nabla w_k + \boldsymbol{\varphi}_k = \gamma_{k_\alpha} \mathbf{i}_\alpha, \tag{12}$$

$$\boldsymbol{\chi}_k \equiv \frac{1}{2}\left(\nabla \boldsymbol{\varphi}_k + (\nabla \boldsymbol{\varphi}_k)^T\right) = \chi_{k_{\alpha\beta}} \mathbf{i}_\alpha \otimes \mathbf{i}_\beta \tag{13}$$

Here $\boldsymbol{\varepsilon}_k$ is the membrane strain tensor, $\mathbf{u}_k$ is the in-plane displacement vector, $\boldsymbol{\gamma}_k$ is the transverse shear strain vector, $\boldsymbol{\varphi}_k$ is the normal rotation vector, w_k is the deflection and $\boldsymbol{\chi}_k$ is the curvature change tensor of the k-th layer. $\mathbf{C}_k$ and $\mathbf{D}_k$ are fourth-order membrane and bending stiffness tensors, while $\boldsymbol{\Gamma}_k$ is the second-order transverse shear stiffness tensor.

For isotropic and homogeneous materials the fourth-order membrane and bending stiffness tensors for the skin layers can be represented as follows

$$\mathbf{C}_k = \frac{E_k \nu_k h_k}{1-\nu_k^2} \mathbf{P} \otimes \mathbf{P} + \frac{E_k h_k}{2(1+\nu_k)} (\mathbf{i}_\alpha \otimes \mathbf{P} \otimes \mathbf{i}_\alpha + \mathbf{i}_\alpha \otimes \mathbf{i}_\beta \otimes \mathbf{i}_\alpha \otimes \mathbf{i}_\beta), \quad k = T, B, \tag{14}$$

$$\mathbf{D}_k = \frac{E_k \nu_k h_k^3}{12(1-\nu_k^2)} \mathbf{P} \otimes \mathbf{P} + \frac{E_k h_k^3}{24(1+\nu_k)} (\mathbf{i}_\alpha \otimes \mathbf{P} \otimes \mathbf{i}_\alpha + \mathbf{i}_\alpha \otimes \mathbf{i}_\beta \otimes \mathbf{i}_\alpha \otimes \mathbf{i}_\beta), \quad k = T, B, \tag{15}$$

where E_k and ν_k are the Young modulus and the Poisson ratio of the kth skin layer, respectively. For the core layer the shear stiffness tensor is given by

$$\boldsymbol{\Gamma}_C = \Gamma_C \mathbf{P}, \quad \Gamma_C = \kappa G_C h_C, \tag{16}$$

where G_C is the shear modulus and κ is the shear correction factor.

2.3 Compatibility Conditions and Further Assumptions

In what follows we imply that the layers are rigidly connected on interfaces. In other words, sliding and delamination between the layers are not allowed. Thus, for $\mathbf{u}_k$ and $\boldsymbol{\varphi}_k$ the following compatibility relations are fulfilled

$$\mathbf{u}_T + \frac{h_T}{2}\boldsymbol{\varphi}_T = \mathbf{u}_C - \frac{h_C}{2}\boldsymbol{\varphi}_C, \quad \mathbf{u}_B - \frac{h_B}{2}\boldsymbol{\varphi}_B = \mathbf{u}_C + \frac{h_C}{2}\boldsymbol{\varphi}_C. \tag{17}$$

Furthermore, we assume that the layers have the same deflections

$$w_B = w_C = w_T = w. \tag{18}$$

Equations (2)–(16) form the system of governing equations for a three-layered plate. These equations can be further simplified with the use of following assumptions [22].

1. The skin layers can be modelled using the Kirchhoff straight normal hypothesis, so we assume that

$$\boldsymbol{\varphi}_T = -\nabla w_T, \quad \boldsymbol{\varphi}_B = -\nabla w_B. \tag{19}$$

Thus, from (18) it follows that

$$\boldsymbol{\varphi}_T = \boldsymbol{\varphi}_B = -\nabla w. \tag{20}$$

We observe that cross section rotations of the skin layers are the same. For the curvature tensors we obtain

$$\chi_T = \chi_B = -\nabla\nabla w. \tag{21}$$

Using $\boldsymbol{\varphi}_T = \boldsymbol{\varphi}_B$ the compatibility conditions can be reduced to

$$\mathbf{u}_T + \mathbf{u}_B - \frac{h_T - h_B}{2}\nabla w = 2\mathbf{u}_C, \quad \mathbf{u}_T - \mathbf{u}_B - \frac{h_T + h_B}{2}\nabla w = -h_C\boldsymbol{\varphi}_C. \tag{22}$$

The assumption 1 is also applied in [3, 8, 9, 20] for the analysis of laminated glass beams and plates.

2. The membrane stiffness of the laminate is primarily determined by the membrane stiffness of the skin layers.

3. The core layer is moments-free. So, in equilibrium equations the tensor $\mathbf{L}_C$ can be neglected. As a result the following relationship can be obtained

$$\mathbf{Q}_C = \frac{h_C}{2}(\mathbf{s}_T + \mathbf{s}_B). \tag{23}$$

2.4 *Differential Equations for Relative In-Plane Displacements and Deflection*

From the engineering point of view a deflection and in-plane displacements play the most important role in the description of deformations. So we consider the relative in-plane displacement vector

$$\mathbf{u}_\delta = \mathbf{u}_T - \mathbf{u}_B \tag{24}$$

and the deflection function of the laminate w as primary variables.

Introducing two scalar potentials Ω and Ψ such that

$$\mathbf{u}_\delta = \nabla\Omega + \nabla \times (\Psi \mathbf{n}) \tag{25}$$

we obtain the following differential equations for Ω and Ψ [22]

$$\Delta\Omega - \frac{2(1-\nu_T^2)}{E_T h_T}\frac{\Gamma_C}{h_C^2}\Omega = \frac{1-\nu_T^2}{E_T h_T}S - 2H\frac{1-\nu_T^2}{E_T h_T}\frac{\Gamma_C}{h_C^2}w, \quad \Delta\Psi - \frac{4(1+\nu_T)}{E_T h_T}\frac{\Gamma_C}{h_C^2}\Psi = 0, \tag{26}$$

where $\Delta = \nabla \cdot \nabla$ is the Laplace operator, $H = h_C + \frac{h_T + h_B}{2}$, and S is a potential of the applied tangential forces

$$\mathbf{s} = -\nabla S. \tag{27}$$

For deflection w we obtain the equation

$$\Delta\Delta w + 3\frac{H}{h_T^2}\Delta\Delta\Omega = \frac{q}{2D_T}, \quad D_T \equiv \frac{E_T h_T^3}{12(1-\nu_T^2)}. \tag{28}$$

Changing variables using relation

$$\tilde{w}\left(1 + 3\frac{H^2}{h_T^2}\right) = w + 3\frac{H}{h_T^2}\Omega$$

we transform (28) and $(26)_1$ into the form

$$\Delta\Delta\tilde{w} = \frac{q}{D}, \quad D \equiv 2D_T\left(1 + 3\frac{H^2}{h_T^2}\right), \tag{29}$$

$$\Delta\Omega - \beta^2\Omega = \frac{1-\nu_T^2}{E_T h_T}S - H\beta^2\tilde{w}, \quad \beta^2 \equiv \frac{2(1-\nu_T^2)}{E_T h_T}\frac{\Gamma_C}{h_C^2}\left(1 + 3\frac{H^2}{h_T^2}\right), \tag{30}$$

where D is the effective bending stiffness of the laminate.

2.5 Maximal Deflection of the Three-Layered Strip

Considering the bending of a three-layered strip of length l and thickness H the maximum deflection $w_{max} = w(l/2)$ is obtained in [22]

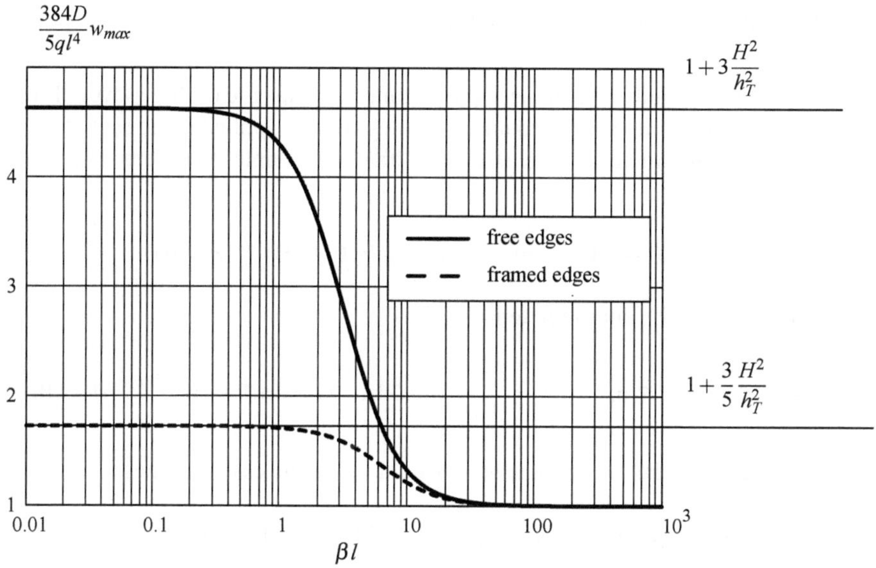

Fig. 1 Normalized maximum deflection vs. shear stiffness parameter for various boundary conditions [22]

$$\frac{D}{ql^2}w_{max} = \frac{5}{384} + \frac{3}{8}\frac{H^2}{\tilde{\beta}^2 h_T^2} - \begin{cases} 3\frac{H^2}{\tilde{\beta}^4 h_T^2}\left(1 - \frac{1}{\cosh\tilde{\beta}}\right), & \text{free edges} \\ \frac{3}{2}\frac{H^2}{\tilde{\beta}^3 h_T^2}\tanh\left(\frac{\tilde{\beta}}{4}\right), & \text{framed edges,} \end{cases} \tag{31}$$

where

$$\tilde{\beta}^2 = \frac{2(1-\nu_T^2)}{E_T h_T}\frac{\Gamma_C l^2}{h_C^2}\left(1 + 3\frac{H^2}{h_T^2}\right).$$

The dimensionless parameter $\tilde{\beta} = \beta l$ characterizes the shear stiffness.

Figure 1 shows the normalized maximum deflection as a function of $\tilde{\beta}$. For $\tilde{\beta} \to \infty$ the maximum deflection tends to deflection calculated using the Kirchhoff theory, see Fig. 1. Analysis shows also that for laminates with $\tilde{\beta} > 20$ the FSDT can be applied instead of more general layer-wise theory. For small $\tilde{\beta}$ the FSDT overestimates the deflection. Moreover, for small $\tilde{\beta}$ there is a significant difference for various boundary conditions.

3 On Layer-Wise Models of Shells

In this section we briefly discuss necessary changes in the case of three-layered shallow shells. In general, all presented un Sect. 2 steps can be applied in the case of shells but with technical difficulties. The stress resultant and couple stress tensors are now given by

$$\mathbf{T}_k = \langle (\mathbf{P} - z\mathbf{B})^{-1} \cdot \boldsymbol{\sigma} \rangle_k, \quad \mathbf{M}_k = -\langle z(\mathbf{P} - z\mathbf{B})^{-1} \cdot \boldsymbol{\sigma} \times \mathbf{n} \rangle_k, \tag{32}$$

where $\mathbf{B} = -\nabla \mathbf{n}$ is the curvature tensor,

$$\mathbf{P} = \mathbf{i}_\alpha \otimes \mathbf{i}^\alpha = \mathbf{I} - \mathbf{n} \otimes \mathbf{n}, \quad \langle \ldots \rangle_k = \int_{-h_k/2}^{h_k/2} (\ldots) G \mathrm{d}z, \quad \mathrm{G} = \det(\mathbf{P} - \mathrm{z}\mathbf{B}), \quad \mathrm{k} = \mathrm{T}, \mathrm{C}, \mathrm{B}.$$

With accuracy of $h\|\mathbf{B}\|$ we can use the formulae as in the case of plates

$$\mathbf{T}_k = \langle \mathbf{P} \cdot \boldsymbol{\sigma} \rangle_k = \mathbf{N}_k + \mathbf{Q}_k \otimes \mathbf{n}, \quad \mathbf{M}_k = -\langle z\mathbf{P} \cdot \boldsymbol{\sigma} \times \mathbf{n} \rangle_k, \tag{33}$$

and using components we have

$$N_{k_{\alpha\beta}} = \langle \sigma_{\alpha\beta} \rangle_k, \quad Q_{k_\alpha} = \langle \sigma_{\alpha 3} \rangle_k, \quad M_{k_{\alpha\beta}} = \langle z\sigma_{\alpha\beta} \rangle_k.$$

For small deformations we use the following constitutive relations (here we omit the index of the layer for brevity)

$$\mathbf{N} = \mathbf{C} \cdot \cdot \boldsymbol{\varepsilon}, \quad \mathbf{Q} = \boldsymbol{\Gamma} \cdot \boldsymbol{\gamma}, \quad \mathbf{M} = -\mathbf{D} \cdot \cdot \boldsymbol{\chi} \times \mathbf{n}, \tag{34}$$

where

$$\boldsymbol{\varepsilon} \equiv \frac{1}{2} \left(\nabla \mathbf{u} \cdot \mathbf{P} + \mathbf{P} \cdot (\nabla \mathbf{u})^T \right) - w\mathbf{B}, \tag{35}$$

$$\boldsymbol{\gamma} \equiv \nabla w + \boldsymbol{\varphi} + \mathbf{B} \cdot \mathbf{u}, \quad \boldsymbol{\chi} \equiv \frac{1}{2} \left(\nabla \boldsymbol{\varphi} \cdot \mathbf{P} + \mathbf{P} \cdot (\nabla \boldsymbol{\varphi})^T \right). \tag{36}$$

Here $\boldsymbol{\varepsilon}$ is the membrane strain tensor, $\mathbf{u}$ is the in-plane displacement vector, $\boldsymbol{\gamma}$ is the transverse shear strain vector, $\boldsymbol{\gamma}_k$ is the normal rotation vector, w is the deflection and $\boldsymbol{\chi}$ is the curvature change tensor. $\mathbf{C}$ and $\mathbf{D}$ are fourth-order membrane and bending stiffness tensors, while $\boldsymbol{\Gamma}$ is the second-order transverse shear stiffness tensor.

For the in-plane displacement vectors $\mathbf{u}_i$ and vectors of cross section rotations $\boldsymbol{\varphi}_i$ we assume the following compatibility relations

$$\mathbf{u}_T + \frac{h_T}{2} \boldsymbol{\varphi}_T = \mathbf{u}_C - \frac{h_C}{2} \boldsymbol{\varphi}_C, \quad \mathbf{u}_B + \frac{h_B}{2} \boldsymbol{\varphi}_B = \mathbf{u}_C + \frac{h_C}{2} \boldsymbol{\varphi}_C. \tag{37}$$

With respect to the deflections we assume the following relation

$$w_B = w_C = w_T = w. \tag{38}$$

We apply the same assumptions as for plates:

1. The skin layers can be assumed to be shear rigid. With the Kirchhoff–Love kinematic hypothesis (straight normal hypothesis) it follows that

$$\boldsymbol{\varphi}_T = -\nabla w_T - \mathbf{B}_T \cdot \mathbf{u}_T, \quad \boldsymbol{\varphi}_B = -\nabla w_B - \mathbf{B}_B \cdot \mathbf{u}_B. \tag{39}$$

   With the kinematical constraints (39) the shear forces $\mathbf{Q}_T$ and $\mathbf{Q}_T$ are not defined by the constitutive equations anymore. They can be computed from equilibrium conditions.
2. The membrane stiffness of the laminated plate is primarily determined by the membrane stiffness of the skin layers.
3. The core layer is moment-free.

The main difference between plates and shells consist of in coupling between in-plane deformations and bending. As a result, the governing equations become coupled and cannot be splitted as in the case of plates. The detailed analysis of shallow three-layered shells will be performed in forthcoming papers.

4 First-Order Shear Deformable Plate Model

Here we consider the first-order shear deformable model of plates called also five-parameter plate theory or Reissner–Mindlin plate model [10, 13, 29, 30]. Let us note that the model is implemented in various commercial software, so the model can be treated as classical one. But for calculations within the model one may need to change properly the stiffness parameters such the transverse shear stiffness.

4.1 Equilibrium Conditions

The equilibrium equations have the following form:

$$\nabla \cdot \mathbf{T} + \mathbf{q} = \mathbf{0}, \quad \nabla \cdot \mathbf{M} + \mathbf{T}_\times + \mathbf{m} = \mathbf{0}, \tag{40}$$

where $\mathbf{T}$, $\mathbf{M}$ are the stress resultant and couple stress tensors, $\mathbf{q}$, $\mathbf{m}$ are the surface force and moment densities, ∇ is the surface nabla operator, $\mathbf{u}$, $\boldsymbol{\varphi}$ are the vectors of the displacements and the rotations, respectively, and ρ is the surface mass density.

The static and kinematic boundary conditions are given by

$$\boldsymbol{\nu} \cdot \mathbf{T} = \mathbf{f}, \quad \boldsymbol{\nu} \cdot \mathbf{M} = \mathbf{l} \quad (\mathbf{l} \cdot \mathbf{n} = 0) \quad \text{or} \quad \mathbf{u} = \mathbf{u}^0, \quad \boldsymbol{\varphi} = \boldsymbol{\varphi}^0 \quad \text{along } S, \tag{41}$$

where $\mathbf{f}$ and $\mathbf{l}$ are external force and couple vectors acting along the contour of the plate S, whereas $\mathbf{u}^0$ and φ^0 are given displacements and rotations, respectively. $\boldsymbol{\nu}$ is the unit normal vector to S ($\nu \cdot \mathbf{n} = 0$). Clearly, mixed types of boundary conditions may be also useful.

4.2 Constitutive Equations

The surface strain energy density is given by the following quadratic form

$$W(\boldsymbol{\mu}, \boldsymbol{\gamma}, \boldsymbol{\kappa}) = \frac{1}{2}\boldsymbol{\mu}\cdot\cdot\mathbf{A}\cdot\cdot\boldsymbol{\mu} + \boldsymbol{\mu}\cdot\cdot\mathbf{B}\cdot\cdot\boldsymbol{\kappa} + \frac{1}{2}\boldsymbol{\kappa}\cdot\cdot\boldsymbol{C}\cdot\cdot\boldsymbol{\kappa} + \frac{1}{2}\boldsymbol{\gamma}\cdot\boldsymbol{\Gamma}\cdot\boldsymbol{\gamma} + \boldsymbol{\gamma}\cdot(\boldsymbol{\Gamma}_1\cdot\cdot\boldsymbol{\mu} + \boldsymbol{\Gamma}_2\cdot\cdot\boldsymbol{\kappa}). \tag{42}$$

Here we introduced strain measures

$$\boldsymbol{\mu} = (\nabla\mathbf{u} \cdot \mathbf{P})^{\text{sym}}, \quad \boldsymbol{\gamma} = \nabla\mathbf{u} \cdot \mathbf{n} + \mathbf{c} \cdot \boldsymbol{\varphi}, \quad \boldsymbol{\kappa} = \nabla\boldsymbol{\varphi}. \tag{43}$$

and stiffness tensors $\mathbf{A}$, $\mathbf{B}$, $\mathbf{C}$, $\boldsymbol{\Gamma}_1$, $\boldsymbol{\Gamma}_2$, and $\boldsymbol{\Gamma}$.

The tensors $\boldsymbol{\mu}$, $\boldsymbol{\gamma}$ and $\boldsymbol{\kappa}$ are the tensor of in-plane strains, vector of transverse shear strains and tensor of the out-of-plane strains, respectively. Here $\mathbf{n}$ is the unit normal vector, $\mathbf{c} = -\mathbf{P} \times \mathbf{n}$ is the discriminant tensor, $(\ldots)^{\text{sym}}$ denotes symmetric part of the tensor. The stiffness tensors $\mathbf{A}$, $\mathbf{B}$, $\mathbf{C}$ are fourth-order tensors, $\boldsymbol{\Gamma}_1$ and $\boldsymbol{\Gamma}_2$ are third-order tensors, and $\boldsymbol{\Gamma}$ is a second-order tensor.

The relations between $\mathbf{T}$, $\mathbf{M}$ and W are

$$\mathbf{T} \cdot \mathbf{P} = \frac{\partial W}{\partial \boldsymbol{\mu}} = \mathbf{A}\cdot\cdot\boldsymbol{\mu} + \mathbf{B}\cdot\cdot\boldsymbol{\kappa} + \boldsymbol{\gamma} \cdot \boldsymbol{\Gamma}_1, \quad \mathbf{T} \cdot \mathbf{n} = \frac{\partial W}{\partial \boldsymbol{\gamma}} = \boldsymbol{\Gamma} \cdot \boldsymbol{\gamma} + \boldsymbol{\Gamma}_1\cdot\cdot\boldsymbol{\mu} + \boldsymbol{\Gamma}_2\cdot\cdot\boldsymbol{\kappa}, \tag{44}$$

$$\mathbf{M}^{\mathrm{T}} = \frac{\partial W}{\partial \kappa} = \mu\cdot\cdot\mathbf{B} + \mathbf{C}\cdot\cdot\kappa + \gamma \cdot \Gamma_2. \tag{45}$$

For isotropic materials and for plates of symmetric structure the stiffness tensors take the form

$$\mathbf{A} = A_{11}\mathbf{a}_1 \otimes \mathbf{a}_1 + A_{22}(\mathbf{a}_2 \otimes \mathbf{a}_2 + \mathbf{a}_4 \otimes \mathbf{a}_4), \quad \mathbf{C} = C_{22}(\mathbf{a}_2 \otimes \mathbf{a}_2 + \mathbf{a}_4 \otimes \mathbf{a}_4) + C_{33}\mathbf{a}_3 \otimes \mathbf{a}_3, \tag{46}$$

$$\boldsymbol{\Gamma} = \Gamma\mathbf{P}, \quad \mathbf{B} = \mathbf{0}, \quad \boldsymbol{\Gamma}_1 = \mathbf{0}, \quad \boldsymbol{\Gamma}_2 = \mathbf{0},$$

where

$$\mathbf{a}_1 = \mathbf{a} = \mathbf{e}_1 \otimes \mathbf{e}_1 + \mathbf{e}_2 \otimes \mathbf{e}_2, \quad \mathbf{a}_2 = \mathbf{e}_1 \otimes \mathbf{e}_1 - \mathbf{e}_2 \otimes \mathbf{e}_2,$$

$$\mathbf{a}_3 = \mathbf{c} = \mathbf{e}_1 \otimes \mathbf{e}_2 - \mathbf{e}_2 \otimes \mathbf{e}_1, \quad \mathbf{a}_4 = \mathbf{e}_1 \otimes \mathbf{e}_2 + \mathbf{e}_2 \otimes \mathbf{e}_1,$$

and $\mathbf{e}_1, \mathbf{e}_2$ are unit basis vectors.

The tangent C and bending D stiffness parameters relate with $\mathbf{A}$ and $\mathbf{C}$ as follows

$$C \equiv A_{11} + A_{22}, \qquad D \equiv C_{33} + C_{22}.$$

4.3 *Stiffness Parameters*

Since the middle layer is soft and thin we can assume that

$$h_c \ll h_f, \qquad E_c \ll E_f, \qquad G_c \ll G_f, \qquad \rho_{c0} \ll \rho_{f0}.$$

Finally, the tangential, bending and transverse shear stiffness take the following values [30]

$$C = \frac{1}{2}\frac{E_f h}{1-\nu_f^2}, \qquad D = \frac{1}{12}\frac{E_f h^3}{1-\nu_f^2}, \quad \Gamma = \frac{1}{3\alpha} G_c h, \quad \text{where} \quad \alpha = \frac{h_c}{h}.$$

Clearly, C and D are entirely determined by thickness and stiffness of faces, whereas Γ depends on the stiffness and thickness of the core. Factor $1/3\alpha$ differs Γ from the Reissner's value $\Gamma_R = G_c h$ which is valid for classical sandwich plates.

5 Conclusions

Recent developments in the manufacturing and design of new thin-walled structures with the increase of efficiency and accuracy of used models require also to develop new nonclassical models of plates and shells or make certain modifications in classical theories. Here we discussed such relatively new structures as laminated glasses, photovoltaic panels and others layered structures with high contrast in material properties. It was shown that depending on the material parameters such as the transverse shear stiffness the first-order theories of plates and shells may lead to the overestimation of deflection. In particular, we also mention that for layer-wise theories the influence of nonclassical boundary conditions may be important.

Acknowledgements V.A.E. acknowledges the support by the Russian Science Foundation (grant number 15-19-10008).

References

1. Aşik, M.Z., Tezcan, S.: Laminated glass beams: strength factor and temperature effect. Comput. Struct. **84**, 364–373 (2006)
2. Chen, S., Zang, M., Xu, W.: A three-dimensional computational framework for impact fracture analysis of automotive laminated glass. Comput. Methods Appl. Mech. Eng. **294**, 72–99 (2015)
3. Ivanov, I.V.: Analysis, modelling, and optimization of laminated glasses as plane beam. Int. J. Solids Struct. **43**, 6887–6907 (2006)
4. Ivanov, I.V., Velchev, D.S., Georgiev, N.G., Ivanov, I.D., Sadowski, T.: A plate finite element for modelling of triplex laminated glass and comparison with other computational models. Meccanica **51**(2), 341–358 (2015)
5. Koutsawa, Y., Daya, E.M.: Static and free vibration analysis of laminated glass beam on viscoelastic supports. Int. J. Solids Struct. **44**, 8735–8750 (2007)
6. Liang, Y., Lancaster, F., Izzuddin, B.A.: Effective modelling of structural glass with laminated shell elements. Composi. Struct. **156**, 47–62 (2016)
7. Eitner, U., Köntges, M., Brendel, R.: Use of digital image correlation technique to determine thermomechanical deformations in photovoltaic laminates: measurements and accuracy. Sol. Energy Mater. Sol. Cells **94**(8), 1346–1351 (2010)
8. Schulze, S., Pander, M., Naumenko, K., Altenbach, H.: Analysis of laminated glass beams for photovoltaic applications. Int. J. Solids Struct. **49**(15–16), 2027–2036 (2012)
9. Weps, M., Naumenko, K., Altenbach, H.: Unsymmetric three-layer laminate with soft core for photovoltaic modules. Compos. Struct. **105**, 332–339 (2013)
10. Altenbach, H., Eremeyev, V.A.: Direct approach based analysis of plates composed of functionally graded materials. Arch. Appl. Mech. **78**(10), 775–794 (2008)
11. Altenbach, H., Eremeyev, V.A.: On the bending of viscoelastic plates made of polymer foams. Acta Mech. **204**(3–4), 137–154 (2009)
12. Brank, B.: On boundary layer in the mindlin plate model: levy plates. Thin-Walled Struct. **46**, 451–465 (2008)
13. Eisenträger, J., Naumenko, K., Altenbach, H., Köppe, H.: Application of the first-order shear deformation theory to the analysis of laminated glasses and photovoltaic panels. Int. J. Mech. Sci. **96–97**, 163–171 (2015)
14. Naumenko, K., Altenbach, J., Altenbach, H., Naumenko, V.K.: Closed and approximate analytical solutions for rectangular Mindlin plates. Acta Mech. **147**, 153–172 (2001)
15. Reddy, J.N., Wang, C.M.: An overview of the relationships between solutions of classical and shear deformation plate theories. Compos. Sci. Technol. **60**, 2327–2335 (2000)
16. Szilard, R.: Theories and Applications of Plate Analysis. Wiley, New Jersey (2004)
17. Wang, C.M., Reddy, J.N., Lee, K.H.: Shear Deformable Beams and Shells. Elsevier, Amsterdam (2000)
18. Carrera, E.: Historical review of Zig-Zag theories for multilayered plates and shells. Appl. Mech. Rev. **56**(2), 287–308 (2003)
19. Carrera, E., Brischetto, S., Nali, P.: Plates and Shells for Smart Structures: Classical and Advanced Theories for Modelling and Analysis. Wiley, Chichester (2011)
20. Foraboschi, P.: Analytical model for laminated-glass plate. Compos. Part B Eng. **43**(5), 2094–2106 (2012)
21. Kulikov, G.M., Plotnikova, S.V.: Exact 3D stress analysis of laminated composite plates by sampling surfaces method. Compos. Struct. **94**(12), 3654–3663 (2012)
22. Naumenko, K., Eremeyev, V.A.: A layer-wise theory for laminated glass and photovoltaic panels. Compos. Struct. **112**, 283–291 (2014)
23. Tornabene, F., Fantuzzi, N., Bacciocchi, M., Viola, E.: Accurate inter-laminar recovery for plates and doubly-curved shells with variable radii of curvature using layer-wise theories. Compos. Struct. **124**, 368–393 (2015)
24. Tornabene, F., Fantuzzi, N., Viola, E., Ferreira, A.: Radial basis function method applied to doubly-curved laminated composite shells and panels with a general higher-order equivalent single layer formulation. Compos. Part B Eng. **55**, 642–659 (2013)

25. Tornabene, F., Francesco, N., Viola, E.: Inter-laminar stress recovery procedure for doubly-curved, singly-curved, revolution shells with variable radii of curvature and plates using generalized higher-order theories and the local gdq method. Mech. Adv. Mater. Struct. **23**(9), 1019–1045 (2016)
26. Aßmus, M., Naumenko, K., Altenbach, H.: A multiscale projection approach for the coupled globallocal structural analysis of photovoltaic modules. Compos. Struct. **158**, 340–358 (2016)
27. Eisenträger, J., Naumenko, K., Altenbach, H., Meenen, J.: A user-defined finite element for laminated glass panels and photovoltaic modules based on a layer-wise theory. Compos. Struct. **133**, 265–277 (2015)
28. Lebedev, L.P., Cloud, M.J., Eremeyev, V.A.: Tensor Analysis with Applications in Mechanics. World Scientific, New Jersey (2010)
29. Altenbach, H., Eremeyev, V.: Eigen-vibrations of plates made of functionally graded material. Comput. Mater. Continua **9**(2), 153–178 (2009)
30. Altenbach, H., Eremeyev, V.A., Naumenko, K.: On the use of the first order shear deformation plate theory for the analysis of three-layer plates with thin soft core layer. ZAMM **95**(10), 1004–1011 (2015)

[illegible]

Ray Tracing Method for a High-Frequency Propagation of the Ultrasonic Wave Through a Triple-Periodic Array of Spheres

Nikolay V. Boyev and M.A. Sumbatyan

Abstract The Ray method is applied to study the propagation of a high-frequency plane wave through a triple-periodic system of the spherical obstacles. The initial plane wave is taken as a superposition of spherical waves, which in discretization are reduced to a system of waves, each of them being studied by the Ray method in a local formulation. On the first step, we calculate the geometric parameters of the trajectory of each ray transmitted through the system of spherical obstacles, which is a spatial broken polyline. On the second step, we calculate the wave characteristics, by using methods of the short-wave diffraction.

Keywords Ray tracing method · Periodic arrays of spheres · Acoustic filter · Ultrasonic impulse · Geometrical diffraction theory

1 Introduction

The Ultrasonic (US) wave propagation through arrays of obstacles in the acoustic and elastic media are nowadays under intensive investigation. As a simplest approximation, one may consider acoustic media where only longitudinal type of US waves can propagate. The most popular obstacles are solids and voids of canonical shapes: rectangles, circles, ellipses—in the two-dimensional (2d) problems, spheres and ellipsoids (3d case), thin scatterers like plates, slabs, etc. One of practically important problems for the acoustic media is the propagation of the impulse through periodic obstacles in the 2d and 3d cases, in both low and high frequency regimes. In the 2d problems about propagation of acoustic and elastic waves through periodic systems of obstacles some analytical and numerical methods have been developed in

N.V. Boyev (✉) · M.A. Sumbatyan
Institute of Mathematics, Mechanics and Computer Science, Southern Federal University, Milchakova Street 8a, Rostov-on-Don 344090, Russia
e-mail: boyev@math.rsu.ru

M.A. Sumbatyan
e-mail: masumbatyan@sfedu.ru

M.A. Sumbatyan (ed.), *Wave Dynamics and Composite Mechanics for Microstructured Materials and Metamaterials*, Advanced Structured Materials 59, DOI 10.1007/978-981-10-3797-9_10

[8–12, 14–18, 20, 22] and some other works, mainly in the low-frequency case. Some practical aspects of the Ultrasonic detection are studied in [4, 13, 21]. In those works the reader can find some further helpful references.

2 Formulation of the Problem

In an infinite acoustic medium there is located a triple-periodic system of equal spheres of radius a located inside a cubic domain. The edge of the cube is $(2a + b)M$. The global coordinate system $OXYZ = OX_1X_2X_3$ is chosen so that the sides of the cube are parallel to the coordinate planes. The period of the grating is $(2a + b)$, along all three Cartesian directions. The case $M = 3$ is shown in Fig. 1, where the spherical obstacles are approximated by a set of plane facets. The minimal distance from the spheres to the cube sides is $b/2$. In this coordinate system the governing equations of the spheres are given by the following equation:

$$[x_1 - m_1(2a + b)]^2 + [x_2 - m_2(2a + b)]^2 + [x_3 - m_3(2a + b)]^2 = a^2,$$
$$m_1, m_2, m_3 = 1, \ldots, M. \qquad (1)$$

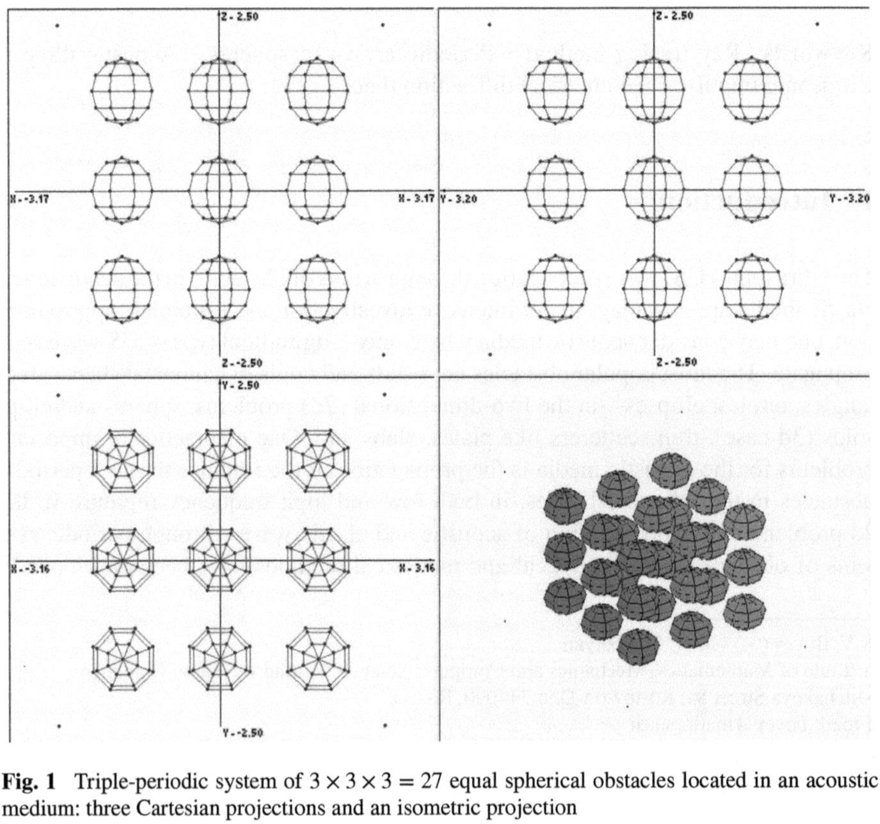

Fig. 1 Triple-periodic system of $3 \times 3 \times 3 = 27$ equal spherical obstacles located in an acoustic medium: three Cartesian projections and an isometric projection

Note that in Fig. 1 axis x for symmetry is shown with a shift.

From one of these sides of the cube there is introduced an impulse into the cube, with a tonal filling by several periods of a plane high-frequency monochromatic acoustic wave, and on the opposite face of the cube a through-transmitted wave is received. Such a structure of the impulse permits calculation of the through-transmitted characteristics in the time-harmonic regime. After the wave has been passed through the system of obstacles, on the opposite side of the cube a modified impulse is received. The problem is to calculate the characteristics of the through-transmitted impulse on the opposite side of the cube and to compare them with the characteristics of the impulse introduced into the cube.

3 The Applied Method of Solution

The receiving face of the considered cube is divided to $N_1 \times N_1$ small quadrates and the plane acoustic wave is changed by a set of point sources of the spherical longitudinal wave, emitted from the centers of these small quadrates. The spatial angle with the vertex at a source directed to the obstacles is divided to $M_1 \times M_1$ small spatial angles, each of them containing respective ray of the propagating acoustic wave. Therefore, the problem is reduced to a short-wave diffraction of the acoustic wave in a local formulation, whose solution is constructed in two steps. The first step is fully geometric—we find the trajectory of each ray arriving at the opposite side. On the second step, by using the found trajectory and the known point of mirror reflection, we determine the acoustic pressure in the through-transmitted wave on the receiving side, on the basis of the Geometrical Diffraction Theory (GDT).

The total wave on the receiving side is a sum of rays passed through the system of spheres, which may be of one of the following types: (1) the rays passed across the system of obstacles without any reflection; (2) the rays reflected from the spheres only once; (3) the rays with multiple reflections. The rays free of reflection are those which propagate from the point sources of the spherical waves in the thin free layer of thickness b between the layers of obstacles. In the case of sufficiently gentle incidence of the ray at the extreme obstacles tangent to the upper and the lower boundary planes of the acoustic layer, a single reflection of the acoustic wave may happen, after which it falls at the face of the receiver. The search of the multiply reflected rays is a more complex diffraction problem. Generally, the trajectory of each such a ray through-transmitted across the system of spheres is a spatial polyline. This takes into account only the directions which form an acute angle with the direction of propagation of the initial plane wave.

4 High-Frequency Analytical Representations

In this section we write out the explicit formulas at the receiving point for the acoustic pressure, for each of three types of waves.

1. For the rays passing inside the cube without any reflection, in the high-frequency regime of oscillations with $k \to \infty$, in the direction of the incident ray, the pressure in the incident wave at the receiving point is

$$p^{inc}(y) = \frac{e^{ik|x_0 - y|}}{|x_0 - y|}, \qquad (2)$$

where $k = \omega/c$, c is the wave number and the speed of the acoustic wave, ω is the frequency of the oscillations.

2. In [2], with the use of the GDT, and in [9] on the basis of the physical Kirchhoff diffraction theory, there have been obtained the expression for the high-frequency acoustic wave once reflected from the rigid obstacle bounded by an arbitrary smooth surface, along the ray $x_0 - y^* - x$. We write out it here in the case of the spherical obstacles. It is a leading asymptotic term of the diffraction integral which is developed in [9] by using the two-dimensional stationary phase method [5, 6]:

$$p(x) = \cos\gamma \, \frac{\exp\left\{ i \left(k(L_0 + L) + \frac{\pi}{4}(\delta_2 - 2) \right) \right\}}{L_0 L \sqrt{|det(D_2)|}},$$

where $\delta_2 = signD_2$ is the difference between the number of positive and negative eigenvalues of the Hessian matrix D_2 of a symmetric structure ($d_{ij} = d_{ji}, i, j = 1, 2$), with the following elements:

$$d_{11} = \left(\frac{1}{L_0} + \frac{1}{L_1} \right) \sin^2\alpha + \frac{2\cos\gamma}{a},$$

$$d_{12} = -\left(\frac{1}{L_0} + \frac{1}{L_1} \right) \cos\alpha \cos\beta,$$

$$d_{22} = \left(\frac{1}{L_0} + \frac{1}{L_1} \right) \sin^2\beta + \frac{2\cos\gamma}{a}.$$

By calculating the determinant in the denominator, taking into account that $d_{21} = d_{12}$, we write out the obtained formula in the explicit form:

$$p(x) = \frac{\exp\left\{ i \left[k(L_0 + L) + \frac{\pi}{4}(\delta_2 - 2) \right] \right\}}{\sqrt{\left| (L_0 + L)^2 + 2L_0 L (L_0 + L) a^{-1} (\cos^{-1}\gamma + \cos\gamma) + 4L_0^2 L^2 a^{-2} \right|}},$$

$$L_0 = |x_0 - y^*|, \qquad L = |y^* - x|. \qquad (3)$$

Here $\{-\cos\alpha, -\cos\beta, -\cos\gamma\}$ is the vector which determines the direction of incidence of the ray $x_0 - y^*$ in the local Cartesian coordinate system, defined by the normal and the tangent to the curvature lines at the point of mirror reflection. Formula (3) is obtained in the case when the high-frequency wave is incident on the convex surface. If the wave is incident on the concave surface then the second term in the denominator should be taken with the negative sign. Formula (3) gives the leading asymptotic term of the acoustic pressure, under the condition that $kL_0 \gg 1$, $kL \gg 1, ka \gg 1$.

3. The most complex problem is the calculation of the acoustic pressure in the multiple reflection from non-plane reflectors. If the shape of the boundary surface of an isolated acoustically hard obstacle, and the geometry and the shape of the array of obstacles located in the acoustic medium admit multiple (N-fold) reflection, then the diffraction of the high-frequency wave can be studied in frames of an integral representation. The basis for this integral representation for the acoustic pressure in N times reflected acoustic wave is given by a modification of the Kirchhoff integral representation [3], realized in the case of double reflection in [2]. The general 2D theory is proposed in [19].

Let the high-frequency spherical wave (2) propagates from point x_0. Let us assume that the ray propagates along the polyline $x_0 - y_1^* - y_2^* - y_3^* - \cdots - y_N^* - x_{N+1}$, where the points of mirror reflection $y_1^*, y_2^*, y_3^*, \ldots, y_N^*$ may belong either to the same reflector or to different surfaces of N reflectors. Also, we admit the cases of sequential location of some mirror reflection points on the surfaces of certain reflectors, while other reflectors contain only one reflection point. The wave is received at point x_{N+1} of the acoustic medium. Further, we consider only the local approach where in the high-frequency regime the pressure at the receiving point is defined by the reflection from small vicinities $S_1^*, S_2^*, \ldots, S_N^*$ of the boundary surfaces at the points of mirror reflection $y_1^*, y_2^*, y_3^*, \ldots, y_N^*$.

Let us describe the representation for the pressure $p(x_{N+1})$ in the reflected wave at the receiving point x_{N+1} in more detail. The pressure in N times reflected wave at point x_{N+1} is sought by integration over surface S_N^* of the vicinity of the last reflection point y_N^*, obtained with the single reflection from the vicinity S_{N-1}^* of the point of mirror reflection y_{N-1}^*. The pressure at the receiving point $p(x_{N+1})$ is given by the following formula:

$$p(x_{N+1}) = \iint_{S_N^*} 2p(y_N) \frac{\partial \Phi(y_N, x_{N+1})}{\partial n_N} \, dS_N. \tag{4}$$

Here $p\left(y_N\right)$ is the pressure in the incident wave at point $y_N \in S_N^*$ of the vicinity of the reflection point y_N^* which is defined after the reflection over the vicinity S_{N-1}^* of point y_{N-1}^*, $\mathbf{n}_N$ is the normal to the surface S_N^* at point y_N, directed to the acoustic medium.

At the same time, pressure $p\left(y_N\right)$ itself is defined as a integral representation in terms of the wave incident to the vicinity S_N^*, arriving after reflection from the vicinity S_{N-1}^*:

$$p(y_N) = \iint_{S_{N-1}^*} 2p(y_{N-1}) \frac{\partial \Phi(y_{N-1}, y_N)}{\partial n_{N-1}} dS_{N-1}. \tag{5}$$

The same approach can be extended, to form the reflected wave from arbitrary vicinity S_n^* along the considered ray. In frames of such a modification the pressure at points of the vicinity $y_n \in S_n^*$, $(n = 2, 3, \dots, N)$ is expressed by the formula

$$p(y_n) = \iint_{S_{n-1}^*} 2p(y_{n-1}) \frac{\partial \Phi(y_{n-1}, y_n)}{\partial n_{n-1}} dS_{n-1}. \qquad (6)$$

in which $p\left(y_n\right)$ is the pressure at point $y_n \in S_n^*$, $p(y_{n-1})$ is the pressure in the incident wave at point $y_{n-1} \in S_{n-1}^*$ of the vicinity y_{n-1}^*, which is defined after the reflection on the vicinity S_{n-2}^* of point y_{n-2}^*, $\mathbf{n}_{n-1}$ is the normal to the surface S_{n-1}^* at point y_{n-1}, directed to the acoustics medium.

Traveling along the re-reflected ray in the inverse direction, i.e. in direction $x_{N+1} - y_N^* - \cdots - y_2^* - y_1^* - x_0$, we come to the forming of the wave on the vicinity S_2^*.

At points y_2 of the vicinity S_2^* of the second mirror reflection y_2^* the integral representation $p(y_2)$ has the form:

$$p(y_2) = \iint_{S_1^*} 2p^{inc}(y_1) \frac{\partial \Phi(y_1, y_2)}{\partial n_1} dS_1, \qquad (7)$$

where

$$p^{inc}(y_1) = \frac{e^{ik|x_0 - y_1|}}{|x_0 - y_1|}$$

defines the incident field, corresponding to the point source x_0 (2).

By collecting together all integral representations in Eqs. (4)–(7), we obtain the following $2N$-fold integral, to determine $p(x_{N+1})$:

$$p(x_{N+1}) = 2^N \iint_{S_N^*} \iint_{S_{N-1}^*} \cdots \iint_{S_2^*} \iint_{S_1^*} \frac{\partial \Phi}{\partial n_1} \frac{\partial \Phi}{\partial n_2} \cdots \frac{\partial \Phi}{\partial n_{N-1}} \frac{\partial \Phi}{\partial n_N} \times$$
$$\times dS_1 dS_2 \dots dS_{N-1} dS_N. \qquad (8)$$

The asymptotic solution constructed below has a local character, giving the leading asymptotic term for the amplitude of the diffracted field in the small vicinity of any ray emitted from point x_0 and reflected from the surfaces of the obstacles sequentially at points $y_1^*, y_2^*, y_3^*, \dots, y_N^*$ and arriving at point x_{N+1}. Obviously, such rays exist only in the case when all reflection points $y_1^*, y_2^*, y_3^*, \dots, y_N^*$, as well as the receiving point x_{N+1}, all are located in the "light" zone.

In order to construct the leading asymptotic term, let us apply the asymptotic representation for the derivative $\partial \Phi\left(y_{n-1}, y_n\right)/\partial n_{n-1}$ of the fundamental potential

$$\Phi(y_{n-1}, y_n) = \frac{e^{ik|y_{n-1} - y_n|}}{4\pi\,|y_{n-1} - y_n|}, \qquad (9)$$

with $k \to \infty$:

$$\frac{\partial \Phi(y_{n-1}, y_n)}{\partial n_{n-1}} = ik\cos\gamma_{n-1}\frac{e^{ik|y_{n-1}-y_n|}}{4\pi\,|y_{n-1}-y_n|}\left[1+O(k^{-1})\right],$$
$$n = 1, 2, \ldots, N+1, \quad y_0 = x_0, \quad y_{N+1} = x_{N+1}, \tag{10}$$

where γ_{n-1} is the angle between the normal $\mathbf{n}_{n-1}$ and the direction of incidence of the ray $y_{n-2} - y_{n-1}$, and $|y_{n-1} - y_n|$ is the distance between the reflection points $y_{n-1} \in S^*_{n-1}$ and $y_n \in S^*_n$. It should be noted that the incident $y^*_{n-1} - y^*_n$ and the reflected $y^*_n - y^*_{n+1}$ rays belong to the same plane with the normal $\mathbf{n}_n$ at point y^*_n.

Let us denote the distances $\left|x_0 - y^*_1\right| = L_0$, $\left|y^*_n - y^*_{n+1}\right| = L_n$, $\left|y^*_N - x_{N+1}\right| = L_N$, $n = 1, \ldots, N-1$. After the slowly varying functions in the asymptotic representations for normal derivatives of Greens functions (10) are taken out of the sign of integral (8), we can write out the following integral representation for the pressure at the receiving point:

$$p(x_{N+1}) = \left(\frac{ik}{2\pi}\right)^N L_0^{-1}\prod_{n=1}^{N} L_n^{-1}\cos\gamma_n \iint_{S^*_N}\iint_{S^*_{N-1}}\cdots\iint_{S^*_1} e^{ik\varphi}\times$$
$$\times\, dS_1 \ldots dS_{N-1} dS_N, \tag{11}$$

$$\varphi = |x_0 - y_1| + |y_1 - y_2| + \cdots + |y_{N-1} - y_N| + |y_N - x_{N+1}|. \tag{12}$$

In the factors in front of the integral there are taken the values of $\cos\gamma_n$ for the ray incident at the point of mirror reflection y^*_n.

Let us relate the vicinities $S^*_1, S^*_2, \ldots, S^*_N$ to the right Cartesian coordinate system, which are defined by the normals $\mathbf{n}_1, \mathbf{n}_2, \ldots, \mathbf{n}_N$ to the surfaces at points $y^*_1, y^*_2, \ldots, y^*_N$ directed to the acoustic medium and by the tangents to the curvature curves. Let us designate as $O_n X_1^{(n)} X_2^{(n)} X_3^{(n)}$, $n = 1, 2, 3, \ldots, N$ the Cartesian coordinate system with the center at point y^*_n.

To introduce the coordinates of the current point $y_n \in S^*_n$, $n = 1, 2, \ldots, N$, on each vicinity $y^*_n \in S^*_N$, we count along the curvature lines the arc lengths $\Delta s_1^{(n)}, \Delta s_2^{(n)}$. Then the current point y_n of the convex surface S_n in the local coordinate system $O_n X_1^{(n)} X_2^{(n)} X_3^{(n)}$ has the following coordinates:

$$y_n = \left(\Delta s_1^{(n)},\ \Delta s_2^{(n)},\ -0.5\left(k_1^{(n)}\left(\Delta s_1^{(n)}\right)^2 + k_2^{(n)}\left(\Delta s_2^{(n)}\right)^2\right)\right),$$

where $k_1^{(n)}, k_2^{(n)}$ are the principal curvatures, and $\left(k_1^{(n)}\left(\Delta s_1^{(n)}\right)^2 + k_2^{(n)}\left(\Delta s_2^{(n)}\right)^2\right)$ is the second quadratic form of the surface S^*_n at point y^*_n.

The forming of the asymptotic representation of the phase φ, up to infinitesimals of the second order in the general case (12), we show on the example of the triple re-reflection of the ray $x_0 - y^*_1 - y^*_2 - y^*_3 - x_4$:

$$\varphi = |x_0 - y_1| + |y_1 - y_2| + |y_2 - x_3| + |y_3 - x_4| = L_0 + L_1 + L_2 + L_3 +$$
$$+0.5d_{11}\left(\Delta s_1^{(1)}\right)^2 + d_{12}\Delta s_1^{(1)}\Delta s_2^{(1)} + d_{13}\Delta s_1^{(1)}\Delta s_1^{(2)} + d_{14}\Delta s_1^{(1)}\Delta s_2^{(2)} +$$
$$+0.5d_{22}\left(\Delta s_2^{(1)}\right)^2 + d_{23}\Delta s_2^{(1)}\Delta s_1^{(2)} + d_{24}\Delta s_2^{(1)}\Delta s_2^{(2)} +$$
$$+0.5d_{33}\left(\Delta s_1^{(2)}\right)^2 + d_{34}\Delta s_1^{(2)}\Delta s_2^{(2)} + d_{35}\Delta s_1^{(2)}\Delta s_1^{(3)} + d_{36}\Delta s_1^{(2)}\Delta s_2^{(3)} +$$
$$+0.5d_{44}\left(\Delta s_2^{(2)}\right)^2 + d_{45}\Delta s_2^{(2)}\Delta s_1^{(3)} + d_{46}\Delta s_2^{(2)}\Delta s_2^{(3)} +$$
$$0.5d_{55}\left(\Delta s_1^{(3)}\right)^2 + d_{56}\Delta s_1^{(3)}\Delta s_2^{(3)} +$$
$$0.5d_{66}\left(\Delta s_2^{(3)}\right)^2. \quad (13)$$

In the general case of the N-fold re-reflection in the phase φ (12) there are absent the terms with the first power of $\Delta s_j^{(n)}$ ($j = 1, 2;\ n = 1, 2, \ldots, N$). This approves that the points $y_n^* \in S_n^*$ of the direct ray reflection correspond to the stationary value of the phase φ.

The leading asymptotic term for the pressure $p(x_{n_1})$ of N times re-reflected wave at the receiving point x_{N+1} can be obtained from expression (11) by applying the multidimensional (2N-fold) stationary phase method [5, 6], as follows:

$$p\left(x_{N+1}\right) = \frac{1}{L_0}\prod_{n=1}^{N}\frac{\cos\gamma_n}{L_n}\frac{\exp\left\{i\left[k\sum_{n=0}^{N}L_n + \frac{\pi}{4}\left(\delta_{2N} - 2N\right)\right]\right\}}{\sqrt{\left|det\left(D_{2N}\right)\right|}}, \quad (14)$$

where $D_{2n} = (d_{ij}), i, j = 1, 2, \ldots, 2N$ is the symmetric Hessian matrix of the bended structure of the width equal to 7. Parameter $\delta_{2N} = signD_{2N}$ is the difference between the number of positive and negative eigenvalues of the matrix D_{2N}.

Let us demonstrate the matrix to indicate non-trivial elements d_{ij}:

$$D_{2N} = \begin{pmatrix}
d_{11} & d_{12} & d_{13} & d_{14} & 0 & 0 & \ldots & 0 & 0 & 0 \\
d_{21} & d_{22} & d_{23} & d_{24} & 0 & 0 & \ldots & 0 & 0 & 0 \\
d_{31} & d_{32} & d_{33} & d_{34} & d_{35} & d_{36} & \ldots & 0 & 0 & 0 \\
d_{41} & d_{42} & d_{43} & d_{44} & d_{45} & d_{46} & \ldots & 0 & 0 & 0 \\
\cdot & \cdot & \cdot & \cdot & \cdot & \cdot & \cdot & \cdot & \cdot & \cdot \\
\cdot & \cdot & \cdot & \cdot & \cdot & \cdot & \cdot & \cdot & \cdot & \cdot \\
0 & 0 & 0 & 0 & \ldots & d_{2N-3,2N-4} & d_{2N-3,2N-3} & d_{2N-3,2N-2} & d_{2N-3,2N-1} & d_{2N-3,2N} \\
0 & 0 & 0 & 0 & \ldots & d_{2N-2,2N-4} & d_{2N-2,2N-3} & d_{2N-2,2N-2} & d_{2N-2,2N-1} & d_{2N-2,2N} \\
0 & 0 & 0 & 0 & \ldots & 0 & d_{2N-1,2N-3} & d_{2N-1,2N-2} & d_{2N-1,2N-1} & d_{2N-1,2N} \\
0 & 0 & 0 & 0 & \ldots & 0 & d_{2N,2N-3} & d_{2N,2N-2} & d_{2N,2N-1} & d_{2N,2N}
\end{pmatrix}$$

The Hessian matrix $D_{2N} = (d_{ij}), i, j = 1, 2, 3, \ldots, 2N$ is symmetric $d_{ij} = d_{ji}$ with the following non-trivial elements $d_{ij}, i \leq j$:

the diagonal elements ($n = 1, \ldots, N$):

$$d_{2n-1,2n-1} = \left(L_{n-1}^{-1} + L_n^{-1}\right)\left(1 - \left(\mathbf{q}_0^{(n)}, \mathbf{e}_\theta^{(n)}\right)^2\right) - 2a^{-1}\left(\mathbf{q}_0^{(n)}, \mathbf{e}_r^{(n)}\right),$$

$$d_{2n,2n} = \left(L_{n-1}^{-1} + L_n^{-1}\right)\left(1 - \left(\mathbf{q}_0^{(n)}, \mathbf{e}_\varphi^{(n)}\right)^2\right) - 2a^{-1}\left(\mathbf{q}_0^{(n)}, \mathbf{e}_r^{(n)}\right), \tag{15}$$

the elements out of the principal diagonal:

$$\begin{aligned}
d_{2n-1,2n} &= \left(L_{n-1}^{-1} + L_n^{-1}\right)\left(\mathbf{q}_0^{(n)}, \mathbf{e}_\theta^{(n)}\right)\left(\mathbf{q}_0^{(n)}, \mathbf{e}_\varphi^{(n)}\right),\\
d_{2n-1,2n+1} &= L_n^{-1}\left(\left(\mathbf{q}_0^{(n)}, \mathbf{e}_\theta^{(n)}\right)\left(\mathbf{q}_0^{(n+1)}, \mathbf{e}_\theta^{(n+1)}\right) - \left(\mathbf{e}_\theta^{(n)}, \mathbf{e}_\theta^{(n+1)}\right)\right),\\
d_{2n-1,2n+2} &= L_n^{-1}\left(\left(\mathbf{q}_0^{(n)}, \mathbf{e}_\theta^{(n)}\right)\left(\mathbf{q}_0^{(n+1)}, \mathbf{e}_\varphi^{(n+1)}\right) - \left(\mathbf{e}_\theta^{(n)}, \mathbf{e}_\varphi^{(n+1)}\right)\right),\\
d_{2n,2n+1} &= L_n^{-1}\left(\left(\mathbf{q}_0^{(n)}, \mathbf{e}_\varphi^{(n)}\right)\left(\mathbf{q}_0^{(n+1)}, \mathbf{e}_\theta^{(n+1)}\right) - \left(\mathbf{e}_\varphi^{(n)}, \mathbf{e}_\theta^{(n+1)}\right)\right),\\
d_{2n,2n+2} &= L_n^{-1}\left(\left(\mathbf{q}_0^{(n)}, \mathbf{e}_\varphi^{(n)}\right)\left(\mathbf{q}_0^{(n+1)}, \mathbf{e}_\varphi^{(n+1)}\right) - \left(\mathbf{e}_\varphi^{(n)}, \mathbf{e}_\varphi^{(n+1)}\right)\right),
\end{aligned} \tag{16}$$

where $n = 1, \ldots, N$ for the first equation in (16), and $n = 1, \ldots, N-1$ for the others.

The elements of the Hessian are written for the considered case of the triple periodic system of spherical obstacles related to a global Cartesian coordinate system, which is defined by the edges of the cube, outgoing from the same vertex. All vector quantities in expressions (15) and (16) are given in this coordinate system. In numerical implementation of the algorithm, when forming the trajectories of the multiple ray re-reflected at the mirror reflection points, it is necessary to consider also the local spherical coordinates. In Eqs. (15) and (16) $\left(\mathbf{e}_r^{(n)}, \mathbf{e}_\theta^{(n)}, \mathbf{e}_\varphi^{(n)}\right)$ is the orthonormal basis of the local spherical coordinate system at the point of mirror reflection $\left(x_n^*, y_n^*, z_n^*\right)$, $\mathbf{q}_0^{(n)} = \{ q_{01}^{(n)}, q_{02}^{(n)}, q_{03}^{(n)} \}$ are the components of the unit vector $\mathbf{q}^{(n)}$, which defines the direction of the incident wave in the global Cartesian coordinates at point $\left(x_n^*, y_n^*, z_n^*\right)$ of the sphere. Let us calculate the local spherical coordinates (a, θ_n, φ_n) of the point of mirror reflection $\left(x_n^*, y_n^*, z_n^*\right)$, located on the boundary surface of respective sphere with certain parameters m_1, m_2, m_3:

$$\begin{aligned}
\theta_n &= \theta_n^{(m_1,m_2,m_3)} = \arccos\left(\left(z_n^* - 2m_3 a\right)/a\right), \qquad 0 \le \theta_n \le \pi,\\
\varphi_n &= \varphi_n^{(m_1,m_2,m_3)} = \begin{cases} \arccos\left((x_n^* - 2m_1 a)/a\right), & y_n^* - 2m_2 a \ge 0,\\ \pi + \arccos\left((x_n^* - 2m_1 a)/a\right), & y_n^* - 2m_2 a < 0, \end{cases} \quad 0 \le \varphi_n \le \pi.
\end{aligned} \tag{17}$$

Let us write out the local basis vectors $\mathbf{e}_r^{(n)}, \mathbf{e}_\theta^{(n)}, \mathbf{e}_\varphi^{(n)}$ in terms of the global basis vectors $\mathbf{i}, \mathbf{j}, \mathbf{k}$ at point $\left(x_n^*, y_n^*, z_n^*\right)$:

$$\begin{aligned}
\mathbf{e}_{\mathbf{r}}^{(\mathbf{n})} &= \mathbf{i}\sin\theta_n\cos\varphi_n + \mathbf{j}\sin\theta_n\sin\varphi_n + \mathbf{k}\cos\theta_n, \\
\mathbf{e}_{\theta}^{(\mathbf{n})} &= \mathbf{i}\cos\theta_n\cos\varphi_n + \mathbf{j}\cos\theta_n\sin\varphi_n - \mathbf{k}\sin\theta_n, \\
\mathbf{e}_{\varphi}^{(\mathbf{n})} &= -\mathbf{i}\sin\varphi_n + \mathbf{j}\cos\varphi_n.
\end{aligned} \tag{18}$$

The local coordinates of the direction of the incident wave at point $\left(x_n^*, y_n^*, z_n^*\right)$ are defined by the following expressions:

$$q_r^{(n)} = \left(\mathbf{q}_0^{(n)}, \mathbf{e}_r^{(n)}\right), \quad q_\theta^{(n)} = \left(\mathbf{q}_0^{(n)}, \mathbf{e}_\theta^{(n)}\right), \quad q_\varphi^{(n)} = \left(\mathbf{q}_0^{(n)}, \mathbf{e}_\varphi^{(n)}\right). \tag{19}$$

For the reflected ray,

$$q_r^{(n+1)} = -q_r^{(n)}, \quad q_\theta^{(n+1)} = q_\theta^{(n)}, \quad q_\varphi^{(n+1)} = q_\varphi^{(n)}.$$

The coordinates of the reflected ray $\mathbf{q}^{(n+1)} = \left\{q_1^{(n+1)}, q_2^{(n+1)}, q_3^{(n+1)}\right\}$ in the global coordinate system,

$$\begin{aligned}
q_1^{(n+1)} &= -q_r^{(n)}\sin\theta_n\cos\varphi_n + q_\theta^{(n)}\cos\theta_n\cos\varphi_n - q_\varphi^{(n)}\sin\varphi_n, \\
q_2^{(n+1)} &= -q_r^{(n)}\sin\theta_n\sin\varphi_n + q_\theta^{(n)}\cos\theta_n\cos\varphi_n + q_\varphi^{(n)}\cos\varphi_n, \\
q_3^{(n+1)} &= -q_r^{(n)}\cos\theta_n - q_\theta^{(n)}\sin\theta_n.
\end{aligned} \tag{20}$$

5 Application of the Ray Tracing Method

For more detailed study in the present section we develop an alternative technique, additionally to the GDT theory. This is the Ray Tracing (RT) method which is founded on another fundamental concept connected with the sound beams, see for example [1, 7, 9]. This method is also valid for high-frequency regimes being applicable, like the GDT, under the condition that the characteristic size of the obstacle is significantly smaller that the wave length.

In the computer realization of the developed RT model, the algorithm uses a huge number of sound rays, up to 500,000, which are irradiated from the source of the sound uniformly along all directions. These rays travel over a closed space, and if the reflecting surfaces possess some sound absorption then the traveling ray loose their energy after each reflection, according to the absorption coefficient of a current reflecting surface. In the problem under consideration we accept the "no absorption" hypothesis. Therefore, the energy of the ray keeps its initial value when flying out from the source; we accept this is of a unit value in a dimensionless form.

In the proposed algorithm the key role is played by a reflection of the rays. Since the traditional treatment is founded upon the approximation of arbitrary reflecting surface by a set of plane facets, we describe this for plane reflectors only.

If the reflection is specular then the parametric representation for the incident ray in the vector form is

$$\mathbf{r} = \mathbf{r_0} + \tau \mathbf{q}, \tag{21}$$

where $\mathbf{r_0} = (x_0, y_0, z_0)$ is the point of the previous reflection, $\mathbf{q}$ is the unit directing vector defining the direction of the incidence. First of all, we find the point of intersection of this ray with the reflecting surface, point (x_0, y_0, z_0), and the distance to this reflecting plane is

$$\tau = \frac{n_x(x_1 - x_0) + n_y(y_1 - y_0) + n_z(z_1 - z_0)}{n_x\, q_x + n_y\, q_y + n_z\, q_z}. \tag{22}$$

Here $\mathbf{n} = (n_x, n_y, n_z)$ is the vector of the unit normal to the current reflecting surface. After that by using formula

$$\mathbf{q_1} = \mathbf{q} - 2(\mathbf{n} \cdot \mathbf{q})\, \mathbf{n} \tag{23}$$

we determine the new directing vector of the ray reflected at point (x_1, y_1, z_1).

Under some conditions, the reflection may happen to be not specular, if the reflecting surface is rough enough. However, we assume that all reflections are strictly specular.

The realization of this method as a computer algorithm consists of the following. The algorithm traces each ray irradiated from the source, when the ray travels over the closed space under consideration, registering all recorded impulses in the passages of the ray across a small ε-vicinity of the receiver, whose radius is taken as 1 mm.

It is very interesting to clarify how the amplitude of the impulse arriving at the receiver can decrease in time, under the condition of zero absorption the energy carried by every ray is equal to the unit value and does not vary in time. This is equivalent to the amplitude decay with distance since the distance increases when a ray travels in the space. In frames of the GDT the decreasing amplitude with distance is clearly seen from formulas like (3) and further ones, prescribing that the amplitude is an inversely proportional function of distance with distance increasing. On the contrary, in frames of the RT method the decreasing amplitude with times at the receiving point is formed because of decreasing number of rays (in our case, all with the same amplitude) per a certain unit of time, say per 1 μs.

A typical diagram of the impulse received at a point located behind the last spherical layer, along the direction of the wave propagation, is shown in Fig. 2. This is constructed in the case when the system of 27 spheres under consideration is located in the open unbounded acoustic space. In the example demonstrated in Fig. 2 the source is located at a distance b from the first spherical layer, if counting along axis x, and the receiver is located at the same distance b behind the last (third) layer. Parameter b is in two times smaller than spheres' radius $a : b = a/2$. For all examples below each of 27 spherical surfaces are approximated by 48 flat reflecting facets. Therefore,

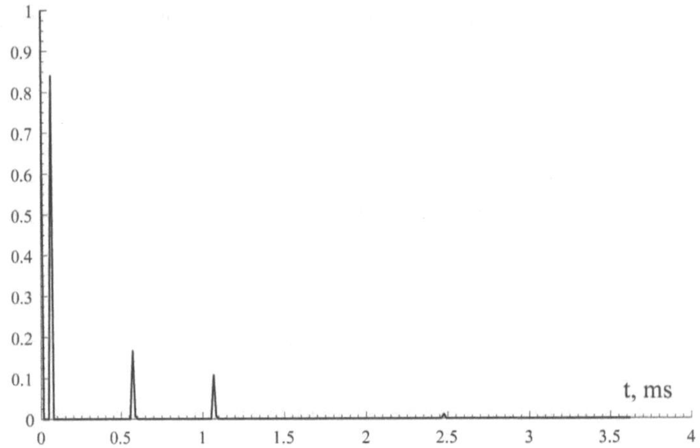

Fig. 2 US impulse through-transmitted along the triple-periodic array of 27 spherical obstacles in the unbounded space and recorded at the receiving point: $b = a$

the total number of the reflecting faces for the RT method is $48 \times 27 = 1296$. It is seen from the diagram that only few rays reach, after some multiple re-reflections from neighbor spheres, the small ε-vicinity of the receiver, for the chosen pair of the source and the receiver.

The next example is related to the same geometry, with the only difference that the array of 27 spherical obstacles, together with the source and the receiver, are placed into the cubic domain. The four lateral faces of the surrounding cube are absolutely reflecting (the absorption is zero), and the back and the front walls are absolutely absorptive. Since the lateral cubic's surfaces are at the distance $b = a/2$ from the nearest sphere, it can be proved that, due to the natural symmetry, the problem becomes equivalent to the same through-propagation of the US impulse in the unbounded space like in the previous example, but for a periodic system of spherical obstacles infinitely repeated in the lateral directions, i.e. along axes y and z. One thus comes to the three layers of the doubly-periodic infinite system of spheres. The diagram in Fig. 3, which shows the US impulse versus time, for this geometry unlike the one in Fig. 2 possesses the absolutely different qualitative behavior. In fact, this demonstrates a number of resonances in the initial period of time. These resonances are connected with a specific interference which predetermines that the periodic structures under consideration possess the properties of acoustic filters. Typically, such filters can provide a cutoff of the through-transmitted energy on some frequency intervals. This is related to the so-called "metamaterials" whose properties are defined by their internal structure rather than by the medium which the solid is made of.

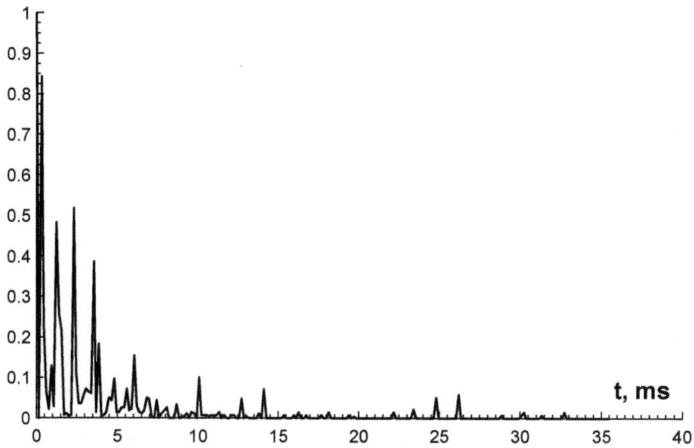

Fig. 3 US impulse through-transmitted along the triple-periodic array of 27 spherical obstacles inside the cubic space with four absolutely reflecting lateral sides and two absolutely absorbing back and front sides, recorded at the receiving point: $b = a$

The property described in the previous paragraph can be translated to the form of the Amplitude-Frequency characteristics. If the full flight of the n-th ray at the chosen point receiver is $L_n, n = 1, \dots, N$ registered at the moment t_n, then in the time-harmonic regime the total amplitude is

$$A(k) = \left| \sum_{n=1}^{N} e^{ikL_n} \right|, \qquad L_n = c\, t_n, \tag{24}$$

where c is the wave speed.

This function is calculated on a dense set of the point receivers distributed over the front side of the cube, as described above. If we model the base of the receiving US transducer as coinciding with the mentioned side of the cube that the amplitude of the full registered impulse is a sum of the complex-valued quantities under the modulus sign in Eq. (24). After that the final true value is the modulus of the obtained complex-valued quantity.

Such a treatment, applied to the two Amplitude-Time relations shown in Figs. 2 and 3, gives the results represented in Fig. 4, by lines 1 and 2, respectively. Line 1 approves that the through-transmission amplitude is a quasi-monotonic decreasing function of frequency. This means that only $3 \times 3 \times 3$ layers cannot provide real filtering. On the contrary, if the three layers of the spherical obstacles are infinitely extended along two coordinate axes, such a structure can guarantee some maxima and minima of the through-transmitted amplitude over certain frequency intervals. Apparently, this property becomes more expressive if the number of doubly-periodic arrays of spherical obstacles, packed along the direction of wave propagation, becomes greater than 3.

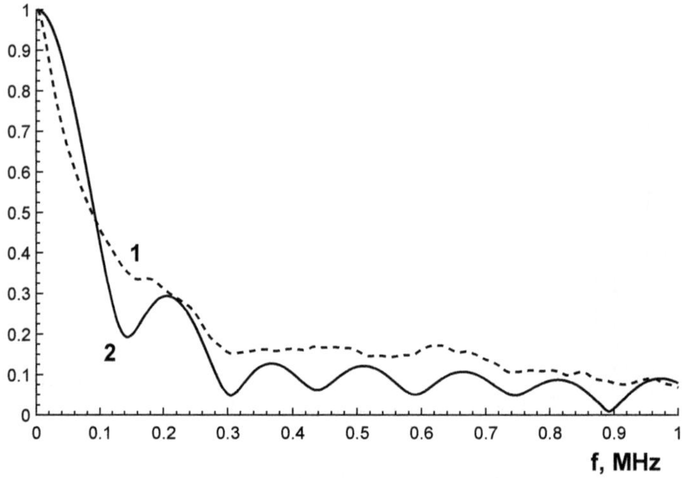

Fig. 4 Amplitude of the through-transmitted US signal versus frequency, $b = a$. *Line 1* $3 \times 3 \times 3 = 27$ spherical obstacles in the open unbounded space; *Line 2* the same geometry inside the cubic shell with absolutely reflecting lateral faces and absolutely absorbing back and front faces

6 Conclusions

In frames of the GDT we write out the explicit expressions for the acoustic pressure, for each type of the acoustic wave through-transmitted across the triple periodic system of spheres: the waves directly passed through the medium, and the waves passed with single and multiple reflections from the obstacles. There is developed the explicit expression (14) for the leading asymptotic term of the diffracted field in the case of multiple reflection of the acoustic wave. The formula (14) for the pressure in the re-reflected wave establishes that this depends upon parameters of the problem, which are the principal curvatures of the surfaces at the points of mirror reflection, the distances between the points of mirror reflection, the distance of the wave source from the first reflection point, and the distance of the receiving point from the last reflection point, as well as depends upon the directions of the incident waves.

We also develop a Ray Tracing algorithm, by approximating each spherical reflector by a number of plane facets, hence the trajectory of each irradiated ray in its travel across the specimen with multiple reflections can be calculated in a simpler way. The resulting signal recorded at the receiving point is a sum of all arriving rays taking with respective phase shift determined by the full time of flight of the current ray. This allows us to construct both the Amplitude-Time and the Amplitude-Frequency diagrams. With so doing, it is approved that the chosen triple-periodic structure under investigation can provide the properties of acoustic filtering, although with not very expressive quality.

Acknowledgements The authors express their gratitude to the Russian Science Foundation (RSCF), for its support by Project 15-19-10008.

References

1. Babich, V.M., Buldyrev, V.S.: Short-Wavelength Diffraction Theory. Springer, Heidelberg (1972)
2. Boyev, N.V., Sumbatyan, M.A.: A short-wave diffraction by bodies, bounded by arbitrary smooth surface. Russian Doklady **392**(5) (2003)
3. Borovikov, V.A., Kinber, B.Y.: Geometrical Theory of Diffraction. The Institution of Electrical Engineers, London (1994)
4. Brigante, M.: On multiple scattering in acoustic media: a deterministic ray tracing method for random structures. Ultrasonics **53**, 652–657 (2013)
5. Fedorjuk, M.V.: Stationary phase method for multiple integrals. J. Comput. Math. Math. Phys. **2**, 152–157 (1962)
6. Fedoryuk, M.V.: The stationary phase method and pseudodifferential operators. Russ. Math. Surv. **26**, 65–115 (1971)
7. Kuttruff, H.: Room Acoustics, 5th edn. Taylor and Francis, New York (2009)
8. Liu, Z., Zhang, X., Mao, Y., Zhu, Y.Y., Yang, Z., Chan, C.T., Sheng, P.: Locally resonant sonic materials. Science **289**(5485), 1734–1736 (2000)
9. McNamara, D.A., Pistorius, C.W.I., Malherbe, I.A.G.: Introduction to the Uniform Geometrical Theory of Diffraction. Artech House, Norwood (1990)
10. Miles, J.W.: On Rayleigh scattering by a grating. Wave Motion **4**, 285–292 (1982)
11. Pompei, A., Sumbatyan, M.A., Boyev, N.V.: Reflection of high-frequency elastic waves from a non-plane boundary surface of the elastic medium. J. Sound Vibr. **302**, 925–935 (2007)
12. Porter, R., Evans, D.V.: Wave scattering by periodic arrays of breakwaters. Wave Motion **23**, 95–120 (1996)
13. Scalia, A., Sumbatyan, M.A.: On efficient quantitative analysis in real-time ultrasonic detection of cracks. Ultrasonics **37**, 239–245 (1999)
14. Scarpetta, E., Sumbatyan, M.A.: On wave propagation in elastic solids with a doubly periodic array of cracks. Wave Motion **25**, 61–72 (1997)
15. Scarpetta, E., Sumbatyan, M.A.: On the oblique wave penetration in elastic solids with a doubly periodic array of cracks. Quart. Appl. Math. **58**, 239–250 (2000)
16. Scarpetta, E., Sumbatyan, M.A.: Wave propagation through elastic solids with a periodic array of arbitrarily shaped defects. Math. Comput. Model. **37**, 19–28 (2003)
17. Scarpetta, E., Tibullo, V.: Explicit results for scattering parameters in three-dimensional wave propagation through a doubly periodic system of arbitrary openings. Acta Mech. **185**, 1–9 (2006)
18. Scarpetta, E., Tibullo, V.: On the three-dimensional wave propagation through cascading screens having a periodic system of arbitrary openings. Int. J. Eng. Sci. **46**, 105–118 (2008)
19. Sumbatyan, M.A., Boyev, N.V.: High-frequency diffraction by nonconvex obstacles. J. Acoust. Soc. Am. **95**, 2347–2353 (1994)
20. Sumbatyan, M.A.: Low-frequency penetration of acoustic waves through a periodic arbitrary shaped grating: the three-dimensional problem. Wave Motion **22**, 133–144 (1995)
21. Sumbatyan, M.A., Brigante, M.: Analysis of strength and wave velocity for micro-damaged elastic media. Eng. Fract. Mech. **145**, 43–43 (2015)
22. Sumbatyan, M.A., Remizov, M.Y.: Low frequency penetration of elastic waves through a triple periodic array of cracks. Springer Proc. Phys. **175**, 459–474 (2016)

Acknowledgements. The authors express their gratitude to the Russian Science Foundation [illegible] Project 15-[illegible].

References

[illegible]

An Experimental Model of the Ultrasonic Wave Propagation Through a Doubly-Periodic Array of Defects

Vladimir V. Zotov, Vitaly V. Popuzin and Alexander E. Tarasov

Abstract In the present chapter we consider both computer and natural experimental approaches for the wave propagation through an elastic material with the doubly-periodic system of holes. The numerical study is performed by applying the Boundary Integral Equation method with further discretization to the algebraic system by the Boundary Element Method. A wide range of numerical experiments is conducted for different setups of the doubly periodic system, varying distances, sizes of the holes and their locations. The influence of hole cross-sections on the wave-transmission coefficient is examined by considering different star-like shapes. Natural experiments are based on the ultrasonic testing performed for the steel and plastic materials with the system of small holes. The experimental data are analyzed from the point of their spectral characteristics as well as the amplitude-time dependence.

Keywords Defects · Periodic array · Numerical modeling · Natural experiment · Acoustic filtering · Acoustic metamaterials

1 Introduction

In the past two decades a great number of studies have been devoted to study the properties in the wave propagation through the materials with internal periodic structures [2, 3, 5, 13, 14]. The wave behavior in such a structured medium reveals new different phenomena which are absent in regular materials. This leads to a great

V.V. Zotov · V.V. Popuzin (✉) · A.E. Tarasov
Institute of Mathematics, Mechanics and Computer Science, Southern Federal University, Milchakova Street 8a, Rostov-on-Don 344090, Russia
e-mail: popuzin@gmail.com

V.V. Zotov
e-mail: zot48@mail.ru

A.E. Tarasov
e-mail: tarasovmech@gmail.com

M.A. Sumbatyan (ed.), *Wave Dynamics and Composite Mechanics for Microstructured Materials and Metamaterials*,
Advanced Structured Materials 59, DOI 10.1007/978-981-10-3797-9_11

number of new interesting applications of such structured materials. Most of the research in this field has focused on the electromagnetic waves and resulted in finding such interesting properties as band-gapping, filtering, cloaking and others [10]. Many of these new features also appear in the case of acoustic and elastic waves. For example, sonic crystals can be considered as an analogue of photonic crystals in electromagnetics because they have similar filtering properties [9].

However, these properties of acoustic meta-materials are not uniform for all frequencies [11]. Thus, the main aim of this chapter is to extensively study various geometrical arrangements of the doubly-periodic array of holes with different cross-sections to attain the most universal wave transmission properties. Since one of the main parameters of interest is the boundary contour of the holes, it is natural to use the Boundary Integral Equation (BIE) approach, to model the problem under consideration. One of the advantages of this method is that when the problem is discretized to the linear algebraic system (LAS) via the Boundary Element Method (BEM), the matrix of the LAS has a smaller size when compared to the Finite Element Method. Nevertheless, this matrix is fully-dense and the solution of such LAS by a direct numerical approach requires cubic number of arithmetic operations. In our numerical implementation we use bi-conjugate gradient method to solve the problem with less computer resources involved.

The natural experiment is described in the next sections. The ultrasonic flaw-detector in conjunction with low-frequency ultrasonic transducers are used to perform a full-scale study.

2 Problem Statement

Let us consider the propagation of the ultrasonic wave through a bounded doubly-periodic system of obstacles demonstrated in Fig. 1. The incident wave is generated by the harmonic oscillations of an ultrasound transducer of length D placed above the system of obstacles, while the receiver of the same size is placed on the opposite side of the array.

For simplicity, we study wave propagation in a steady-state regime, when the dependence on the time parameter for all quantities implies the multiplication by exponential term e^{-iwt}, where w is the angular frequency of the wave, i is the imaginary unit and t is time. Moreover, we restrict the discussion to the acoustic case considering only propagation of longitudinal waves. Then, the wave behavior in the medium is described by the Helmholtz equation for the full pressure:

$$\Delta \mathbf{p} + k^2 \mathbf{p} = 0, \tag{1}$$

where k is the wavenumber and full pressure $\mathbf{p}$ is a sum of the incident and reflected wave fields:

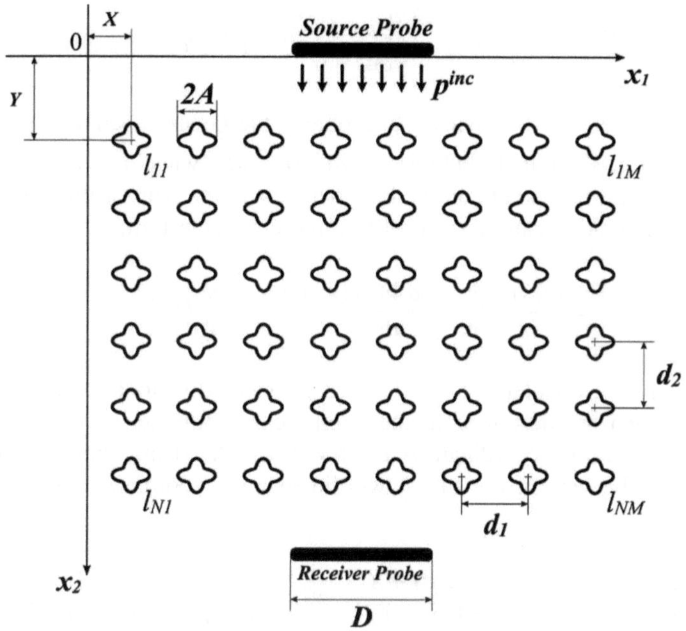

Fig. 1 Arrangement of the doubly-periodic array of cylindrical obstacles with the star-like shape of cross-section areas

$$\mathbf{p} = \mathbf{p}^{inc} + \mathbf{p}^{sc}. \tag{2}$$

In the case of the normal incidence of the plane wave $\mathbf{p}^{inc} = \mathrm{e}^{\mathrm{i}kx_2}$ the problem can be considered in frames of a two-dimensional approach. Thus, each nm-th obstacle is described by its center and the geometry of its contour l_{nm}. As shown in Fig. 1, we use matrix notation to the obstacle subscripts where the first index $n = 1, \ldots, N$ corresponds to the number of the horizontal line of the array and the second one—to the column number $m = 1, \ldots, M$ If we denote by L the sum of all boundary contours l located in the medium $L = \sum_{n=1}^{N} \sum_{m=1}^{M} l_{nm}$, then the presence of the holes can be described by the impermeability condition (Neumann-type boundary conditions for the case of the acoustically hard boundary)

$$v_n|_L = 0 \sim \left.\frac{\partial \mathbf{p}}{\partial \mathbf{n}}\right|_L = 0 \sim \left.\frac{\partial \mathbf{p}^{inc}}{\partial \mathbf{n}}\right|_L = -\left.\frac{\partial \mathbf{p}^{sc}}{\partial \mathbf{n}}\right|_L, \tag{3}$$

Here, the second and third conditions simply follow from the evident relationship between the normal component of the velocity vector and the normal derivative of the pressure $v_n = 1/i\rho w * \partial \mathbf{p}/\partial \mathbf{n}$, where ρ is the mass density of the medium and $\mathbf{n}$ represents the outward unit normal vector to the boundary contour.

The application of the BIE method [15] to the system (1)–(3) reduces it to the linear second kind Fredholm integral equation over the total boundary L

$$\frac{\mathbf{p}(y_0)}{2} - \int_L \mathbf{p}(y)\frac{\partial\Phi(|y_0 - y|)}{\partial\mathbf{n}_y}dL = \mathbf{p}^{inc}(y_0), \quad y_0 \in L \tag{4}$$

In the considered two-dimensional problem, the Green function is given by the Hankel function of the first kind $\Phi(r) = i/4 * H_0^{(1)}(kr)$, where $r = |\mathbf{r}| = |y_0 - y|$ is the distance between two points on the boundary. With the help of this representation the normal derivative can be reduced to a more suitable form

$$\frac{\partial\Phi(r)}{\partial\mathbf{n}_y} = \frac{\partial\Phi(r)}{\partial r}\frac{\partial r}{\partial\mathbf{n}_y} = \frac{\partial\Phi(r)}{\partial r}\frac{(\mathbf{r},\mathbf{n}_y)}{r} = -\frac{ik}{4}H_1^{(1)}(kr)\frac{(\mathbf{r},\mathbf{n}_y)}{r}, \tag{5}$$

In the present chapter we restrict our consideration to the case of star-like obstacles which are described in the cylindrical coordinate system by the polar equation of its radial coordinate $\rho(\phi)$, $\phi = 0, \ldots, 2\pi$. Thus, "inner" and "outer" variables can be written out in its two-dimensional coordinate representation as follows:

$$y^{nm} = \begin{cases} \rho(\theta)\cos(\theta) + (m-1)d_1 + X \\ \rho(\theta)\sin(\theta) + (n-1)d_2 + Y \end{cases}, \quad y_0^{uw} = \begin{cases} \rho(\phi)\cos(\phi) + (w-1)d_1 + X \\ \rho(\phi)\sin(\phi) + (u-1)d_2 + Y \end{cases}$$

where d_1 and d_2 are the horizontal and vertical distances between the centers of two neighbor obstacles, X is the x_1-coordinate and Y is the x_2-coordinate of the first column of the doubly-periodic array (see Fig. 1). Thus, the coordinate of the outward normal vector is expressed by

$$\mathbf{n}_y = \frac{1}{|\mathbf{n}_y^{nm}|}\frac{\partial}{\partial\theta}\{y_2^{nm}, -y_1^{nm}\} = \frac{\{\rho(\theta)\cos(\theta) + \rho'(\theta)\sin(\theta), \rho(\theta)\sin(\theta) - \rho'(\theta)\cos(\theta)\}}{\sqrt{\rho^2(\theta) + \rho'^2(\theta)}}$$

where $\left|\mathbf{n}_y^{nm}\right| = \sqrt{\rho^2(\theta) + \rho'^2(\theta)}$.

The radius-vector between two points has the form

$$\mathbf{r} = y_0^{uw} - y^{nm} = \begin{cases} \rho(\phi)\cos(\phi) - \rho(\theta)\cos(\theta) + (w-m)d_1 \\ \rho(\phi)\sin(\phi) - \rho(\theta)\sin(\theta) + (u-n)d_2 \end{cases}$$

By using the trigonometric identity for an arbitrary phase shift, we arrive to the following representation for the scalar product

$$(\mathbf{r},\mathbf{n}_y) = -\frac{\rho^2(\theta)}{\left|\mathbf{n}_y^{nm}\right|} + \frac{R}{\left|\mathbf{n}_y^{nm}\right|}(\rho(\theta)\sin(\theta + \psi_R) - \rho'(\theta)\cos(\theta + \psi_R))$$
$$+ \rho(\phi)\sin(\theta - \phi + \psi_n)$$

where the phase shift is calculated as follows:

$$\psi_n: \begin{cases} \sin(\psi_n) = \frac{\rho(\theta)}{\left|n_y^{nm}\right|}, \\ \cos(\psi_n) = \frac{\rho'(\theta)}{\left|n_y^{nm}\right|} \end{cases}; \quad \phi_R: \begin{cases} \sin(\psi_R) = \frac{(w-m)d_1}{R}, \\ \cos(\psi_R) = \frac{(u-n)d_2}{R} \end{cases}$$

and $R = \sqrt{(w-m)^2 d_1^2 + (u-n)^2 d_2^2}$ is the distance between the centers of nm-th and uw-th obstacles.

Taking this into account, we can rewrite the integral Eq. (4) in a more explicit form

$$\frac{\mathbf{p}(y_0^{uw})}{2} + \frac{ik}{4} \sum_{n=1}^{N} \sum_{m=1}^{M} \int_0^{2\pi} \mathbf{p}(y^{nm}) H_1^{(1)}(kr) \frac{(\mathbf{r}, \mathbf{n}_y)}{r} \left|\mathbf{n}_y^{nm}\right| d\theta = \mathbf{p}^{inc}(y_0^{uw}), \tag{6}$$

where $y_0^{uw} \in l_{uw}$.

The solution of the integral Eq. (4) gives the value of the full pressure field on the boundary contour of each obstacle. In order to obtain a scattered wave at an arbitrary point $z = \{z_1, z_2\} \notin L$ inside the medium, the following integral formula should be used

$$p^{sc}(z) = \int_L p(y) \frac{\partial \Phi(|z-y|)}{\partial \mathbf{n}_y} dL_y = -\frac{ik}{4} \sum_{n=1}^{N} \sum_{m=1}^{M} \int_{l_{nm}} \mathrm{p}(y) H_1^{(1)}(kr) \frac{(\mathbf{r}, \mathbf{n}_y)}{r} dl \tag{7}$$

3 Numerical Treatment and the Solution

The series of the numerical experiments has been performed with the use of the BIE approach, which is described in the previous section for the system of star-like contours. In order to represent the integral notation in the form adapted to the computer calculations, we use the Boundary Element Method [4]. Under this approach we divide each contour l_{nm} into Q small arc-intervals with a uniform step over the angle coordinate $h_\theta = 2\pi/Q$. The nodes of the grid are placed in accordance with the collocation technique equally for "internal" and "external" variables $y_{0_q}^{nm} = y_q^{nm} = y^{nm}(\theta_q)$ at angles $\theta_q = (q-0.5) * h_\theta, q = 1, \ldots, Q$. Then, assuming that the integrand function is a constant on each elementary arc-interval, we can substitute the integration by its approximate sum, which leads to the following LAS

$$\frac{\mathbf{p}(y_{0_j}^{uw})}{2} + \frac{ik}{4} \sum_{n=1}^{N} \sum_{m=1}^{M} \sum_{q=1}^{Q} \mathbf{p}\left(y_q^{nm}\right) H_1^{(1)}(kr) \frac{(\mathbf{r}, \mathbf{n}_y)}{r} \left|\mathbf{n}_y^{nm}\right| h_\theta = \mathbf{p}^{inc}\left(y_{0_j}^{uw}\right) \tag{8}$$

where $j = 1, \ldots, Q$.

However, when the "external" and "internal" variables belong to the same point, the singularity appears in the argument of Hankel function [1]. This term can be calculated by the explicit integration over corresponding small arc-interval. Nevertheless, its contribution to the integral equation is relatively small in comparison with the contribution ½ of the first term and can be neglected in the approximate numerical calculations.

It is convenient to rewrite Eq. (8) in the matrix notation. For this purpose we introduce the following indexes through: index $a=j+(w-1+(u-1)*M)*Q$, corresponds to the "inner" variable $y_{0_g}^{uw}$, $g=1,\ldots,Q, w=1,\ldots,M, u=1,\ldots,N$, while for the "outer" variable we use index $b=q+(m-1+(n-1)*M)*Q, q=1,\ldots,Q$, $m=1,\ldots,M$, $n=1,\ldots,N$. This notation leads to the matrix form of Eq. (8)

$$\mathbf{Gp}=\mathbf{p}^{inc} \tag{9}$$

with matrix elements $\{g\}_{ab}=\frac{ik}{4}H_1^{(1)}(kr)\frac{(\mathbf{r},\mathbf{n}_y)}{r}\left|\mathbf{n}_y^{nm}\right|h_\theta$.

Due to the nature of BEM, the matrix G of system (9) is fully-dense with $(N*M*Q)^2$ non-zero elements. Thus, the solution of such a system by applying classical direct approaches requires $(N*M*Q)^3$ arithmetic operations. In the case of high-frequency regime, a dense numerical grid on the boundary of each contour should be taken, in order to adequately describe the wave-like nature of the problem, at least 10 nodes per each wave-length. This leads to the huge size of the algebraic system (9) and as a result, the direct solution requires too much computational time even on a modern PC. It becomes even more critical in performing a wide range of numerical experiments which involve varying boundary geometry and obstacles dispositions.

In our current study we use the iterative bi-conjugate gradient method, to solve the problem (9) with less computational cost than it requires when using direct approaches. This method can be written out as simple relations:

$$\begin{aligned}
&\alpha_j=(\hat{\mathbf{e}}_{j-1},\mathbf{e}_{j-1})/(\hat{\mathbf{s}}_j,\mathbf{G}\mathbf{s}_j),\\
&\mathbf{p}_j=\mathbf{p}_{j-1}+\alpha_j\mathbf{s}_j,\\
&\mathbf{e}_j=\mathbf{e}_{j-1}-\alpha_j\mathbf{G}\mathbf{s}_j,\\
&\hat{\mathbf{e}}_j=\hat{\mathbf{e}}_{j-1}-\alpha_j\mathbf{G}^T\hat{\mathbf{s}}_j,\\
&\beta_j=(\hat{\mathbf{e}}_j,\mathbf{e}_j)/(\hat{\mathbf{e}}_{j-1},\mathbf{e}_{j-1}),\\
&\mathbf{s}_{j+1}=\mathbf{e}_j+\beta_j\mathbf{s}_j,\\
&\hat{\mathbf{s}}_{j+1}=\hat{\mathbf{e}}_j+\beta_j\hat{\mathbf{s}}_j.
\end{aligned} \tag{10}$$

For the first iteration we set $\mathbf{e}_0=\mathbf{p}^{inc}-\mathbf{G}\mathbf{p}_0$ where initial guess p_0 can be set zero or taken from some physical assumption. Further reduction of the computational time can be attained by application of the modern fast algorithms [6–8, 12], to perform the fast matrix-vector multiplication in the right-hand sides of relations (10).

4 Numerical Results

In order to test the influence of the obstacle geometry on the wave propagation, a series of numeric computations has been performed on the basis of the numerical representation of the problem described above. For the purpose of a more general analysis we suppose the normal incidence of the plane wave with a unit amplitude $\mathbf{p}^{inc}(x_2) = e^{ikx_2}$. Physically, it corresponds to the far-field of the transducer with its length D significantly larger than the total horizontal area of the periodic array of obstacles.

In the accomplished calculations one of the main investigated parameters is the transmission coefficient. However, the distribution of the wave pressure transmitted through the material is non-uniform and the plane-wave structure is not retained at the receiver. Thus, we consider the average value of the full field on the +interval $x_1 = [-0.003; 0.031]$ m at the receiver position $x_2 = -0.026$ m which is placed below the periodic structure.

The first set of experiments is devoted to different geometries of the hole boundaries, starting from a simple circle shape to more complex polar-rose shapes with two, three and four leaves. This is adjusted by putting the polar distance of the contour equal to $\rho(\theta) = A((1 - Ratio) + Ratio * \cos(leaf * \theta))$, where the parameter *leaf* corresponds to the number of leaves in the contour, and *Ratio* represents the impact of the polar-rose shape on the contour and A is used for an appropriate scale on the contour (see Fig. 1). Obviously, the case $Ratio = 0$ gives the circular shape of the contour with the radius equal to A.

The transmission coefficient for various values of these parameters is given in Fig. 2, where on the right-hand side of each diagram the shapes and their corresponding *Ratio* are presented. The computations have been made for the following conditions: the speed of the longitudinal wave in the medium $c = 2200$ m/s, the doubly periodic array has $M = 8$ number of rows and $N = 6$ number of columns with $Q = 2^6 = 64$ points of the numerical grid on each boundary contour, the holes' radius is $A = 1$ mm and they are placed at distances $d_1 = d_2 = 2A$, the first row starts at the x_2-coordinate $Y = -3A$, while x_1-coordinate $\mathrm{X} = 0$.

From the presented figures one can observe that the transmission coefficient in a low-frequency range (up to 0.4 MHz) weakly depends on the contour shape and significant changes appear only at higher frequencies. Parameters $leaf = 2$ and $Ratio = 0.3$ give the widest band gap (dashed line on the top diagram in Fig. 2) on the interval $\approx [0.49, 0.91]$ MHz when compared to other boundary contours.

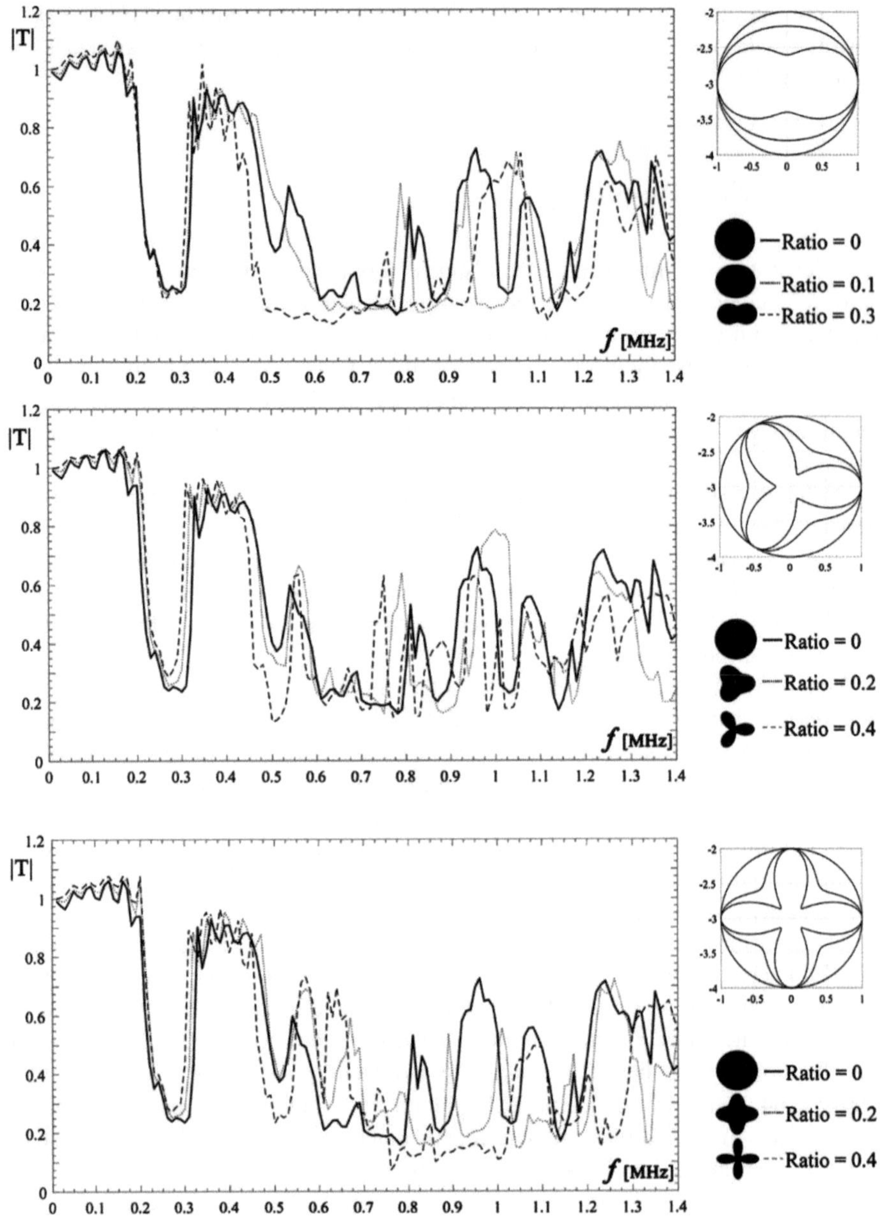

Fig. 2 Band-gapping at low frequencies: transmission coefficient $|T|$ versus frequency f (MHz) for various cross-section contours l of holes in doubly-periodic arrays

The lowest transmission coefficient (≈ 0.1) is attained for the four-leaf rose with *Ratio* = 0.4 in the range $\approx [0.75, 1]$ MHz. The three-leaf rose is the only non-symmetrical contour which is placed at an angle with respect to the incident wave. As a result, it is difficult to provide a uniform distribution of the transmission coefficient for the wide frequency range.

The distribution of the amplitudes of the full pressure in the current three configurations is presented in Fig. 3. Predictably, we can notice the symmetry of the pressure field for the four-leaf and two-leaf contours, while the wave transmitted through the array of holes with the three-leaf contours is asymmetric. In this case we can also notice a high concentration of the transmitted wave energy on the left side of the figure. An interesting beam-forming of the wave is shown for the case **3** with parameters $\rho(\theta) = 0.6 + 0.4 * \cos(2 * \theta)$ at $f = 2.49$ MHz. However, from cases **4** and **5** we can conclude that such behavior is not retained for other frequencies or other *Ratio* values.

Another series of experiments have been performed in order to test the influence of the hole positions on the wave propagation. For these purpose, parameters d_1 and d_2 are varied and the results are presented in Figs. 4 and 5. The transmission coefficient $|T|$ in this cases is calculated as the average value of the full pressure on the interval $x_1 = [-0.003; 0.031 * d_1/0.04]$ m at the receiver position $x_2 = -0.026 + 0.02 * (1 - d_2/0.04)$ m. It is noticeable that such changes in the arrangement of the array can affect the low-frequency range. The horizontal parameter d_1 can extend the first band gap and the vertical distance d_2 makes higher the value of the lower frequency of this band gap. As expected, the decreasing of the horizontal distance between the nearest neighbor holes reduces the transmitted wave energy.

It is also interesting to test the effect of the horizontal shift of each second row in the doubly-periodic array. Corresponding calculations are reflected in Fig. 6 where X denotes the horizontal coordinate of the center of the first hole in each second row in the doubly-periodic array. Surprisingly, wave behavior for this arrangement in the low-frequency range is identical to the array without such a shift. In the range $f = [0.4; 1.1]$ MHz the array with the shift gives the highest transmission coefficient for all performed tests demonstrated above.

Another set of experiments are demonstrated in Fig. 7 for the material with sound speed $c = 6000$ m/s, where the impact of the number of rows in the doubly-periodic array of cylindrical holes to the transmission properties of the material is evaluated. It is clear from this figure that with increasing number of rows a smaller amount of the wave energy passes through the array of holes. However, the array with shift $X = 2$ mm of each second row is less sensitive to this effect than the array with normal arrangement. Nevertheless, at low frequencies the coefficient $|T|$ demonstrates only small perturbations around the unit value for both the arrays.

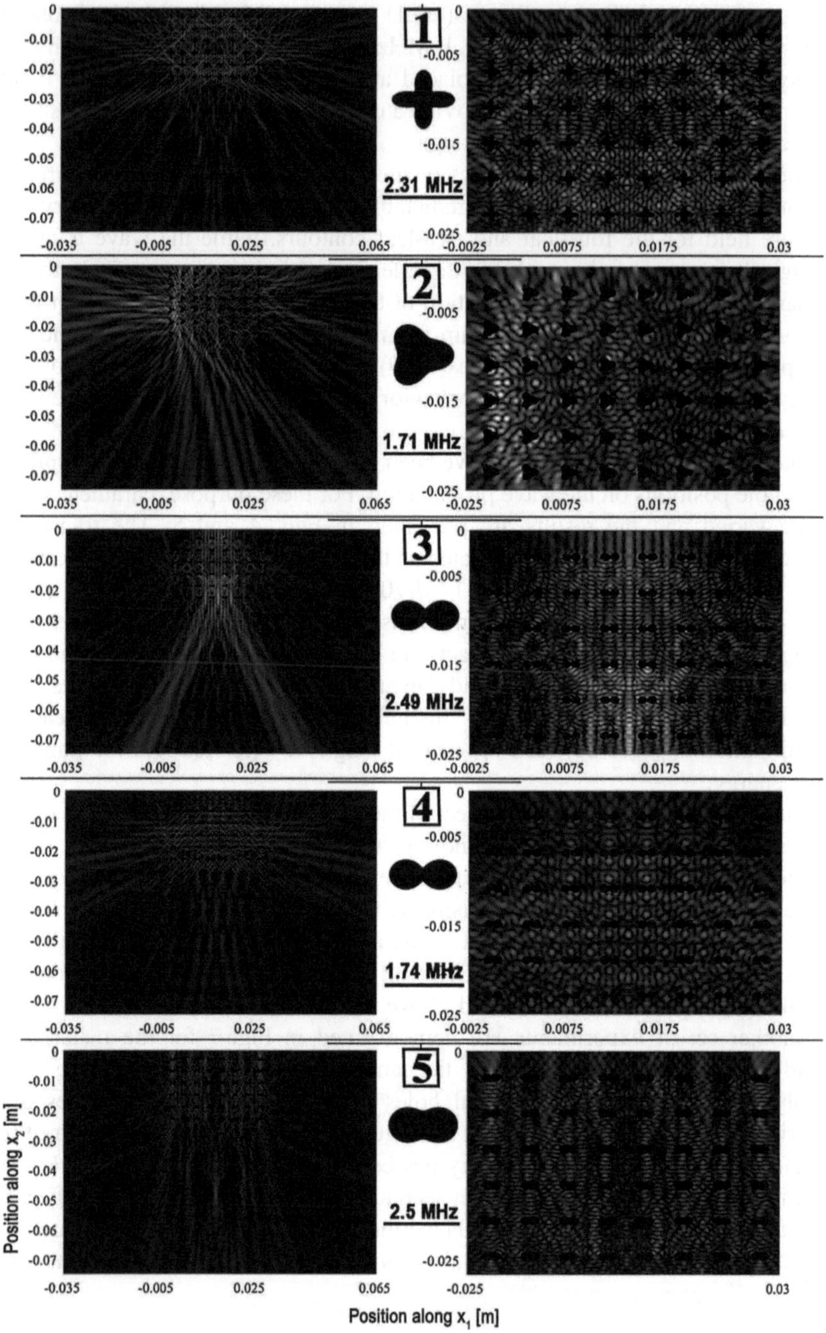

Fig. 3 Full pressure magnitude $|p|$ outside (*left*) and inside (*right*) doubly-periodic array of holes. **Array 1**—$\rho(\theta) = 0.7 + 0.3 * \cos(4 * \theta)$ at $f = 2.31$ MHz; **Array 2**—$\rho(\theta) = 0.8 + 0.2 * \cos(3 * \theta)$ at $f = 1.71$ MHz; **Array 3**—$\rho(\theta) = 0.6 + 0.4 * \cos(2 * \theta)$ at $f = 2.49$ MHz; **Array 4**—$\rho(\theta) = 0.6 + 0.4 * \cos(2 * \theta)$ at $f = 1.74$ MHz; **Array 5**—$\rho(\theta) = 0.7 + 0.3 * \cos(2 * \theta)$ at $f = 2.5$ MHz

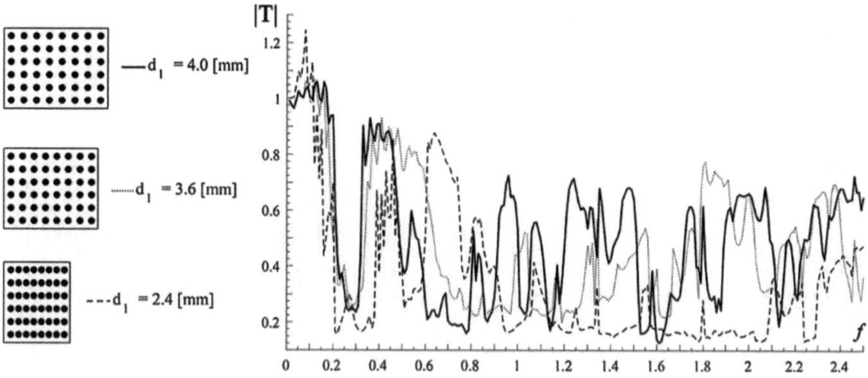

Fig. 4 Transmission coefficient $|T|$ versus frequency $f\,(MHz)$ for various horizontal distances between the centers of the neighbor holes in the doubly-periodic array

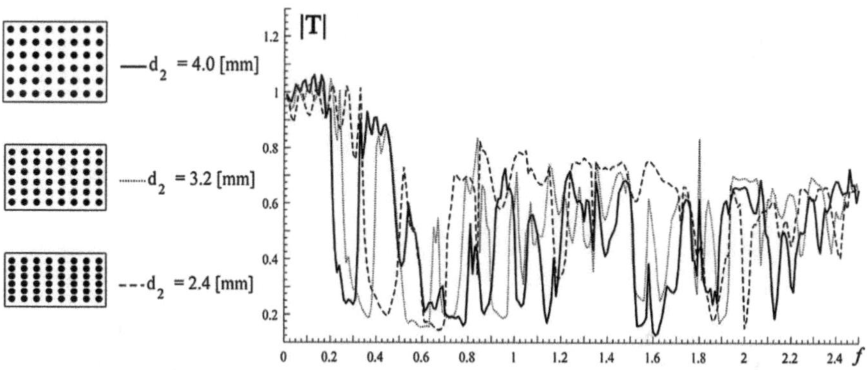

Fig. 5 Transmission coefficient $|T|$ versus frequency $f\,(MHz)$ for various vertical distances between the centers of the neighbor holes in the doubly-periodic array

5 Full-Scale Experiment

The natural experiments have been performed using industrial low-frequency ultrasonic flaw detector UDC-60 N with the working frequency range from 0.04 to 2.5 MHz (see Fig. 8). Several samples from different materials have been prepared by drilling doubly-periodic arrays of holes. We restrict our experimental study to the case of simple cylindrical holes, since it is difficult to produce more complex contour shapes with appropriate accuracy. Moreover, it should be noted that the energy transition of the wave energy is lower at high-frequencies due to the damping properties of the material. This is especially noticeable for the plastic materials with high damping. Nevertheless, some characteristic features predicted by numerical calculations can be approved experimentally with the use of the through transmission scanning pattern demonstrated in Fig. 8.

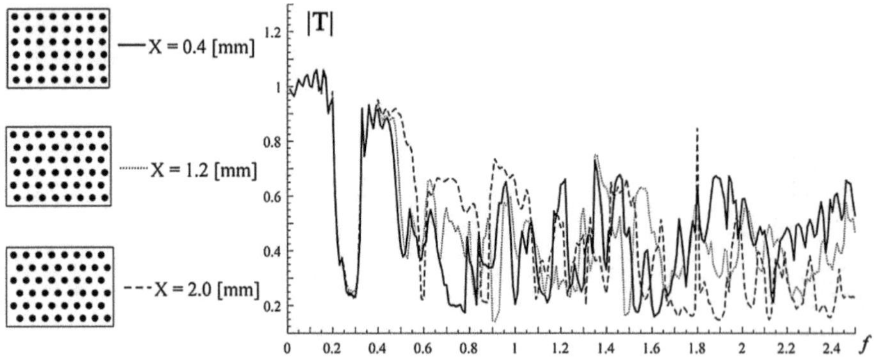

Fig. 6 Comparison between transmission coefficient $|T|$ for doubly-periodic arrays with different horizontal shift of each second row

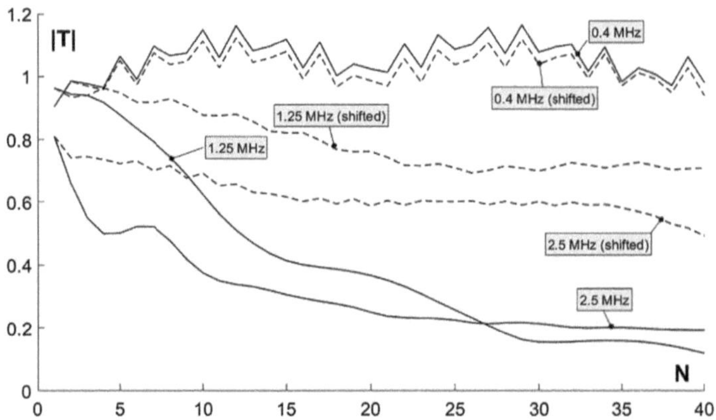

Fig. 7 Transmission coefficient $|T|$ versus number of rows N in the doubly-periodic array of cylindrical holes without shift (*solid line*) and with shift $X = 2$ mm (*dashed line*)

Fig. 8 Measurement equipment and through transmission scanning pattern

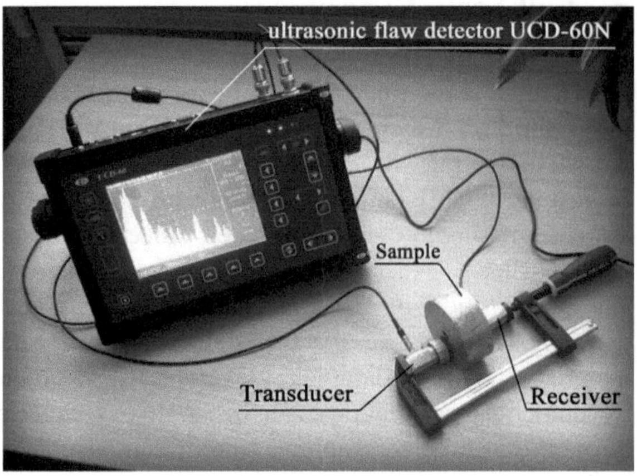

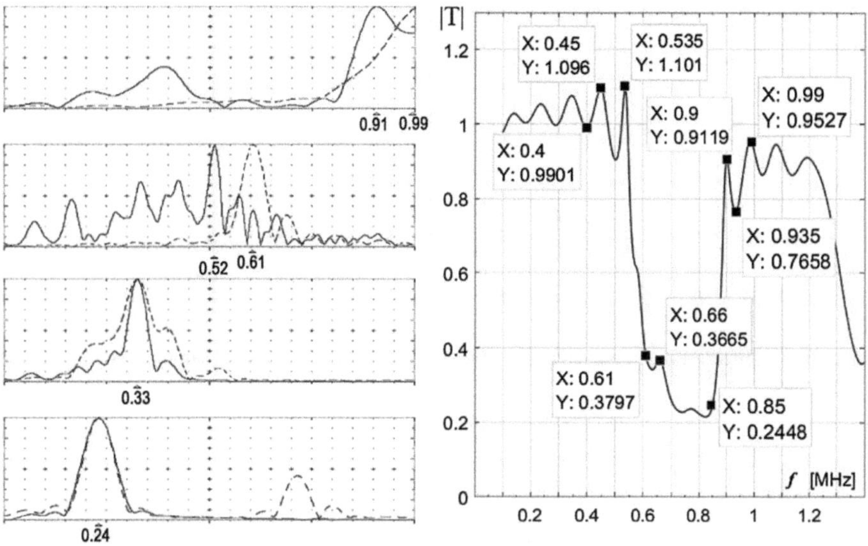

Fig. 9 Comparison of the natural (at the *left*) and numerical (at the *right*) experimental data for the sample with array of cylindrical holes with parameters $a = 1\,\text{mm}, d_1 = d_2 = 4\,\text{mm}, c = 6000\,\text{m/s}$. At the *left* normalized spectral characteristic of the wave measured at the receiver position for the sample without array of holes (*dashed line*) and with array of holes (*solid line*) in the frequency range $[0, 1]$ MHz. At the *right* transition coefficient predicted by the numerical method

In Fig. 9 the comparison of the numerical calculations (right-hand side of the figure) and full-scale experiment data is presented for the steel sample with doubly-periodic array of cylindrical holes of radius $A = 1\,\text{mm}$ and distances between two neighbor holes $d_1 = d_2 = 4A$, where sound speed is equal to $c = 6000\,\text{m/s}$. At the left-hand side of the figure the normalized spectrum characteristics of the transmitted wave measured at the receiver is shown over the range $[0, 1]$ MHz. The sample with holes is marked by the solid line, while the dashed line represents the sample without any defects. The experiment has been performed for four different probes with the leading frequencies $1, 0.61, 0.33, 0.24$ MHz, respectively. Detailed analysis of this comparison shows that the true behavior predicted by the numerical calculations is predicted with a small error.

Another set of experiments has been performed to examine the influence of the vertical distance d_2 between rows in the periodic array. For this purpose, two samples have been prepared from plastic materials with the sound speed $c \approx 2200\,\text{m/s}$. The measured amplitude-time characteristics is reflected in Fig. 10 where the first row presents the wave transmitted through the material free of holes, the second row demonstrates the data for the material with the array arrangement $A = 1\,\text{mm}, d_1 = d_2 = 4\,\text{mm}$, and the bottom row shows the measurements for the sample with $A = 1\,\text{mm}, d_1 = 4\,\text{mm}, d_2 = 2.4\,\text{mm}$. This should be compared with Fig. 5 from the previous section. As it is predicted numerically, the transmission in the last sample is higher than in the shown low-frequency region because its first

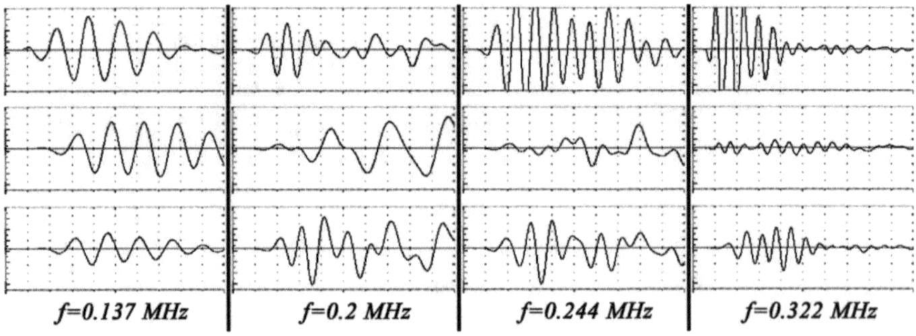

Fig. 10 Amplitude-time characteristics measured for three plastic samples with different frequencies

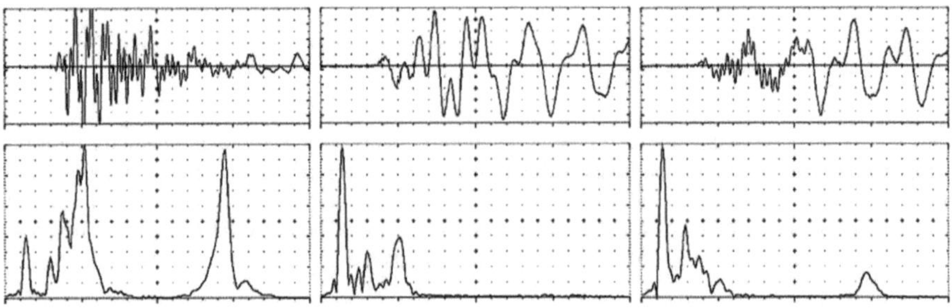

Fig. 11 Comparison of the impulse and the spectrum characteristics for the wave transmitted through a continuous sample (at the *left*), the sample with the array of holes without any shift (at the *center*), the sample with the shift of each second row in the hole array (at the *right*)

band-gap starts with a higher frequency. However, the second sample demonstrates better transmission before this band-gap, which is in good agreement with the numerical experiments.

In order to verify the prediction which is reflected in Fig. 5, two samples have been prepared made of plastic material with $c = 2200\,\mathrm{m/s}$. The first one contains a doubly periodic array of holes with parameters $A = 1$ mm, $d_1 = d_2 = 4$ mm with no shift. The second sample is made with the same parameters as the previous one but with the shift of each second row equal to $X = 2$ mm. These samples are scanned by the through transmission ultrasonic technique with the ultrasonic transducer which has two dominant frequencies, one at ≈0.25 MHz and the other at ≈0.7 MHz. The results is reflected in Fig. 11, where the graph at the left side represents the signal transmitted through the material free of voids, the central graph shows the wave transmitted through the sample with the array of holes without any shift, and the one at the right demonstrates the transmission through the sample with the shift of each second row in the array. Note, that for demonstration purposes, the last two graphs are presented with double amplification. As can be sees from the central diagram, the propagation of the wave through the sample without shift “cleans” the signal from its high-frequency components, which is reflected both on impulse response and on the spectrum. From the right-hand side of the diagram we can conclude that the second sample permits transmission of the high-frequency component, as it is predicted numerically (see Fig. 6). However, the impact of the high-frequency component is small, due to the high damping of the used material.

6 Conclusions

- The boundary elements method is a powerful tool, which gives accurate prediction for the wave behavior in the medium with doubly-periodic array of holes and can be successfully applied to the study of the properties of the modern acoustic meta-materials.
- As numerical experiments show, small changes in the hole cross-section can significantly effect the transmission properties of the material. Such adjustments can give high amplification on certain frequencies, concentrating and redirecting the wave energy in a specific way. However, such resonance properties are not maintained even for small changes of the hole boundaries. Thus, the low-frequency range is not sensitive to the contours shape.
- The horizontal and vertical distance between the neighbor holes in the doubly-periodic array effect the low-frequency range of the transmission coefficient, namely the variation of the horizontal distance changes the width of the first band gap and the variation of the vertical distance can translate this band-gap to higher frequency range.
- The transition coefficient in the frequency range before the first band-gap is weakly sensitive to the number of rows in the doubly-periodic array, while at the higher frequencies the increasing number of rows decreases considerably the transmitted wave energy. The array with the horizontal shift of each second row demonstrates more resistances in the wave process.
- The Ultrasonic non-destructive testing technology with the through transmission scanning pattern can verify some properties predicted by the calculations. Nevertheless, high damping of some materials constricts verification on higher frequencies. Thus, it is more convenient to use immersion ultrasonic tests in such cases.

Acknowledgements The work is performed within the framework of the Project № 15-19-10008 of the Russian Science Foundation (RSCF).

References

1. Abramowitz, M., Stegun, I.: Handbook of Mathematical Functions. Dover, New York (1965)
2. Achenbach, J.D., Li, Z.L.: Reflection and transmission of scalar waves by a periodic array of screens. Wave Motion **8**, 225–234 (1986)
3. Achenbach, J.D., Kitahara, M.: Harmonic waves in a solid with a periodic distribution of spherical cavities. J. Acoust. Soc. Am. **81**, 595–598 (1987)
4. Banerjee, P.K.: The Boundary Element Methods in Engineering, 2nd edn. McGraw-Hill, London (1994)
5. Banerjee, B.: An Introduction to Metamaterials and Waves in Composites. CRC Press, Boca Raton (2011)
6. Bron, S.: Efficient numerical methods for non-local operators. EMS Tracts Math. **14** (2010)
7. Brunner, D., Junge, M., Rapp, P., Bebendorf, M., Gaul, L.: Comparison of the fast multipole method with hierarchical matrices for the helmholtz-BEM. Comput. Model. Eng. Sci. **58**, 131–158 (2010)
8. Cheng, H., Crutchfield, W.Y., Gimbutas, Z., et al.: A wideband fast multipole method for the Helmholtz equation in three dimensions. J. Comput. Phys. **216**, 300–325 (2006)
9. Deymier, P.A.: Acoustic Metamaterials and Phononic Crystals. Springer, Berlin (2013)
10. Guenneau, S., Craster, R.V.: Acoustic Metamaterials Negative Refraction, Imaging, Lensing and Cloaking. Springer, Netherlands (2013)
11. Liu, Z., Zhang, X., Mao, Y., et al.: Locally resonant sonic materials. Science **289**, 1734–1736 (2000)
12. Cho, M.H., Cai, W.: A wideband fast multipole method for the two-dimensional complex Helmholtz equation. Comput. Phys. Commun. **181**, 2086 (2010)
13. Scarpetta, E., Sumbatyan, M.A.: On wave propagation in elastic solids with a doubly periodic array of cracks. Wave Motion **25**, 61–72 (1997)
14. Scarpetta, E., Sumbatyan, M.A.: Wave propagation through a periodic array of inclined cracks. Eur. J. Mech. A/Solids. **19**, 949–959 (2000)
15. Sumbatyan, M.A., Scalia, A.: Equations of Mathematical Diffraction Theory. CRC Press, Boca Raton (2005)

Finite Element Simulation of Thermoelastic Effective Properties of Periodic Masonry with Porous Bricks

A.V. Nasedkin, A.A. Nasedkina and A. Rajagopal

Abstract In this work an integrated approach has been proposed for the determination of the effective mechanical and temperature properties of thermoelastic periodic brick masonry wall with various porous structures. According to the classical method of determining effective moduli of composites, in order to describe internal micro- or macrostructure, we consider a representative volume cell, which enables us to describe effective properties of the equivalent homogeneous anisotropic material. The problems for representative cells are simulated and analyzed as thermoelastic boundary value problems, using special programs in APDL language for ANSYS finite element package. The post processing of the solution gives averaged characteristics of the stress–strain state and thermal flux fields that allow computing the effective moduli of the composite. The proposed method has been applied to several examples of periodic masonry with porous, hollow and porous–hollow bricks. A periodic part of masonry with porous, hollow and porous–hollow bricks was chosen as a representative cell with thermoelastic tetrahedral and hexahedral finite elements. In order to take into account the porosity of the bricks in the masonry, using similar approaches we have preliminary solved at the microlevel the problems of the effective moduli detection for the porous bricks as thermoelastic composite bodies with random porosity structures. After that at the macrolevel the material of porous brick was considered as a homogeneous body with its own effective properties. The results of numerical experiments showed that the structures of the representative cells and porosity could significantly affect the values of the effective moduli for the considered brick masonry walls.

A.V. Nasedkin (✉) · A.A. Nasedkina
Institute of Mathematics, Mechanics and Computer Science, Southern Federal University, Milchakova Street 8a, Rostov-on-Don 344090, Russia
e-mail: nasedkin@math.sfedu.ru

A. Rajagopal
Department of Civil Engineering, Indian Institute of Technology Hyderabad, Kandi, Hyderabad 502285, Telangana, India
e-mail: rajagopal@iith.ac.in

M.A. Sumbatyan (ed.), *Wave Dynamics and Composite Mechanics for Microstructured Materials and Metamaterials*,
Advanced Structured Materials 59, DOI 10.1007/978-981-10-3797-9_12

Keywords Periodic composite · Masonry · Porous brick · Hollow brick · Homogenization · Thermoelastic effective moduli · Basic unit cell · Representative volume · Finite element method

1 Introduction

Masonry is a composite material consisting of bricks and mortar which are normally arranged in a periodic manner. Global behaviour of masonry structure is determined by the behaviour of its phases, namely, the brick units and the joints between the units. However, modeling each unit and joint in the structure would be impractical for a large masonry structure.

The composite behaviour of masonry in terms of average stresses and strains can be described using the homogenization theory so that the compound material can be considered as an elastic homogeneous material with effective properties. Homogenization procedure is a very important step in masonry modelling. There are different approaches to the homogenization techniques described in literature, which can be analytical, semi-analytical and numerical. These methods fall into two general categories of one-step and two-step methods.

The two-step methods are analytical and are aimed at representing the complex geometry of the basic cell by a simplified geometry. When the homogenisation is being performed in two steps, head joints and bed joints are introduced successively, so that the masonry can be considered as a layered material. Some examples of such techniques can be found in [17, 18]. In [17] a multilayer system with alternating mortar joints was considered. In a two-step technique of [18] for the first step the homogenized matrix with head joints as elliptical inclusions was considered and the second step consisted in the consideration of a laminate structure with continuous bed joints. The final result of the two-step homogenization methods depends on the step order. Also in these approximate methods of homogenization the unit-mortar interface, bond patterns and thickness of masonry wall are often not taken into account.

A one-step fully analytical homogenization technique was proposed in [5] were parallelepiped-shaped units where approximated by elliptic cylinders which were considered as inclusions. However, most of the one-step homegenization methods are semi-analytical by nature. They imply rigorous application of the homogenisation theory for the basic cell with adequate boundary conditions and exact geometry. Because of the complexity of masonry structure, these methods include the numerical solution of the problem with the help of the finite element method. For example, a mixed one-step homogenization by using asymptotic method in combination with the finite element method was considered in [8]. Later in [7] an analytical multi-step and a numerical single-step technique were compared to determine mechanical characteristics of CFRP reinforced masonry walls. Homogenization process of stratified masonry was discussed in [6] where a three-dimensional problem was studied by asymptotic techniques using variational approach. In [3] in-plane elastic characteristics of masonry were derived through homogenization theory with periodic boundary conditions and the numerical solution was obtained by the finite element method.

A strain energy approach was used in [23] for a one-step periodic eigenstrain homogenization method, where the periodicity of field quantities was imposed by Fourier series, and the Eshelby tensor for the unit cell was derived to relate the disturbance field with the eigenstrain field. This approach was also applied in [11] to calculate effective properties of masonry by finite element method using stress-prescribed or strain-prescribed analysis.

The question of whether plane stress, plane strain or three-dimensional modelling should be used for the simulation of periodic masonry was discussed in [4]. This study took into account the thickness of masonry wall and concluded that two-dimensional assumptions had little influence on the mascorscopic elastic behaviour of masonry in the linear range, but could affect its non-linear response and therefore three-dimensional computations were unavoidable when the characteristics of masonry varyied through the masonry wall thickness.

A nonlinear homogenization procedure for periodic masonry has been discussed in several papers. For example, linear constitutive relationship for the bricks and nonlinear constitutive law for mortar joints was proposed in [20]. Nonlinear homogenization technique to describe the damage of the masonry structure subjected to alternating tension-compression cyclic loads was applied in [19] where a linear elastic constitutive law was considered for the bricks and the unilateral damage model was implemented for the mortar. In [24] homogenization technique and damage mechanics theory were used to model a three-dimensional masonry basic cell and numerically derive the equivalent elastic properties, strength envelope, and failure characteristics of masonry material. The basic cell was modelled with distinctive consideration of non-linear material properties of mortar and brick. Recent non-linear methodology for the analysis of masonry was implemented [1] where a macro-element based on the equivalent frame approach was used. An improved micro-mechanical model for the masonry homogenisation in the non-linear domain was proposed and validated by comparison with experimental and numerical results in [25–27].

Most of the papers concerning masonry structures deal with periodic masonry. However, some of real masonry structures, such as old masonry walls made up of stones, are non-periodic. For example, the concept of representative volume was used instead of the concept of periodic cell in case of quasi-periodic masonry [10] where the representative volume element was obtained by using test-window method and by increasing its finite dimensions.

Meanwhile, the thermoelastic properties of masonry porous structure have not been studied enough [14, 22].

In this chapter we apply the effective moduli method for thermoelastic composites to calculate the effective properties of various masonry structures. The paper has the following structure. In Sect. 2 we describe the mathematical statement of standard effective moduli method for thermoelastic composites of arbitrary anisotropy class. We formulate six static elastic problems for the calculation of the effective stiffnesses, one problem of thermal stresses for the calculation of the effective thermal stress coefficients and three heat conduction problems for finding the effective thermal conductivities. We also note that in the case of isotropic medium we can solve only three problems: one elastic problem, one problem of thermal stresses and one heat conduction problem.

In Sect. 3 we consider a periodic masonry basic structure as a representative volume of masonry wall one brick thick. We investigate the mechanical and thermal properties of masonry wall made of different porous and hollow bricks with the same total porosity. For the case of porous brick with only stochastic microporosity at first we calculate the effective moduli of the isotropic thermoelastic material with microporosity of stochastic structure. Then we consider the porous brick material as a homogeneous isotropic medium with effective properties. We describe solid and finite element models for the case of porous bricks, for the case of bricks with hollows made of dense brick material, and for the case of porous–hollow bricks made of porous material with the hollows. In the end of Sect. 3 we present the computation results for the effective moduli obtained in ANSYS finite element package.

In Sect. 4 we discuss the results. We note that for all cases the masonry basic structure has orthotropic class of anisotropy for the mechanical and thermal moduli. We discuss the values of the various moduli and the influence effects of porosity and hollows on the masonry strength and on its thermal insulation properties.

2 Effective Moduli Method for Linear Thermoelastic Composite Materials

Let us consider heterogeneous anisotropic thermoelastic material in a representative volume Ω of a composite body. Then in the frames of static linear theory of thermoelasticity [13] we have the following system of equations

$$\sigma_{ij,j} = 0, \quad \sigma_{ij} = c_{ijkl}\varepsilon_{kl} - \beta_{ij}\,\theta, \quad \varepsilon_{ij} = (u_{i,j} + u_{j,i})/2\,, \tag{1}$$

$$q_{j,j} = 0, \quad q_i = -k_{ij}\theta_{,j}\,, \tag{2}$$

where σ_{ij} are the components of the second rank stress tensor $\boldsymbol{\sigma}$; ε_{ij} are the components of the second rank strain tensor $\boldsymbol{\varepsilon}$; u_i are the components of the displacement vector $\mathbf{u}$; θ is the temperature increment from natural state, c_{ijkl} are the forth rank tensor $\mathbf{c}$ of elastic stiffness moduli; β_{ij} are the thermal stress coefficients; q_i are the components of the heat flux vector $\mathbf{q}$; k_{ij} are the components of the tensor $\mathbf{k}$ of thermal conductivities.

In vector-matrix symbols in R^3 the formulas (1), (2) can be presented in the form

$$\mathbf{L}^*(\nabla) \cdot \mathbf{T} = 0, \quad \mathbf{T} = \mathbf{c} \cdot \mathbf{S} - \boldsymbol{\beta}\,\theta, \quad \mathbf{S} = \mathbf{L}(\nabla) \cdot \mathbf{u}\,, \tag{3}$$

$$\nabla^* \cdot \mathbf{q} = 0, \quad \mathbf{q} = -\mathbf{k} \cdot \nabla\theta\,, \tag{4}$$

where

$$\mathbf{L}^*(\nabla) = \begin{bmatrix} \partial_1 & 0 & 0 & 0 & \partial_3 & \partial_2 \\ 0 & \partial_2 & 0 & \partial_3 & 0 & \partial_1 \\ 0 & 0 & \partial_3 & \partial_2 & \partial_1 & 0 \end{bmatrix}, \quad \nabla = \begin{Bmatrix} \partial_1 \\ \partial_2 \\ \partial_3 \end{Bmatrix}, \tag{5}$$

$\mathbf{T} = \{\sigma_{11}, \sigma_{22}, \sigma_{33}, \sigma_{23}, \sigma_{13}, \sigma_{12}\}$ is the stress array; $\mathbf{S} = \{\varepsilon_{11}, \varepsilon_{22}, \varepsilon_{33}, 2\varepsilon_{23}, 2\varepsilon_{13}, 2\varepsilon_{12}\}$ is the strain array; $\mathbf{c}$ is the 6×6 matrix of elastic moduli; $c_{\gamma\zeta} = c_{ijkl}$, $\gamma, \zeta = 1, ..., 6$, $i, j = 1, 2, 3$ with correspondence law $\gamma \leftrightarrow (ij)$, $\zeta \leftrightarrow (kl)$, $1 \leftrightarrow (11)$, $2 \leftrightarrow (22)$, $3 \leftrightarrow (33)$, $4 \leftrightarrow (23) = (32)$, $5 \leftrightarrow (13) = (31)$, $6 \leftrightarrow (12) = (21)$; $\boldsymbol{\beta} = \{\beta_{11}, \beta_{22}, \beta_{33}, \beta_{23}, \beta_{13}, \beta_{12}\}$; $(...)^*$ is the transpose operation; and $(...) \cdot (...)$ is the scalar product operation.

Setting the appropriate principal boundary conditions at $\Gamma = \partial\Omega$, we can find the solutions of the problems (1), (2) or (3)–(5) for heterogeneous medium with $\mathbf{c} = \mathbf{c}(\mathbf{x})$, $\boldsymbol{\beta} = \boldsymbol{\beta}(\mathbf{x})$ and $\mathbf{k} = \mathbf{k}(\mathbf{x})$ in the representative volume Ω. Then the comparison of the averaged over Ω stresses and heat fluxes with analogous values for homogeneous medium (the comparison medium) similarly to [12, 14, 15] will permit to determine the effective moduli $c_{\gamma\zeta}^{\text{eff}}$, $\beta_{\gamma}^{\text{eff}}$ and k_{ij}^{eff} for the composite material.

Thus, for finding the effective stiffness moduli $c_{\gamma\zeta}^{\text{eff}}$ and the effective thermal stress coefficients $\beta_{\gamma}^{\text{eff}}$ we assume the following boundary condition

$$\mathbf{u} = \mathbf{L}^*(\mathbf{x}) \cdot \mathbf{S}_0, \quad \theta = \theta_0, \quad \mathbf{x} \in \Gamma, \tag{6}$$

where $\mathbf{S}_0 = \{\varepsilon_{011}, \varepsilon_{022}, \varepsilon_{033}, 2\varepsilon_{023}, 2\varepsilon_{013}, 2\varepsilon_{012}\}$; ε_{0ij}, θ_0 are some constant values that do not depend on $\mathbf{x}$.

Note, that in the region Ω for $\theta = \theta_0$ the Eq. (4) are satisfied identically, and therefore we can solve the elastic problem with thermal stresses with respect only to the displacement vector $\mathbf{u}$

$$\mathbf{L}^*(\nabla) \cdot \mathbf{T} = 0, \quad \mathbf{T} = \mathbf{c} \cdot \mathbf{S} - \boldsymbol{\beta}\,\theta_0, \quad \mathbf{S} = \mathbf{L}(\nabla) \cdot \mathbf{u}, \tag{7}$$

$$\mathbf{u} = \mathbf{L}^*(\mathbf{x}) \cdot \mathbf{S}_0, \quad \mathbf{x} \in \Gamma. \tag{8}$$

Thus, setting in (7), (8)

$$\theta_0 = 0, \quad \mathbf{S}_0 = \varepsilon_0 \mathbf{h}_\zeta, \quad \zeta = 1, 2, ..., 6, \tag{9}$$

where $\varepsilon_0 = \text{const}$, ζ is some fixed index mark ($\mathbf{h}_\zeta$ is the vector from six-dimensional basic set for the components of the strain tensor basic set; $\mathbf{h}_j = \mathbf{e}_j\mathbf{e}_j$, $j = 1, 2, 3$; $\mathbf{h}_4 = (\mathbf{e}_2\mathbf{e}_3 + \mathbf{e}_3\mathbf{e}_2)/2$; $\mathbf{h}_5 = (\mathbf{e}_1\mathbf{e}_3 + \mathbf{e}_3\mathbf{e}_1)/2$; $\mathbf{h}_6 = (\mathbf{e}_1\mathbf{e}_2 + \mathbf{e}_2\mathbf{e}_1)/2$; $\mathbf{e}_j$ are the orts of the Cartesian coordinate system), we obtain the computation formulas for the elastic moduli $c_{\gamma\zeta}^{\text{eff}}$:

$$c_{\gamma\zeta}^{\text{eff}} = \langle T_\gamma \rangle / \varepsilon_0. \tag{10}$$

Here, the angle brackets $\langle(...)\rangle$ denote the averaged by the volume values

$$\langle(...)\rangle = (1/|\Omega|) \int_{\Omega} (...)\, d\Omega\,.$$

If we assume

$$\theta_0 \neq 0, \quad \mathbf{S}_0 = 0\,, \tag{11}$$

then from the boundary problem (7), (8) we find the thermal stress effective moduli:

$$\beta_{\gamma}^{\text{eff}} = -\langle T_{\gamma}\rangle/\theta_0\,. \tag{12}$$

To determine the effective thermal conductivities k_{ij} we solve uncoupled thermal problems (4) with the boundary conditions

$$\theta = G_0 \mathbf{x}^* \cdot \mathbf{e}_l = G_0 x_l, \quad G_0 = \text{const}, \quad l = 1,2,3, \quad \mathbf{x} \in \Gamma\,, \tag{13}$$

and from the results we can calculate the effective thermal conductivity coefficients

$$k_{il}^{\text{eff}} = -\langle q_i\rangle/G_0\,. \tag{14}$$

In the case of material of arbitrary anisotropy class, in order to determine the complete set of effective moduli we have to solve six elastic problems (7)–(9) for the calculation of the effective stiffnesses by (10), one problem (7), (8), (11) of thermal stresses for the calculation of the effective thermal stress coefficients by (12) and three heat conduction problems (4), (13) for finding the effective thermal conductivities by (14).

For certain crystalline composites systems the number of problems can be reduced, since some problems will provide the same (or nearly the same) values of effective moduli. The type of the composite crystalline system is defined by the both crystalline systems of the initial phases and the geometrical and structural features of the representative volume of the composite.

If the initial phase is made of isotropic materials and the geometry of the composite has no peculiarities in different directions, then the homogenised material will also be isotropic. In this case, a module structure of the initial phases and of the homogenized material will have the form

$$\mathbf{c} = \begin{bmatrix} c_{11} & c_{12} & c_{12} & 0 & 0 & 0 \\ c_{12} & c_{11} & c_{12} & 0 & 0 & 0 \\ c_{12} & c_{12} & c_{11} & 0 & 0 & 0 \\ 0 & 0 & 0 & c_{66} & 0 & 0 \\ 0 & 0 & 0 & 0 & c_{66} & 0 \\ 0 & 0 & 0 & 0 & 0 & c_{66} \end{bmatrix}, \quad \boldsymbol{\beta} = \begin{Bmatrix} \beta \\ \beta \\ \beta \\ 0 \\ 0 \\ 0 \end{Bmatrix}, \quad \mathbf{k} = \begin{bmatrix} k & 0 & 0 \\ 0 & k & 0 \\ 0 & 0 & k \end{bmatrix}, \tag{15}$$

where $c_{66} = (c_{11} - c_{12})/2$.

Thus, an isotropic material is characterized by four moduli from (15): two elastic stiffness moduli c_{11} and c_{12}, thermal stress coefficient β and thermal conductivity k. In practice we usually know other elastic and thermal moduli are obtained, for example, the Young's modulus E, the Poisson's ratio ν, or the bulk modulus K, the shear modulus G, and the thermal expansion coefficient α. The main moduli E, ν and α are related to c_{11}, c_{12} and β for isotropic materials by formulae

$$c_{11} = \frac{(1-\nu)E}{(1+\nu)(1-2\nu)}, \quad c_{12} = \frac{\nu E}{(1+\nu)(1-2\nu)}, \quad \beta = \alpha\frac{E}{(1-2\nu)}, \tag{16}$$

and $G = c_{66} = E/[2(1+\nu)]$, $K = E/[3(1-2\nu)]$.

Then for isotropic case we can solve the problems (7), (8) with two variants of boundary conditions, such as

$$\text{iI)} \quad \theta_0 = 0, \quad \mathbf{S}_0 = \varepsilon_0 \mathbf{h}_1 \quad \Rightarrow \quad c_{11}^{\text{eff}} = \langle\sigma_{11}\rangle/\varepsilon_0, \quad c_{12}^{\text{eff}} = \langle\sigma_{22}\rangle/\varepsilon_0, \tag{17}$$

$$\text{iII)} \quad \theta_0 \neq 0, \quad \mathbf{S}_0 = 0 \quad \Rightarrow \quad \beta^{\text{eff}} = -\langle\sigma_{11}\rangle/\theta_0, \tag{18}$$

and one thermal problem (4) with the boundary condition (13), for example, with $j = 1$

$$\text{iIII)} \quad \theta = G_0 x_1 \quad \Rightarrow \quad k^{\text{eff}} = -\langle q_1\rangle/G_0. \tag{19}$$

Next, we will model the periodicity cell of masonry, consisting of the elements of bricks and mortar. In turn, the brick can be regarded as solid, porous, hollow, porous and hollow.

On the macro level we will consider the porous material of brick as a homogeneous body with effective moduli, depending on the porosity. This material may be considered isotropic, and its moduli can be found using the above-described procedures from solutions of three problems for representative volumes with stochastic porosity. Here, after finding the effective moduli c_{11}^{eff}, c_{12}^{eff}, β^{eff}, k^{eff} from (17)–(19), we can also calculate the effective moduli E^{eff}, ν^{eff} and α^{eff} in sequence from (16) by formulae

$$E = \frac{(c_{11}-c_{12})(c_{11}+2c_{12})}{(c_{11}+c_{12})}, \quad \nu = \frac{c_{12}}{(c_{11}+c_{12})}, \quad \alpha = \frac{\beta}{(c_{11}+2c_{12})}.$$

We will also model the mortar as an isotropic material. Thus, the periodicity cell of brick masonry will consist of two isotropic materials and macrocavities in bricks. Despite the fact that all the phases of the composite materials are isotropic, the effective characteristics of the composite will have the module structure of anisotropic material, since the geometry of the masonry and arrangement of macrocavities differ in various directions.

As a result, to determine the effective thermoelastic constants of brick masonry all ten boundary value problems should be solved as in the case of the material of an arbitrary anisotropic class (six problems (7)–(9), one problem (7), (8), (11) and the three problems (4), (13)).

3 Computer Design of the Effective Properties of Masonry Made of Porous, Hollow and Porous–Hollow Bricks

For numerical example we will consider a periodic masonry basic structure as a representative volume of masonry wall one brick thick (Fig. 1) that represents the cell of periodicity in horizontal and vertical directions [11, 16]. The basic cell illustrated in Figs. 1 and 2a, contains two half-sized in length brick units, two half-sized in height brick units, and mortar joints between the bricks. Such choice of a representative volume enables to describe the microstructure of masonry in exact way, and some numerical experiments have shown that the extension of periodic masonry structure does not essentially improve the results.

We will investigate the mechanical and thermal properties of masonry wall for different types of bricks with the same total porosity $p = V_p/V$, where V_p is the volume of pores and voids (hollows), $V = abc$, a is the length, b is the width and c is the height of brick.

For hollow brick we will consider eight identical parallelepiped hollows located on the top surface of the brick with the length a_h, width b_h and height c_h, where all these dimensions are related to the corresponding dimensions of brick (a, b and c, respectively). We denote by p_m microporosity of the brick material, $V_h = 8a_hb_hc_h$ is the total volume of hollows.

In the case of porous–hollow brick, in order to obtaine the same porosity p we will select the dimensions a_h, b_h and c_h of hollows so that their total volume V_h can be represented by formula: $V_h = V(p - p_m)/(1 - p_m)$.

We will examine the case of porous brick with only stochastic microporosity (P), the case of brick with hollows made of dense brick material (H), and the case of porous–hollow brick made of porous material with the hollows (PH). Basic data for all considered cases is summarized in Table 1.

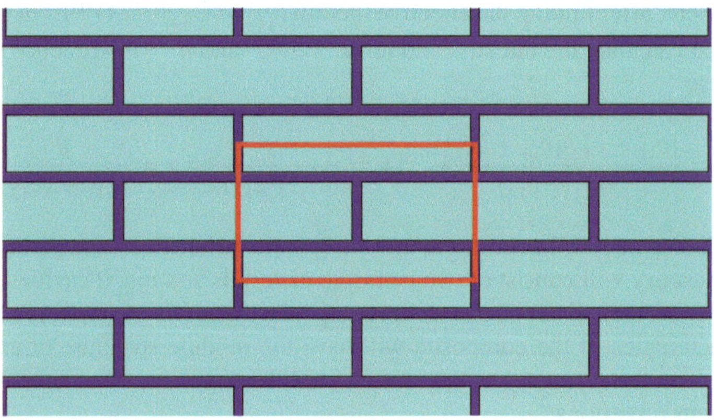

Fig. 1 Brick masonry wall with periodic structure

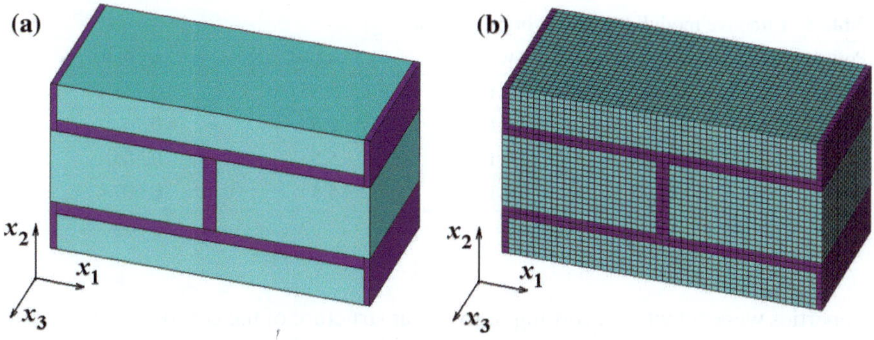

Fig. 2 Case 1*P* of basic cell: solid model (**a**) and finite element model (**b**)

Table 1 Data for different cases of computation basic cell

Case	p	p_m	a_h (m)	b_h (m)	c_h (m)
1*P*	0.1	0.1	–	–	–
1*H*	0.1	–	0.03	0.02	0.041
1*PH*	0.1	0.05	0.03	0.02	0.021
2*P*	0.2	0.2	–	–	–
2*H*	0.2	–	0.042	0.028	0.041
2*PHa*	0.2	0.05	0.038	0.025	0.041
2*PHb*	0.2	0.1	0.03	0.02	0.045
2*PHc*	0.2	0.15	0.03	0.02	0.024

Table 2 Material properties

Material	E (GPa)	ν	α (10^{-6} K^{-1})	k (W m^{-1}K^{-1})
Non-porous brick	16.7	0.15	5.5	0.8
Mortar	1.0	0.15	10.0	1.0

For all these cases, we consider the following standard input data: brick length $a = 0.25$ (m), brick width $b = 0.12$ (m), brick height $c = 0.065$ (m), and thickness of mortar layer $h = 0.01$ (m). We also adopt the moduli for dense brick and mortar materials presented in Table 2.

At the beginning we will calculate the effective moduli of the brick with microporosity of stochastic structure. For this case the representative volume Ω was constructed in ANSYS finite element package in a shape of cube consist from in $n \times n \times n$ small geometrically equal cubes which were 20-node hexahedral finite elements SOLID226 with KEYOPT(1) = 11 [2]. In such two-phase composite the finite elements of the first phase had material properties of the non-porous brick from Table 2, and the pores had negligibly small elastic stiffness moduli and thermal conductivities of air: $k_p = 0.025$ (W m^{-1}K^{-1}). Then the elements with the pore

Table 3 Effective moduli for porous brick materials

Porosity p_m	E (GPa)	ν	α (10^{-6} K^{-1})	k (W m^{-1}K^{-1})
0.05	15.49	0.15	5.5	0.75
0.10	14.313	0.15	5.5	0.7
0.15	13.093	0.15	5.5	0.654
0.20	11.973	0.15	5.5	0.607

properties were selected according to irregular structure of the composite. Stochastic structure was simulated so that for the given value p_m of porosity, $[p_m n^3]$ of the finite elements by random algorithm were modified to pore material properties ([...] is the number integer part). For the obtained representative volume we applied different boundary conditions from (8), (17)–(19), solved the problems (7), (8), (17), (19) and (4), (19), and in the result we have found the effective moduli of the brick material with porous stochastic structure given in Table 3.

Further the porous brick material was consider as a homogeneous isotropic medium with effective properties from Table 3.

For masonry wall from bricks without hollows we mesh the solid model of the representative unit periodic cell (Fig. 2a) with 3-D 20-node hexahedral structural elements SOLID95 for elastic and thermal stress problems or with 20-node hexahedral thermal elements SOLID90 for steady-state thermal problems with the element size of 5 mm. The corresponding finite element model of the representative unit cell is shown in Fig. 2b. The resulting regular finite element mesh consists of 38688 elements and 165779 nodes.

The step of solid model construction is more complicated for masonry wall made of bricks with hollows (Figs. 3 and 4a). Here, the resulting models differ for the case $c_s \geq c$ (Fig. 3a) and for the case $c_s \leq c$ (Fig. 4a), because the periodic masonry structure consists of two half-sized in height brick units. For these complicated periodic masonry structures we create the finite element meshes with 3-D 10-node tetrahedral structural elements SOLID92 for elastic problems and for thermal stress problems or with 10-node tetrahedral thermal elements SOLID87 for steady-state thermal problems with the same element size of 5 mm (Figs. 3 and 4b). For these cases the sizes of resulting tetrahedral finite element meshes depend on the dimensions of hollows. For example, tetrahedral finite element mesh for problem 1H (Fig. 3b) consists of 272809 elements and 387949 nodes, and the finite element mesh for problem 1PH (Fig. 4b) consists of 293847 elements and 413862 nodes.

Note that we choose sufficiently small value of the element size with two elements along mortar thickness. Because we also use the accurate midside node serendipity elements, we obtain high-order finite element system with approximately one million equations for elastic problems. The coarser meshes or less precise finite elements give qualitatively similar data, but of course less accurate. We remind that for each case in order to determine effective moduli we need to solve six boundary elastic problems, one problem of thermal stresses and three heat conduction problems.

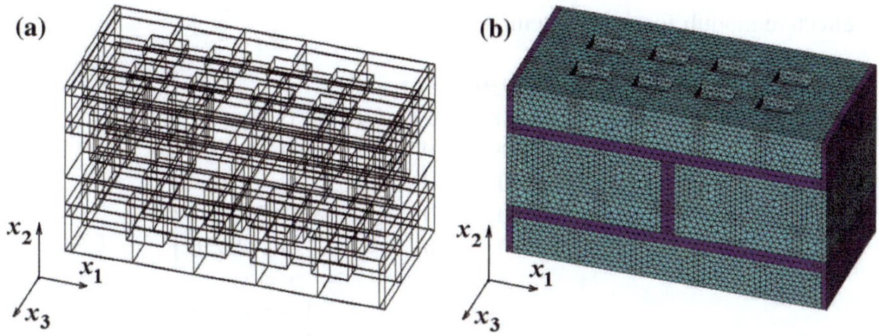

Fig. 3 Case 1*H* of basic cell: solid model in transparent image (**a**) and finite element model (**b**)

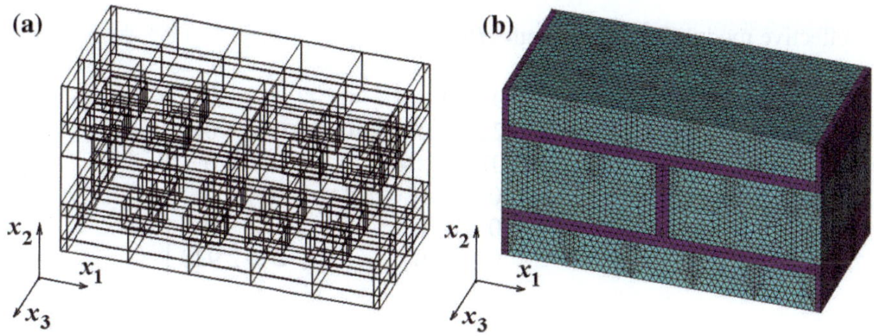

Fig. 4 Case 1*PH* of basic cell: solid model in transparent image (**a**) and finite element model (**b**)

In the results of finite element calculation in ANSYS we obtain the effective elastic stiffness moduli $c_{\gamma\zeta}^{\text{eff},TV}$ (GPa), thermal stress coefficients $\beta_{\zeta}^{\text{eff},TV}$ (MPa/K) and thermal conductivities $k_{il}^{\text{eff},TV}$ (W m^{-1}K^{-1}) for eight cases of bricks in masonry wall from Table 1. We present these results by the following formulas, where superscript "TV" indicate types of the cases (1*P*, ..., 2*PHc*):
— effective moduli for the problem 1*P*

$$\mathbf{c}^{\text{eff},1P} = \begin{bmatrix} 10.775 & 1.020 & 1.769 & 0 & 0 & 0 \\ 1.020 & 6.511 & 1.130 & 0 & 0 & 0 \\ 1.769 & 1.130 & 12.529 & 0 & 0 & 0 \\ 0 & 0 & 0 & 3.357 & 0 & 0 \\ 0 & 0 & 0 & 0 & 4.819 & 0 \\ 0 & 0 & 0 & 0 & 0 & 2.987 \end{bmatrix}, \tag{20}$$

$$\boldsymbol{\beta}^{\text{eff},1P} = \{0.077, 0.052, 0.086, 0, 0, 0\}, \quad \mathbf{k}^{\text{eff},1P} = \begin{bmatrix} 0.748 & 0 & 0 \\ 0 & 0.741 & 0 \\ 0 & 0 & 0.750 \end{bmatrix}, \tag{21}$$

— effective moduli for the problem $1H$

$$\mathbf{c}^{\mathrm{eff},1H} = \begin{bmatrix} 10.498 & 0.959 & 1.511 & 0 & 0 & 0 \\ 0.959 & 6.384 & 0.990 & 0 & 0 & 0 \\ 1.511 & 0.990 & 11.439 & 0 & 0 & 0 \\ 0 & 0 & 0 & 3.163 & 0 & 0 \\ 0 & 0 & 0 & 0 & 4.042 & 0 \\ 0 & 0 & 0 & 0 & 0 & 2.956 \end{bmatrix}, \tag{22}$$

$$\boldsymbol{\beta}^{\mathrm{eff},1H} = \{0.073, 0.049, 0.078, 0, 0, 0\}, \quad \mathbf{k}^{\mathrm{eff},1H} = \begin{bmatrix} 0.739 & 0 & 0 \\ 0 & 0.735 & 0 \\ 0 & 0 & 0.718 \end{bmatrix}, \tag{23}$$

— effective moduli for the problem $1PH$

$$\mathbf{c}^{\mathrm{eff},1PH} = \begin{bmatrix} 10.665 & 0.973 & 1.651 & 0 & 0 & 0 \\ 0.973 & 6.228 & 1.070 & 0 & 0 & 0 \\ 1.651 & 1.070 & 12.048 & 0 & 0 & 0 \\ 0 & 0 & 0 & 3.241 & 0 & 0 \\ 0 & 0 & 0 & 0 & 4.482 & 0 \\ 0 & 0 & 0 & 0 & 0 & 2.928 \end{bmatrix}, \tag{24}$$

$$\boldsymbol{\beta}^{\mathrm{eff},1PH} = \{0.075, 0.049, 0.083, 0, 0, 0\}, \quad \mathbf{k}^{\mathrm{eff},1PH} = \begin{bmatrix} 0.746 & 0 & 0 \\ 0 & 0.729 & 0 \\ 0 & 0 & 0.739 \end{bmatrix}, \tag{25}$$

— effective moduli for the problem $2P$

$$\mathbf{c}^{\mathrm{eff},2P} = \begin{bmatrix} 9.209 & 0.937 & 1.522 & 0 & 0 & 0 \\ 0.937 & 5.899 & 1.025 & 0 & 0 & 0 \\ 1.522 & 1.025 & 10.526 & 0 & 0 & 0 \\ 0 & 0 & 0 & 2.934 & 0 & 0 \\ 0 & 0 & 0 & 0 & 4.071 & 0 \\ 0 & 0 & 0 & 0 & 0 & 2.647 \end{bmatrix}, \tag{26}$$

$$\boldsymbol{\beta}^{\mathrm{eff},2P} = \{0.066, 0.047, 0.073, 0, 0, 0\}, \quad \mathbf{k}^{\mathrm{eff},2P} = \begin{bmatrix} 0.669 & 0 & 0 \\ 0 & 0.657 & 0 \\ 0 & 0 & 0.673 \end{bmatrix}, \tag{27}$$

— effective moduli for the problem $2H$

$$\mathbf{c}^{\text{eff},2H} = \begin{bmatrix} 9.352 & 0.833 & 1.195 & 0 & 0 & 0 \\ 0.833 & 5.553 & 0.790 & 0 & 0 & 0 \\ 1.195 & 0.790 & 9.510 & 0 & 0 & 0 \\ 0 & 0 & 0 & 2.515 & 0 & 0 \\ 0 & 0 & 0 & 0 & 3.302 & 0 \\ 0 & 0 & 0 & 0 & 0 & 2.586 \end{bmatrix}, \quad (28)$$

$$\boldsymbol{\beta}^{\text{eff},2H} = \{0.065, 0.042, 0.065, 0, 0, 0\}, \quad \mathbf{k}^{\text{eff},2H} = \begin{bmatrix} 0.671 & 0 & 0 \\ 0 & 0.628 & 0 \\ 0 & 0 & 0.634 \end{bmatrix}, \quad (29)$$

— effective moduli for the problem $2PHa$

$$\mathbf{c}^{\text{eff},2PHa} = \begin{bmatrix} 9.142 & 0.848 & 1.222 & 0 & 0 & 0 \\ 0.848 & 5.633 & 0.830 & 0 & 0 & 0 \\ 1.222 & 0.790 & 9.469 & 0 & 0 & 0 \\ 0 & 0 & 0 & 2.612 & 0 & 0 \\ 0 & 0 & 0 & 0 & 3.294 & 0 \\ 0 & 0 & 0 & 0 & 0 & 2.588 \end{bmatrix}, \quad (30)$$

$$\boldsymbol{\beta}^{\text{eff},2PHa} = \{0.064, 0.042, 0.065, 0, 0, 0\}, \quad \mathbf{k}^{\text{eff},2PHa} = \begin{bmatrix} 0.662 & 0 & 0 \\ 0 & 0.636 & 0 \\ 0 & 0 & 0.634 \end{bmatrix}, \quad (31)$$

— effective moduli for the problem $2PHb$

$$\mathbf{c}^{\text{eff},2PHb} = \begin{bmatrix} 9.001 & 0.878 & 1.286 & 0 & 0 & 0 \\ 0.878 & 5.823 & 0.888 & 0 & 0 & 0 \\ 1.286 & 0.888 & 9.577 & 0 & 0 & 0 \\ 0 & 0 & 0 & 2.762 & 0 & 0 \\ 0 & 0 & 0 & 0 & 3.371 & 0 \\ 0 & 0 & 0 & 0 & 0 & 2.629 \end{bmatrix}, \quad (32)$$

$$\boldsymbol{\beta}^{\text{eff},2PHb} = \{0.064, 0.045, 0.066, 0, 0, 0\}, \quad \mathbf{k}^{\text{eff},2PHb} = \begin{bmatrix} 0.658 & 0 & 0 \\ 0 & 0.651 & 0 \\ 0 & 0 & 0.639 \end{bmatrix}, \quad (33)$$

— effective moduli for the problem $2PHc$

$$\mathbf{c}^{\mathrm{eff},2PHc} = \begin{bmatrix} 9.080 & 0.887 & 1.402 & 0 & 0 & 0 \\ 0.887 & 5.650 & 0.955 & 0 & 0 & 0 \\ 1.402 & 0.955 & 10.030 & 0 & 0 & 0 \\ 0 & 0 & 0 & 2.810 & 0 & 0 \\ 0 & 0 & 0 & 0 & 3.726 & 0 \\ 0 & 0 & 0 & 0 & 0 & 2.586 \end{bmatrix}, \tag{34}$$

$$\boldsymbol{\beta}^{\mathrm{eff},2PHc} = \{0.065, 0.045, 0.070, 0, 0, 0\}\,,\ \mathbf{k}^{\mathrm{eff},2PHc} = \begin{bmatrix} 0.664 & 0 & 0 \\ 0 & 0.645 & 0 \\ 0 & 0 & 0.658 \end{bmatrix}. \tag{35}$$

4 Discussion and Conclusions

As it can be seen from (20)–(35), the effective elastic stiffnesses, thermal stress coefficients and thermal conductivities of the composite have significantly different values in the directions x_1 (along the wall length), x_2 (along the wall height) and x_3 (along the wall width). The obtained orthotropic class of anisotropy for the mechanical and temperature moduli is due only to the geometric structure of the masonry wall, as all phases of masonry are taken to be isotropic.

As the Young's moduli of brick and mortar differ by one order of magnitude, the brick material will be stiffer in the directions where there is less mortar material. Hence, the overall stiffness of the structure has the largest value in the direction of the masonry wall thickness, i.e. in the axis x_3. In this case under the extension along x_3 in the major part of the composite the directions of tension along the thickness go only along the brick material, which is stiffer. In the case of the extension along the length, i.e. in the axis x_1, the directions of tension cross both materials of brick and mortar. However, the zones filled by the mortar go along the whole basic unit cell, therefore the whole structure remains almost as stiff as in the case of the tension along x_3. The masonry composite is softer along the direction of the height of the masonry wall, i.e. in the axis x_2, because here under the tension along the whole layer the soft mortar is undergoing uniform tension.

The moduli $c_{\gamma\zeta}$ at $\gamma \neq \zeta$, $\gamma, \zeta = 1, 2, 3$ are considerably smaller than the axial tension-compression moduli, because both brick and mortar Poisson's ratios are small for the considered basic unit cell. The shear moduli are smaller than the axial moduli because of the same reason, and also because the shear strains entirely affect the soft layers of mortar for all homogenization shear problems. Note that these trends for the elastic moduli previously have been mentioned for other types of masonry in [16].

The thermal stress coefficients $\beta_{\gamma}^{\mathrm{eff},TV}$ with $\gamma = 1, 2, 3$ are related to each other in the same manner as the stiffness moduli $c_{\gamma\gamma}^{\mathrm{eff},TV}$. The reason for this is that the thermal stress coefficients are proportional to the stiffness moduli from the corresponding matrix row ($\beta_{\gamma} = c_{\gamma\zeta}\alpha_{\zeta}$, where α_{ζ} are the thermal expansion coefficients).

Thermal conductivities $k_{ii}^{\mathrm{eff},TV}$ for each fixed type of brick differ from each other to a lesser extent. Since we accept that the thermal conductivity of the mortar is greater that the thermal conductivity of solid brick, then, depending on the porosity and on the direction in masonry wall, there may be cases when different coefficients $k_{ii}^{\mathrm{eff},TV}$ dominate.

If we compare the results for the laying of bricks of different types, we can make the following conclusions.

In the cases $1P$, $1H$, $1PH$, i.e. for low porosity, elastic stiffness moduli, thermal stress coefficients and thermal conductivities have the largest value in the case of porous brick and have the smallest values in the case of bricks with hollows, because $c_{\gamma\zeta}^{\mathrm{eff},1P} \geq c_{\gamma\zeta}^{\mathrm{eff},1PH} \geq c_{\gamma\zeta}^{\mathrm{eff},1H}$, $\beta_{\zeta}^{\mathrm{eff},1P} \geq \beta_{\zeta}^{\mathrm{eff},1PH} \geq \beta_{\zeta}^{\mathrm{eff},1H}$, $k_{ii}^{\mathrm{eff},1P} \geq k_{ii}^{\mathrm{eff},1PH} \geq k_{ii}^{\mathrm{eff},1H}$, except $k_{22}^{\mathrm{eff},1PH}$.

Thus, among the cases $1P$, $1H$, $1PH$ the masonry made of porous bricks is more stable, but has worse thermal insulation properties.

The moduli of masonry from (20)–(35) decrease with an increase of total porosity. However, for total porosity $p = 0.2$ the comparative values of moduli for the cases $2P$, $2H$, $2PHa$, $2PHb$ and $2PHc$ depend on the size of the hollows in different directions, and we cannot make simple conclusions about all moduli. Nevertheless, these results allow us to draw conclusions about the advantages and disadvantages for each case of the brick structure.

Acknowledgements The first author is grateful to the Russian Scientific Foundation (RSCF) for its support by Project 15-19-10008.

References

1. Addessi, D., Mastrandrea, A., Sacco, E.: An equilibrated macro-element for nonlinear analysis of masonry structures. Eng. Struct. **70**, 82–93 (2014)
2. ANSYS Rel. 11.0. Elements Reference. SAS IP Inc. Canonsburg (2007)
3. Anthoine, A.: Derivation of the in-plane elastic characteristics of masonry through homogenization theory. Int. J. Solids Struct. **32**(2), 137–163 (1995)
4. Anthoine, A.: Homogenization of periodic masonry: plane stress, generalized plane strain or 3D modelling? Commun. Numer. Methods Eng. **13**, 319–326 (1997)
5. Bati, S.B., Ranocchiai, G., Rovero, L.: Suitability of micro- mechnical model for elastic analysis of masonry. J. Eng. Mech. **125**(8), 922–929 (1999)
6. Carbone, V.I., Codegone, M.: Homogenization process of stratified masonry. Math. Comput. Model. **42**, 375–380 (2005)
7. Cecchi, A., Milani, G., Tralli, A.: Out-of-plane loaded CFRP reinforced masonry walls: mechanical characteristics by homogenization procedures. Compos. Sci. Technol. **65**, 1480–1500 (2005)
8. Cecchi, A., Rizzi, N.L.: Heterogeneous elastic solids: a mixed homogenization-rigidification technique. Int. J. Solids Struct. **38**(1), 29–36 (2001)
9. Christensen, R.M.: Mechanics of Composite Materials, pp. 1–348. Krieger Publishing Company (1991)
10. Cluni, F., Gusella, V.: Homogenization of non-periodic masonry structures. Int. J. Solids Struct. **41**, 1911–1923 (2004)

11. Kumar, N., Lambadi, H., Pandey, M., Rajagopal, A.: Homogenization of periodic masonry using self-consistent scheme and finite element method. Int. J. Comput. Methods Eng. Sci. Mech. **17**(1), 7–21 (2016)
12. Nasedkin, A.V., Nasedkina, A.A.: Finite element modeling and computer design of porous composites. In: Hellmich, C., Pichler, B., Adam, D. (eds.) Poromechanics V. Proceedings of the Fifth Biot Conference on Poromechanics, July 10–12, 2013, pp. 608–617. Publ. ASCE, Vienna, Austria (2013)
13. Nasedkin, A.V., Nasedkina, A.A.: Finite element modeling of coupled problems: textbook, pp. 1–174. Southern Federal University Publ., Rostov-on-Don (2015)
14. Nasedkin, A.V., Nasedkina, A.A., Rajagopal, A., Remizov, V.V.: Some finite element approaches for modeling of anisotropic thermoelastic mixture and periodic composites with internal microstructure. In: Pelekasis, N., Stavroulakis, G. (eds.) Proceedings VIII GRACM International Congress Computational Mechanics, 12–15 July 2015, pp. 1–10. University Thessaly Publ, Volos, Greece (2015)
15. Nasedkin, A.V., Nasedkina, A.A., Remizov, V.V.: Finite element modeling of porous thermoelastic composites with account for their microstructure. Vycisl. meh. splos. sred—Comput. Contin. Mech. **7**(1), 100–109 (2014)
16. Nasedkina, A.A., Rajagopal, A.: Finite element homogenization of periodic block masonry by the effective moduli method. In: Parinov, I.A., Chang, S.-H., Jani, M.A. (eds.) Advanced Materials—Techniques, Physics, Mechanics and Applications, Series Springer Proceedings in Physics. Springer, Heidelberg, New York, Dordrecht, London (2017) [submitted for publication]
17. Pande, G.N., Liang, J.X., Middleton, J.: Equivalent elastic moduli for brick masonry. Comput. Geotech. **8**(3), 243–265 (1989)
18. Pietruszczak, S., Niu, X.: A mathematical description of macroscopic behaviour of brick masonry. Int. J. Solids Struct. **29**(5), 531–546 (1992)
19. Quinteros, R.D., Oller, S., Nallim, L.G.: Nonlinear homogenization techniques to solve masonry structures problems. Compos. Struct. **94**(2), 724–730 (2012)
20. Sacco, E.: A nonlinear homogenization procedure for periodic masonry. Eur. J. Mech. A Solids. **28**(2), 209–222 (2009)
21. Shieh-Beygi, B., Pietruszczak, S.: Numerical analysis of structural masonry: mesoscale approach. Comput. Struct. **86**, 1958–1973 (2008)
22. Sýkora, J.: Krejčí T., Kruis J., Šejnoha M. Computational homogenization of non-stationary transport processes in masonry structures. J. Comput. Appl. Math. **236**, 4745–4755 (2012)
23. Wang, G., Li, S., Nguyen, H., Sitar, N.: Effective elastic stiffness for periodic masonry structures via eigenstrain homogenization. J. Mater. Civ. Eng. **19**(3), 269–277 (2007)
24. Wu, C., Hao, H.: Derivation of 3D masonry properties using numerical homogenization technique. Int. J. Numer. Methods Eng. **66**(11), 1717–1737 (2006)
25. Zucchini, A., Lourenço, P.B.: A micro-mechanical model for the homogenisation of masonry. Int. J. Solids Struct. **39**(12), 3233–3255 (2002)
26. Zucchini, A., Lourenço, P.B.: Mechanics of masonry in compression: results from a homogenisation approach. Comput. Struct. **85**, 193–204 (2007)
27. Zucchini, A., Lourenço, P.B.: A micro-mechanical homogenisation model for masonry: application to shear walls. Int. J. Solids Struct. **46**, 871–886 (2009)

Spectral Properties of Nanodimensional Piezoelectric Bodies with Voids and Surface Effects

G. Iovane and A.V. Nasedkin

Abstract This chapter considers the eigenvalue problems for nanodimensional piezoelectric bodies with voids and with account for uncoupled mechanical and electric surface effect. The piezoelectric body is examined in frictionless contact with massive rigid plane punches and covered by the system of open-circuited and short-circuited electrodes. The linear theory of piezoelectric materials with voids for porosity change properties according to Cowin-Nunziato model is used. For modelling the nanodimensional effects the theory of uncoupled surface stresses and dielectric films is applied. The weak statements for considered eigenvalue problem are given in the extended and reduced forms. By using methods of functional analysis, the discreteness of the spectrum, completeness of the eigenfunctions and orthogonality relations are proved. A minimax principle for natural frequencies is constructed which has the properties of minimality, similar to the well-known minimax principle for problems with pure elastic media. As a consequence of the general principles, the properties of an increase or a decrease in the natural frequencies, when the mechanical, electric and "porous" boundary conditions and the moduli of piezoelectric body with voids change, are established. All of these results have been determined for both problems with and without account for surface effects.

Keywords Piezoelectricity · Nanomechanics · Surface effect · Porous material · Voids · Eigenfrequency

---

G. Iovane (✉)
Department of Computer Science, University of Salerno, 84084 Fisciano (SA), Italy
e-mail: giovane@unisa.it

A.V. Nasedkin
Institute of Mathematics, Mechanics and Computer Science, Southern Federal University, Milchakova Street 8a, Rostov-on-Don 344090, Russia
e-mail: nasedkin@math.sfedu.ru

M.A. Sumbatyan (ed.), *Wave Dynamics and Composite Mechanics for Microstructured Materials and Metamaterials*,
Advanced Structured Materials 59, DOI 10.1007/978-981-10-3797-9_13

1 Introduction

As it is well known, a several nanomaterials have anomalous physical properties the differ considerably from usual analogous macromaterials. One of the factors that are responsible of the behavior of nanomaterials can be the surface effects. Numerous recent investigations (see reviews [23, 24]) show, that for nanosize bodies the surface stresses play a considerable role and affect to the change of mechanical stiffness. Lately, in several works the theory of elasticity with surface stresses was extended for piezoelectric solids, see [7, 9–11, 13, 20, 21, 25] among others. Following these paper, similar to the elastic bodies, when analyzing the piezoelectric nanosize media one can introduce uncoupled surface stresses and distributed electric charges (dielectric films). This approach is used in the present work for investigation of the natural oscillations of the piezoelectric nanosize bodies with voids.

The present paper is performed in the context of the linear theory of piezoelectric materials with surface effects and with voids in accordance with porosity change properties in the Cowin-Nunziato's theory [4–6, 14–17]. Here, we restrict our attention to eigenvalue problems for piezoelectric media with surface effects and voids. The investigated problem also becomes complicated if we consider non-classical contact mechanical boundary conditions with rigid stamps (punches) and electric boundary conditions for the system of open-circuited and short-circuited electrodes. Note that the boundary conditions for covered electrodes surfaces are required for modelling the real piezoelectric devices. The similar statments for the correspondent electroelasticity problem has been considered in [2], but only for pure piezoelectric body without taking into account surface effects and mass characteristics of punches, in [14] for elastic body with voids without taking into account surface effects and the piezoelectric properties of material, in [15] for piezoelectric body with voids without taking into account surface effects, and also in [8, 19] for piezoelectric body with surface effects, but without taking into account porosity change properties.

After having formulated the usual statement of eigenvalue problem for piezoelectric media with voids and surface effects, we obtain the weak or generalized settings of these problems. For the displacement function and for electric potential function by analogy with [2, 14, 15] we introduce special spaces of displacement functions conformable to possible motions of rigid stamps and special spaces of electric potential functions satisfying homogeneous boundaries conditions for the system of open-circuited and short-circuited electrodes, but with modifications related to tacking into account the uncoupled mechanical and electric surface effects [8, 19]. We also introduce the usual Sobolev space of porosity change function [14, 15].

Reducing the porosity function and the electric potential function from quasi-static equations we get a generalized statement of eigenvalue problem only for the functions of mechanical displacements. The properties of this problem turn out to be analogous to the properties of the problem for purely elastic medium. This allows to easily determine its spectral properties and minimax principle for natural frequencies of the piezoelectric body with voids and surface effects.

Then, following the concept in [2, 8, 14, 15, 19], we investigate the properties of eigenvalues changes depending on the modifications of the surface effects, the material properties or the types of mechanical, electric and "porous" boundary conditions.

2 Classical Formulation of Eigenvalue Problems for Piezoelectric Media with Voids and Surface Effects

Let Ω be a region occupied by the piezoelectric material with voids; $\Gamma = \partial\Omega$ is the boundary of the region, $\mathbf{n}$ is the vector of the external unit normal to Γ. We will assume that the region Ω and its boundary Γ are subjected to the following conditions: Ω is the sum of a finite number of sets, star-shaped with respect to any spheres contained in them, while Γ is a Lipschitz boundary of class C^1.

We consider linear piezoelectric material with voids where the porosity is simulated under Cowin-Nunziato model with memory effect for the intrinsic equilibrated body force [4, 6]. Confining ourselves to investigate the harmonic vibration $\exp(j\omega t)$ with a circular frequency ω, we will only use the amplitude values for all physical quantities without additional conditions. Let $\mathbf{u} = \mathbf{u}(\mathbf{x})$ denote the vector of mechanical displacements; $\varphi = \varphi(\mathbf{x})$ is the electric potential; $\psi = \psi(\mathbf{x})$ is the porosity change function (the change in the volume fraction from the reference configuration); $(\mathbf{x}) = \{x_1, x_2, x_3\} = \{x, y, z\}$. By the first two functions one can determine the second-order strain tensor $\boldsymbol{\varepsilon} = \boldsymbol{\varepsilon}(\mathbf{u})$ and the electric field vector $\mathbf{E} = \mathbf{E}(\varphi)$

$$\boldsymbol{\varepsilon} = (\nabla\mathbf{u} + (\nabla\mathbf{u})^T)/2\,, \quad \mathbf{E} = -\nabla\varphi\,, \tag{1}$$

where by $(...)^T$ we define the transpose operation.

In the context of the linear theory, we assume the constitutive equations for piezoelectric bodies similarly to [5] with missing some terms [15]

$$\boldsymbol{\sigma} = \mathbf{c} : \boldsymbol{\varepsilon} - \mathbf{e}^T \cdot \mathbf{E} + \mathbf{B}\psi\,, \tag{2}$$

$$\mathbf{D} = \mathbf{e} : \boldsymbol{\varepsilon} + \boldsymbol{\kappa} \cdot \mathbf{E} - \mathbf{g}\psi - \mathbf{G} \cdot \nabla\psi\,, \tag{3}$$

$$\mathbf{h} = \mathbf{A}\nabla\psi + \mathbf{G} \cdot \mathbf{E}\,, \quad g = -\mathbf{B} : \boldsymbol{\varepsilon} - \mathbf{g} \cdot \mathbf{E} - \xi\psi\,, \tag{4}$$

where $\boldsymbol{\sigma}$ is the second-order stress tensor; $\mathbf{D}$ is the electric displacement vector; $\mathbf{h}$ is the equilibrated stress vector; g is the intrinsic equilibrated body force; $\mathbf{c} = \mathbf{c}^E$ is the forth-order tensor of elastic modules, measured at constant electric field; $\mathbf{e}$ is the third-order tensor of piezomoduli; $\boldsymbol{\kappa} = \boldsymbol{\kappa}^S = \boldsymbol{\varepsilon}^S$ is the second-order tensor of dielectric permittivities, measured at constant strain; $\mathbf{A}$, $\mathbf{B}$, $\mathbf{g}$, $\mathbf{G}$ and ξ are the constitutive constants characterizing the properties of porosity changes ($\mathbf{A}$, $\mathbf{B}$, $\mathbf{G}$ are the second-order tensors, and $\mathbf{g}$ is the vector; and the symbol ":" denotes the double scalar product operation between two tensors). Note that in comparison with [5] we use some different definitions and neglect dependencies from the temperature. We also omitted some terms so that when considering materials without piezoelectric connectivity we could get the relationships adopted in [4] for the case of pure elastic materials with voids.

In the case of homogeneous harmonic problem for piezoelectric medium with voids without damping and inertia effect, caused from porosity, we have the following system of field equations

$$-\nabla\cdot\boldsymbol{\sigma} = \rho\,\omega^2\mathbf{u}\,, \tag{5}$$

$$\nabla\cdot\mathbf{D} = 0\,, \tag{6}$$

$$\nabla\cdot\mathbf{h} + g = 0\,, \tag{7}$$

where ρ is the density of material.

We suppose that the density function $\rho(\mathbf{x})$ is piecewise-continuous and $\exists\rho_0 > 0$: $\rho(\mathbf{x}) \geq \rho_0$. The material moduli of the medium in (2)–(4) are piecewise-continuous together with their first derivatives by $\mathbf{x}$ with usual symmetry conditions ($c_{ijkl} = c_{jikl} = c_{klij}$, $e_{ikl} = e_{ilk}$, $\kappa_{kl} = \kappa_{lk}$, $A_{kl} = A_{lk}$, $B_{kl} = B_{lk}$, $G_{kl} = G_{lk}$), and for the positive definiteness of intrinsic volumetric energy $\Pi_\Omega(\boldsymbol{\varepsilon}, \mathbf{E}, \psi, \mathbf{b})$ the following inequality should satisfy $\forall\,\boldsymbol{\varepsilon} = \boldsymbol{\varepsilon}^T$, $\mathbf{E}$

$$\exists\, c_\Omega > 0 : \quad \Pi_\Omega(\boldsymbol{\varepsilon}, \mathbf{E}, \psi, \mathbf{b}) \geq c_\Omega(\boldsymbol{\varepsilon}^T : \boldsymbol{\varepsilon} + \psi^2 + \mathbf{b}^T\cdot\mathbf{b} + \mathbf{E}^T\cdot\mathbf{E})\,, \tag{8}$$

where

$$\Pi_\Omega(\boldsymbol{\varepsilon}, \mathbf{E}, \psi, \mathbf{b}) = \frac{1}{2}(\boldsymbol{\varepsilon}^T : \mathbf{c} : \boldsymbol{\varepsilon} + \xi\psi^2 + 2\mathbf{B}^T : \boldsymbol{\varepsilon}\psi + \mathbf{b}^T\cdot\mathbf{A}\cdot\mathbf{b} + \mathbf{E}^T\cdot\boldsymbol{\kappa}\cdot\mathbf{E})\,. \tag{9}$$

The formulas (1)–(7) give the coupled system of equations for piezoelectric body with voids relatively to the components of the vector of mechanical displacements $\mathbf{u}$, the functions of electric potential φ and the porosity change function ψ. For eigenvalue problem we must add to the system (1)–(7) the homogeneous boundary conditions. The boundary conditions are of three types: mechanical, electric and "porous".

To formulate the mechanical boundary conditions we will assume that the boundary Γ can be split into two subsets: Γ_σ and Γ_u ($\Gamma = \Gamma_\sigma \cup \Gamma_u$). We suggest that at the boundary Γ_σ there are surface stresses $\boldsymbol{\tau}^s$ in accordance with the Gurtin – Murdoch model [12], i.e.

$$\mathbf{n}\cdot\boldsymbol{\sigma} = \nabla^s\cdot\boldsymbol{\tau}^s\,, \quad \mathbf{x}\in\Gamma_\sigma\,, \tag{10}$$

where ∇^s is the surface gradient operator, associated with spatial nabla-operator by the formula $\nabla^s = \nabla - \mathbf{n}(\partial/\partial r)$, where r is the coordinate, measured by the normal Γ_σ; $\boldsymbol{\tau}^s$ is the second-order tensor of surface stresses.

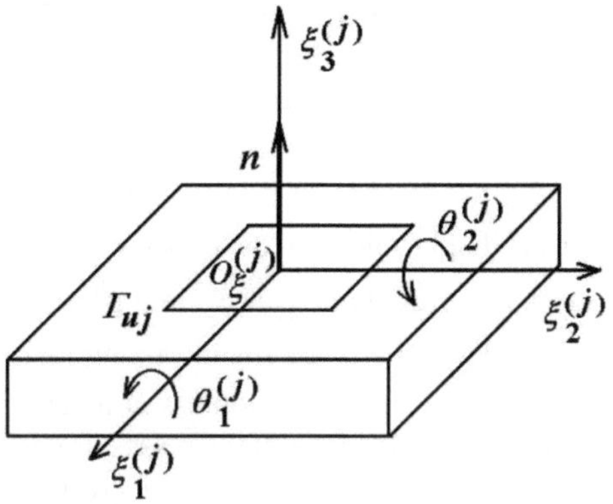

Fig. 1 Contact boundary with rigid plane punch

As for purely elastic body, when taking into account the surface stresses we adopt that the surface stresses $\boldsymbol{\tau}^s$ are jointed only to the surface strains $\boldsymbol{\varepsilon}^s$ by the uncoupled mechanical constitutive relations

$$\boldsymbol{\tau}^s = \mathbf{c}^s : \boldsymbol{\varepsilon}^s, \quad \boldsymbol{\varepsilon}^s = (\nabla^s \mathbf{u}^s + (\nabla^s \mathbf{u}^s)^T)/2, \quad \mathbf{u}^s = \mathbf{A} \cdot \mathbf{u}, \tag{11}$$

where $\mathbf{c}^s$ is the forth-order tensor of surface elastic moduli; $\mathbf{A} = \mathbf{I} - \mathbf{n} \otimes \mathbf{n}$, $\mathbf{I}$ is the unit tensor in R^3.

The properties of the tensor of surface elastic moduli $\mathbf{c}^s$ are analogous to the corresponding properties of the tensor $\mathbf{c}$, i.e. $c^s_{ijkl} = c^s_{jikl} = c^s_{klij}$, and for all $\boldsymbol{\varepsilon}^s$ with structure (11)

$$\exists\, c_\Gamma > 0 \ : \qquad \Pi_{\Gamma m}(\boldsymbol{\varepsilon}^s) = \frac{1}{2}\boldsymbol{\varepsilon}^{sT} : \mathbf{c}^s : \boldsymbol{\varepsilon}^s \geq c_\Gamma \boldsymbol{\varepsilon}^{sT} : \boldsymbol{\varepsilon}^s, \tag{12}$$

that follow from the condition of the positive definiteness of the surface energy density $\Pi_{\Gamma m}(\boldsymbol{\varepsilon}^s)$.

Suppose that $\Gamma_u = \cup \Gamma_{uj}$; $j = 0, 1, ..., L$; $\Gamma_{u0} \neq \emptyset$, Γ_{uj} do not border one another, while among Γ_{uj} there are $L + 1 - l$ rigidly clamped surfaces ($j \in J_r = \{0, l+1, l+2, ..., L\}$) and l plane regions ($j \in J_p = \{1, 2, ..., l\}$), in contact with rigid massive punches. We will connect with region Γ_{uj}, $j \in J_p$ the local coordinate system $O_\xi^{(j)}\xi_1^{(j)}\xi_2^{(j)}\xi_3^{(j)}$ so that the axis $\xi_3^{(j)}$ coincides in direction with direction of external normal $\mathbf{n}$ in the point $O_\xi^{(j)}$ and the axes $\xi_1^{(j)}$ and $\xi_2^{(j)}$ will be the main axes of inertia for the punch j (see Fig. 1). Then, we can assume the following boundary conditions on Γ_{uj}

$$\mathbf{n}^T \cdot \mathbf{u} = \sum_{k=0}^{2} \alpha^u_{jk} \xi_k^{(j)}, \quad (\xi_0^{(j)} = 1), \quad \mathbf{x} \in \Gamma_{uj}, \quad j \in J_p, \tag{13}$$

$$\int_{\Gamma_{uj}} \xi_p^{(j)} \mathbf{n}^T \cdot \boldsymbol{\sigma} \cdot \mathbf{n}\, d\Gamma = \omega^2 \alpha^u_{jp} M_p^{(j)}, \quad p = 0, 1, 2, \quad j \in J_p, \tag{14}$$

$$\mathbf{n}\cdot\boldsymbol{\sigma} - (\mathbf{n}^T\cdot\boldsymbol{\sigma}\cdot\mathbf{n})\mathbf{n} = \nabla^s\cdot\boldsymbol{\tau}^s, \quad \mathbf{x}\in\Gamma_{uj}, \quad j\in J_p, \tag{15}$$

$$\mathbf{u} = 0, \quad \mathbf{x}\in\Gamma_{uj}, \quad j\in J_r, \quad \Gamma_{u0}\neq\emptyset, \tag{16}$$

where in (13), (14) the summation by repeating index j and p is missing; α^u_{j0} is the normal displacement of the punch with number j; $\alpha^{ju}_{j1} = -\theta^{(j)}_2$, $\alpha^u_{j2} = \theta^{(j)}_1$ are the punch rotation angles about axes $\xi^{(j)}_2$ and $\xi^{(j)}_1$ respectively; $M^{(j)}_0$ is the mass of punch; $M^{(j)}_1 = J_{\xi^{(j)}_2\xi^{(j)}_2}$, $M^{(j)}_2 = J_{\xi^{(j)}_1\xi^{(j)}_1}$ are the inertia moments of punch.

Note that, in the frictionless contact of the nanodimensional body with a rigid plane punch, the tangential stresses are not zero but, according to (15), are counterbalanced by the action of surface stresses, just as in (10).

To set the electric boundary conditions we assume that the surface Γ is also divided in two subsets: Γ_D and Γ_φ $(\Gamma = \Gamma_D\cup\Gamma_\varphi)$.

The region Γ_D does not contain electrodes, and the following uncoupled electrical constitutive conditions are satisfied on them:

$$\mathbf{n}^T\cdot\mathbf{D} = \nabla^s\cdot\mathbf{d}^s, \qquad \mathbf{x}\in\Gamma_D. \tag{17}$$

where $\mathbf{d}^s = \mathbf{A}\cdot\boldsymbol{\kappa}^s\cdot\mathbf{A}\cdot\mathbf{E}^s$, $\mathbf{E}^s = -\nabla^s\varphi$, $\boldsymbol{\kappa}^s$ is the second-order dielectric permittivity tensor that is symmetrical positive definite relatively to the vectors $\mathbf{E}^s$.

The subset Γ_φ itself is subdivided into $M+1$ subdomains $\Gamma_{\varphi j}$ ($j\in J_Q\cup J_V$), $J_Q = \{1, 2, ..., m\}$, $J_V = \{0, m+1, ..., M\}$), which are not adjacent to one another, coated by infinitely thin electrodes. We will specify the following boundary conditions on these areas

$$\varphi = \Phi_j, \quad \mathbf{x}\in\Gamma_{\varphi j}, \quad j\in J_Q, \quad \Phi_j = \text{const}, \tag{18}$$

$$\int_{\Gamma_{\varphi j}}\mathbf{n}^T\cdot\mathbf{D}\,d\Gamma = 0, \quad \mathbf{x}\in\Gamma_{\varphi j}, \quad j\in J_Q, \tag{19}$$

$$\varphi = 0, \qquad \mathbf{x}\in\Gamma_{\varphi j}, \quad j\in J_V, \quad \Gamma_{j0}\neq\emptyset. \tag{20}$$

By (18), (19) there are m open-circuited electrodes on which the electric potential Φ_j are initially unknown, but the overall electric charges on each electrode are equal to zero. The remaining $M+1-m$ electrodes are assumed to be short-circuited with zero value of the potential. The cases $m = 0$ and $m = M$ are not ignored. In the case of $m = 0$ all the electrodes are short-circuit. With $m = M$ the electric potential appears to be defined up to an arbitrary constant, and without loss of generality for the uniqueness of the solution we can adopt the condition (20) for one of the electrodes, for example, for Γ_{j0}.

For clearness, a variant of mechanical boundary conditions (13)–(16) for $L = 2$, $l = 1$ and a variant of electric boundary conditions (18)–(20) for $M = 3$, $m = 2$ are shown on Fig. 2, (*a*) and (*b*), respectively.

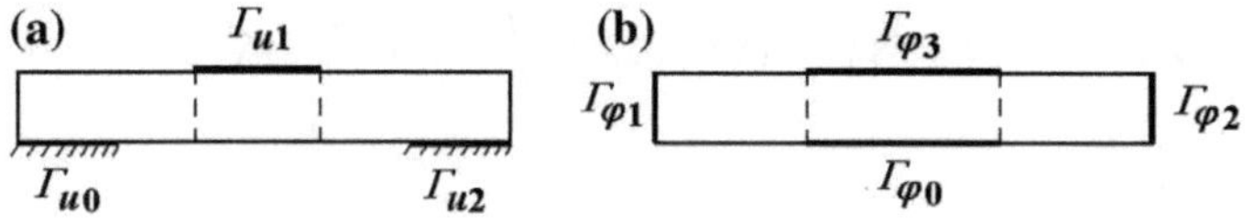

Fig. 2 Examples of boundary conditions: mechanical boundary conditions for $L = 2$, $l = 1$ (*a*); electric boundary conditions for $M = 3$, $m = 2$ (*b*)

To specify the "porous" boundary condition we suppose that the boundary Γ is also split into two parts: $\Gamma_{\psi 0}$ and $\Gamma_{\psi n}$ $(\Gamma = \Gamma_{\psi 0} \cup \Gamma_{\psi n})$. We take the principal condition (Dirichlet condition) on the surface $\Gamma_{\psi 0}$

$$\psi = 0, \quad \mathbf{x} \in \Gamma_{\psi 0}, \tag{21}$$

and we accept the Neumann condition on $\Gamma_{\psi n}$

$$\mathbf{n}^T \cdot \mathbf{h} = 0, \quad \mathbf{x} \in \Gamma_{\psi n}. \tag{22}$$

Note that in real situations the boundary condition (21) is usually absent.

Here we assume that all the areas Γ_σ, Γ_{uj}, Γ_D, $\Gamma_{\varphi j}$, $\Gamma_{\psi 0}$ and $\Gamma_{\psi n}$ have the Lipschitz boundaries of the class C^1.

Problem (1)–(22) is the eigenvalue problem of natural oscillations of piezoelectric body with voids and with surface effects and this problem consists in the determination of natural frequencies ω (or eigenvalues $\lambda = \omega^2$) and corresponding eigenfunctions $\mathbf{u}$, φ, ψ, which give non-trivial solutions of the homogeneous boundary problem.

The spectral properties of this problem will be set using the approaches applied in [1, 2, 8, 14, 15, 19, 20].

3 Generalized Problem Formulations

We transfer from the formulations (1)–(22) of the eigenvalue problems for piezoelectric bodies with voids and surface effects to their generalized or weak settings.

Previously we will introduce the space of functions φ, ψ and of vector functions $\mathbf{u}$, defined on Ω, which we will need later.

We denote by H^0_ρ the space of vector functions $\mathbf{u} \in L_2$ with scalar product

$$(\mathbf{v}, \mathbf{u})_{\mathrm{H}^0_\rho} = \int_\Omega \rho \mathbf{v}^T \cdot \mathbf{u} \, d\Omega. \tag{23}$$

On the set of vector functions $\mathbf{u} \in C^1$ which satisfy (16) on Γ_{uj}, $j \in J_r$, and (13) for arbitrary α^u_{jk} on Γ_{uj}, $j \in J_p$, we introduce the scalar product

$$(\mathbf{v},\mathbf{u})_{\mathrm{H}^1_{ul}} = \int_\Omega (\nabla \mathbf{v})^T : (\nabla \mathbf{u})\, d\Omega + \int_{\Gamma_\tau} (\nabla^s \mathbf{v}^s)^T : (\nabla^s \mathbf{u}^s)\, d\Gamma\,, \tag{24}$$

where $\Gamma_\tau = \Gamma_\sigma \cup (\cup_{j\in J_p} \Gamma_{uj})$, $\mathbf{v}^s = \mathbf{A}\cdot\mathbf{v}$.

The closure of this set of vector functions $\mathbf{u}$ in the norm generated by scalar product (24) will be denoted by H^1_{ul}, where l is the number of rigid punches.

On the set of functions $\varphi \in C^1$ which satisfy (20) on $\Gamma_{\varphi j}$, $j \in J_V$, and (18) for arbitrary $\boldsymbol{\Phi}_j$ on $\Gamma_{\varphi j}$, $j \in J_Q$, we will introduce the scalar product

$$(\chi,\varphi)_{\mathrm{H}^1_{\varphi m}} = \int_\Omega (\nabla \chi)^T \cdot \nabla\varphi\, d\Omega + \int_{\Gamma_D} (\nabla^s \chi)^T \cdot \nabla^s \varphi\, d\Gamma\,. \tag{25}$$

The closure of this set of functions φ in the norm generated by scalar product (25) will be denoted by $\mathrm{H}^1_{\varphi m}$, where m is the number of open-circuite electrodes.

We will introduce the scalar product on the set of functions $\psi \in C^1$ which satisfy (21)

$$(\eta,\psi)_{\mathrm{H}^1_\psi} = \int_\Omega (\nabla \eta)^T \cdot \nabla\psi\, d\Omega\,. \tag{26}$$

The closure of this set in the norm generated by scalar product (26) will be denoted by H^1_ψ.

In order to formulate the generalized or weak solution of eigenvalue problem for piezoelectric media with voids and surface effects we multiply equations (5) by arbitrary vector function $\mathbf{v} \in \mathrm{H}^1_u$, equation (6) by some function $\chi \in \mathrm{H}^1_{\varphi m}$, and Eq. (7) by function $\eta \in \mathrm{H}^1_\psi$. By integrating the obtained equations on Ω and by using the standard technique of the integration by parts, with Eqs. (1)–(4) and formulated boundary conditions, we obtain the following integral relations

$$c(\mathbf{v},\mathbf{u}) + e(\varphi,\mathbf{v}) + b(\mathbf{v},\psi) = \omega^2 \tilde{\rho}(\mathbf{v},\mathbf{u})\,, \tag{27}$$

$$-\, e(\chi,\mathbf{u}) + \kappa(\chi,\varphi) + g(\chi,\psi) = 0\,, \tag{28}$$

$$b(\mathbf{u},\eta) - g(\varphi,\eta) + a(\eta,\psi) = 0\,, \tag{29}$$

where

$$\tilde{\rho}(\mathbf{v},\mathbf{u}) = \rho(\mathbf{v},\mathbf{u}) + \sum_{j=1}^{l}\sum_{k=0}^{2} \alpha^v_{jk}\alpha^u_{jk} M^{(j)}_k\,, \quad \rho(\mathbf{v},\mathbf{u}) = (\mathbf{v},\mathbf{u})_{\mathrm{H}^0_\rho}\,, \tag{30}$$

$$c(\mathbf{v},\mathbf{u}) = c_\Omega(\mathbf{v},\mathbf{u}) + c_\Gamma(\mathbf{v},\mathbf{u})\,, \quad \kappa(\chi,\varphi) = \kappa_\Omega(\chi,\varphi) + \kappa_\Gamma(\chi,\varphi)\,, \tag{31}$$

$$c_\Omega(\mathbf{v},\mathbf{u}) = \int_\Omega \boldsymbol{\varepsilon}(\mathbf{v}) : \mathbf{c} : \boldsymbol{\varepsilon}(\mathbf{u})\, d\Omega, \quad c_\Gamma(\mathbf{v},\mathbf{u}) = \int_{\Gamma_\tau} \boldsymbol{\varepsilon}^s(\mathbf{v}) : \mathbf{c}^s : \boldsymbol{\varepsilon}^s(\mathbf{u})\, d\Gamma\,, \tag{32}$$

$$e(\varphi, \mathbf{v}) = \int_{\Omega} \mathbf{E}(\varphi) \cdot \mathbf{e} : \boldsymbol{\varepsilon}(\mathbf{v})\, d\Omega\,, \quad b(\mathbf{v}, \psi) = \int_{\Omega} \mathbf{B} : \boldsymbol{\varepsilon}(\mathbf{v})\, \psi\, d\Omega\,, \tag{33}$$

$$\kappa_{\Omega}(\chi, \varphi) = \int_{\Omega} \mathbf{E}(\chi) \cdot \boldsymbol{\kappa} \cdot \mathbf{E}(\varphi)\, d\Omega\,, \quad \kappa_{\Gamma}(\chi, \varphi) = \int_{\Gamma_D} \mathbf{E}^s(\chi) \cdot \boldsymbol{\kappa}^s \cdot \mathbf{E}^s(\varphi)\, d\Gamma\,, \tag{34}$$

$$g(\chi, \psi) = \int_{\Omega} (\nabla \chi \cdot \mathbf{G} \cdot \nabla \psi + \mathbf{g}^T \cdot \nabla \chi\, \psi)\, d\Omega\,, \tag{35}$$

$$a(\eta, \psi) = \int_{\Omega} (\nabla \eta \cdot \mathbf{A} \cdot \nabla \psi + \xi \eta \psi)\, d\Omega\,. \tag{36}$$

As it can be easily noted, an account for surface effects for piezoelectric bodies in relations (27)–(36) is reduced to adding the forms $c_{\Gamma}(\mathbf{v}, \mathbf{u})$ and $\kappa_{\Gamma}(\chi, \varphi)$. Therefore we can use the approaches from [2, 8, 14, 15] for conventional eigenvalue problems for elastic and piezoelectric bodies only with voids or only with surface effects.

By virtue of the properties assumed earlier, the forms $\tilde{\rho}(\mathbf{v}, \mathbf{u})$, $c(\mathbf{v}, \mathbf{u})$, $\kappa(\chi, \varphi)$ and $a(\eta, \psi)$ are symmetrical, bilinear and positive defined in L_2, H^1_{ul}, $\mathrm{H}^1_{\varphi m}$ and H^1_{ψ} respectively, while $e(\varphi, \mathbf{v})$, $b(\mathbf{v}, \psi)$ and $g(\chi, \psi)$ are bilinear forms.

We will transform the system (27)–(29) by consecutive elimination of the functions ψ and φ.

Since for fixed $\mathbf{u} \in \mathrm{H}^1_{ul}$, $\varphi \in \mathrm{H}^1_{\varphi m}$ and $\psi \in \mathrm{H}^1_{\psi}$ the forms $b(\mathbf{u}, \eta)$, $g(\varphi, \eta)$ and $a(\eta, \psi)$ are linear-bounded functionals in H^1_{ψ}, by Riesz' theorem [22] the elements $b\mathbf{u}$, $g\varphi$, $a\psi \in \mathrm{H}^1_{\psi}$ exist and are unique so for that for $\eta \in \mathrm{H}^1_{\psi}$

$$b(\mathbf{u}, \eta) = (b\mathbf{u}, \eta)_{\mathrm{H}^1_{\psi}}\,, \quad g(\varphi, \eta) = (g\varphi, \eta)_{\mathrm{H}^1_{\psi}}\,, \quad a(\eta, \psi) = (\eta, a\psi)_{\mathrm{H}^1_{\psi}}\,. \tag{37}$$

It is obvious that $b\mathbf{u}$, $g\varphi$ and $a\psi$ are linear operators acting from H^1_{ul} into H^1_{ψ}, from $\mathrm{H}^1_{\varphi m}$ into H^1_{ψ} and from H^1_{ψ} into H^1_{ψ}, respectively, and an inverse exists for the operator $a\psi$. Then it follows from (29) and (37) that

$$a\psi = -b\,\mathbf{u} + g\varphi, \quad \psi = -B\,\mathbf{u} + G\varphi, \quad B = a^{-1}b, \quad G = a^{-1}g\,, \tag{38}$$

where the operators B and G act from H^1_{ul} into H^1_{ψ} and from $\mathrm{H}^1_{\varphi m}$ into H^1_{ψ}, respectively, and are linear and bounded.

Using (37), (38) we can reduce system (27)–(29) to the form

$$\tilde{c}(\mathbf{v}, \mathbf{u}) + \tilde{e}(\varphi, \mathbf{v}) = \omega^2 \tilde{\rho}(\mathbf{v}, \mathbf{u})\,, \tag{39}$$

$$-\,\tilde{e}(\chi, \mathbf{u}) + \tilde{\kappa}(\chi, \varphi) = 0\,, \tag{40}$$

where

$$\tilde{c}(\mathbf{v}, \mathbf{u}) = c(\mathbf{v}, \mathbf{u}) - a(B\mathbf{v}, B\mathbf{u})\,, \tag{41}$$

$$\tilde{e}(\varphi, \mathbf{v}) = e(\varphi, \mathbf{v}) + a(B\mathbf{v}, G\varphi), \quad \tilde{\kappa}(\chi, \varphi) = \kappa(\chi, \varphi) + a(G\varphi, G\chi). \tag{42}$$

Note that for obtaining the system (39), (40) with (41), (42) we used the transform $b(\mathbf{v}, \psi) = (b\mathbf{v}, \psi)_{H^1_\psi} = (aa^{-1}b\mathbf{v}, \psi)_{H^1_\psi} = a(B\mathbf{v}, \psi)$ and similar chains of equalities.

By excluding the porosity change function ψ we obtain the system (39), (40), which describe the natural vibrations for porous nanosized electroelastic medium with modified piezoelectric properties. By virtue of positive definited potential volumetric and surface energies of material (9), (12) and similar for electric surface energy $\Pi_{\Gamma e}(\mathbf{E}^s) = (1/2)\mathbf{d}^{sT} \cdot \mathbf{E}^s$, the forms $\tilde{c}(\mathbf{v}, \mathbf{u})$ and $\tilde{\kappa}(\chi, \varphi)$ are positive defined in H^1_{ul} and $H^1_{\varphi m}$, respectively. Therefore we can proceed with reduction of system (39), (40) by eliminating the function of electric potential φ.

Analogously, since for fixed $\mathbf{u} \in \mathrm{H}^1_{ul}$ and $\varphi \in \mathrm{H}^1_{\varphi m}$ $\tilde{e}(\chi, \mathbf{u})$ and $\tilde{\kappa}(\chi, \varphi)$ are linear-bounded functions in $\mathrm{H}^1_{\varphi m}$, by Riesz' theorem the elements $\tilde{e}\mathbf{u}$, $\tilde{\kappa}\varphi \in \mathrm{H}^1_{\varphi m}$ exist and are unique so for that for $\chi \in \mathrm{H}^1_{\varphi m}$

$$\tilde{e}(\chi, \mathbf{u}) = (\chi, \tilde{e}\mathbf{u})_{\mathrm{H}^1_{\varphi m}}, \quad \tilde{\kappa}(\chi, \varphi) = (\chi, \tilde{\kappa}\varphi)_{\mathrm{H}^1_\psi}. \tag{43}$$

It is obvious that $\tilde{e}\mathbf{u}$ and $\tilde{\varepsilon}\varphi$ are linear operators acting from H^1_{ul} into $\mathrm{H}^1_{\varphi m}$ and from $\mathrm{H}^1_{\varphi m}$ into $\mathrm{H}^1_{\varphi m}$, respectively, and an inverse exists for the operator $\tilde{\kappa}\varphi$.

From (40) and (43) we obtain that

$$\tilde{\kappa}\varphi = \tilde{e}\mathbf{u}, \quad \varphi = A_{lm}\mathbf{u}, \quad A_{lm} = \tilde{\kappa}^{-1}\tilde{e}, \tag{44}$$

where the operator A_{lm} acting from H^1_{ul} into $\mathrm{H}^1_{\varphi m}$, is linear and bounded and the subscripts l, m indicate to the problem with l rigid punches and m open-circuited electrodes.

Using (43), (44) we can represent system (39), (40) in the final form

$$\tilde{c}_{lm}(\mathbf{v}, \mathbf{u}) = \omega^2 \tilde{\rho}(\mathbf{v}, \mathbf{u}), \tag{45}$$

where

$$\tilde{c}_{lm}(\mathbf{v}, \mathbf{u}) = \tilde{c}(\mathbf{v}, \mathbf{u}) + \tilde{\kappa}(A_{lm}\mathbf{v}, A_{lm}\mathbf{u}). \tag{46}$$

Definition. *We will call the set of quantities* (ω^2, $\mathbf{u} \in \mathrm{H}^1_{ul}$, $\varphi \in \mathrm{H}^1_{\varphi m}$, $\psi \in \mathrm{H}^1_\psi$) *which satisfies* (45) *for arbitrary vector function* $\mathbf{v} \in \mathrm{H}^1_{ul}$ *or, which is equivalent* (39), (40) *for arbitrary* $\mathbf{v} \in \mathrm{H}^1_{ul}$, $\chi \in \mathrm{H}^1_{\varphi m}$ *or* (27)–(29) *for arbitrary* $\mathbf{v} \in \mathrm{H}^1_{ul}$, $\chi \in \mathrm{H}^1_{\varphi m}$, $\eta \in \mathrm{H}^1_\psi$, *a generalized solution of eigenvalue problem for piezoelectric body with voids and with uncoupled surface mechanical and electric effects.*

By repeating the topics presented in [3], we can show that the space H^1_c, which is the closure of the set of vector function $\mathbf{u} \in C^1$, satisfying (13) and (16) in the norm generated by the scalar product (46), is the equivalent to H^1_{ul}, and the next two theorems follow from the complete continuity of the embedding operator from H^1_{ul} into H^0_ρ, as in the general situation [18] (see, also [8, 15, 19]).

Theorem 1 *The operator equation (45) for piezoelectric body with voids and with surface effects has a discrete spectrum* $0 < \omega_1^2 \leq \omega_2^2 \leq ... \leq \omega_k^2 \leq ...; \ \omega_k^2 \to \infty$ *as* $k \to \infty$, *and the corresponding eigenfunctions* $\mathbf{u}^{(k)}$ *form a system that is orthogonal and complete in the spaces* H^0_ρ *and* H^1_c.

Theorem 2 *(The Courant–Fisher minimax principle).*

$$\omega_k^2 = \max_{\mathbf{w}_1,\mathbf{w}_2,...,\mathbf{w}_{k-1}\in \mathrm{H}^1_{ul}} \left[\min_{\substack{\mathbf{v}\neq 0,\ \mathbf{v}\in \mathrm{H}^1_{ul} \\ \tilde{\rho}(\mathbf{v},\mathbf{w}_j)=0;\, j=1,2,...,k-1}} R_{lm}(\mathbf{v}) \right]$$

where $R_{lm}(\mathbf{v})$ is the Rayleigh quotient

$$R_{lm}(\mathbf{v}) = \frac{\tilde{c}_{lm}(\mathbf{v},\mathbf{v})}{\tilde{\rho}(\mathbf{v},\mathbf{v})}\,. \tag{47}$$

Note that these theorems are important for the justification of the mode superposition method of solving harmonic and transient problems for piezoelectric media with voids and with surface effects.

We observe that the orthogonality conditions in Theorem 1 can be presented in the forms

$$(\mathbf{u}^{(i)},\mathbf{u}^{(j)})_{\mathrm{H}^0_\rho} = 0, \quad (\mathbf{u}^{(i)},\mathbf{u}^{(j)})_{\mathrm{H}^1_c} = 0, \quad i \neq j, \tag{48}$$

and also in the extended writing $(i \neq j)$

$$c(\mathbf{u}^{(i)},\mathbf{u}^{(j)}) + e(\varphi^{(j)},\mathbf{u}^{(i)}) + b(\mathbf{u}^{(i)},\psi^{(j)}) = 0\,, \tag{49}$$

$$-\,e(\varphi^{(i)},\mathbf{u}^{(j)}) + \kappa(\varphi^{(i)},\varphi^{(j)}) + g(\varphi^{(i)},\psi^{(j)}) = 0\,, \tag{50}$$

$$b(\mathbf{u}^{(j)},\psi^{(i)}) - g(\varphi^{(j)},\psi^{(i)}) + a(\psi^{(j)},\psi^{(i)}) = 0\,, \tag{51}$$

where $\varphi^{(i)} = A_{lm}\mathbf{u}^{(i)}$, $\psi^{(i)} = -B\mathbf{u}^{(i)} + G\varphi^{(i)}$.

There are another reduced forms of the orthogonality relations between $(\mathbf{u}^{(i)},\varphi^{(j)})$ *and* $(\mathbf{u}^{(i)},\psi^{(j)})$ *similarly to stated in [15].*

4 Theorems about Changes of Eigenfrequencies

We establish some consequences from variation descriptions of natural frequencies of piezoelectric bodies with voids and surface effects. We will formulate these consequences in the form of the theorems about the change of natural frequencies under the changes of boundary conditions and material parameters of the medium. These changes will be indicated explicitly in the formulations of the following theorems, and for the initial and modified problems not specified in formulations of theorems the determining parameters are assumed to be identical. The proofs of further theorems will be omitted as they can be carried out analogously to the corresponding theorems for elastic and piezoelectric bodies with voids and without voids [2, 14, 15] and for elastic and piezoelectric bodies with surface effects [1, 8, 19].

At the beginning we note the influence of the surface effects. Along with the problem formulated we will also consider the corresponding problems without account of surface stresses (mechanical surface effects) or without account of surface electric charges (electric surface effects). We note different variables for the problems without surface stresses by subscripts "*fu*", and the variables for the problems without surface dielectric films by subscripts "$f\varphi$".

In the problem without surface stresses instead of the boundary condition (10) we use conventional natural condition $\mathbf{n}\cdot\boldsymbol{\sigma} = 0\,,\quad \mathbf{x}\in\Gamma_\sigma$, and instead of the boundary condition (15) we use natural condition $\mathbf{n}\cdot\boldsymbol{\sigma} - (\mathbf{n}^T\cdot\boldsymbol{\sigma}\cdot\mathbf{n})\mathbf{n} = \nabla^s\cdot\boldsymbol{\tau}^s\,,\quad \mathbf{x}\in\Gamma_{uj}\,,\quad j\in J_p$.

Thus, for the solution $\mathbf{u}$ and the projection functions $\mathbf{v}$ in the problem without surface stress we can introduce the functional space H^1_{ful} with the norm

$$(\mathbf{v},\mathbf{u})_{\mathrm{H}^1_{ful}} = \int_\Omega (\nabla\mathbf{v})^T : \nabla\mathbf{u}\,d\Omega\,.$$

By analogy, in the problem without account for surface electric charges (without surface dielectric films) we change the boundary condition (17), setting $\mathbf{d}_\mathrm{s} = 0$, and also introduce the functional space $\mathrm{H}^1_{f\varphi m}$ with the norm

$$(\chi,\varphi)_{\mathrm{H}^1_{f\varphi m}} = \int_\Omega (\nabla\chi)^T\cdot\nabla\varphi\,d\Omega\,.$$

From comparison of the space H^1_{ul}, $\mathrm{H}^1_{\varphi m}$ H^1_{ful}, $\mathrm{H}^1_{f\varphi m}$, it can be concluded that

$$\mathrm{H}^1_{ul}\subset\mathrm{H}^1_{ful},\quad \mathrm{H}^1_{\varphi m}\subset\mathrm{H}^1_{f\varphi m}\,, \tag{52}$$

as for the functions from H^1_{ul} $\mathrm{H}^1_{\varphi m}$ additional smoothness at the boundaries Γ_τ and Γ_D is required.

Obviously, the following inequalities take place

$$c_{fu}(\mathbf{v},\mathbf{v}) = c_\Omega(\mathbf{v},\mathbf{v}) \leq c(\mathbf{v},\mathbf{v}), \quad \forall\, \mathbf{v} \in \mathrm{H}^1_{ul}\,, \tag{53}$$

$$\kappa_{f\varphi}(\chi,\chi) = \kappa_\Omega(\chi,\chi) \leq \kappa(\chi,\chi), \quad \forall\, \chi \in \mathrm{H}^1_{\varphi}\,. \tag{54}$$

Using the relations –, we can conclude, that similarly [8, 19], for the problem of natural vibrations for piezoelectric bodies with voids and with surface effects such theorems hold.

Theorem 3 *The natural frequencies ω_k for the problem of oscillations with account for surface stresses are not less than the corresponding natural frequencies ω_{fuk} for the problem without account for surface stresses, i.e. $\omega^2_{fuk} \leq \omega^2_k$ for all k.*

Theorem 4 *The natural frequencies ω_k for the problem of oscillations with account for surface electric charges are not greater than the corresponding natural frequencies $\omega_{f\varphi k}$ for the problem without account for surface electric charges, i.e. $\omega^2_k \leq \omega^2_{f\varphi k}$ for all k.*

We will also call problem (1)–(22) the *lm*-problem, emphasizing by this the presence of l areas Γ_{uj}, $j = 1, 2, ..., l$ in contact with rigid plane punches and m open-circuited electrodes $\Gamma_{\varphi j}$, $j = 1, 2, ..., m$.

We will consider two similar *lm*- and *pm*-problems, which solely differ in the number l and p of contacting surfaces of Γ_{uj} in (13)–(15). All the remaining input data from (1)–(22) in the *lm*- and *pm*-problems are assumed to be the same. Thus, the following theorems take place.

Theorem 5 *If $0 \leq l < p \leq L$, for any k the kth natural frequency ω_{lmk} of lm-problem is no less than kth natural frequency ω_{pmk} of pm-problem, i.e. $\omega^2_{lmk} \geq \omega^2_{pmk}$.*

We note that in conditions of Theorem 5 we do not change the boundary Γ_u. When passing from *lm*-problem to *pm*-problem we only change conditions of fixed boundary by the conditions of contact with punches on the parts of Γ_{uj}.

Theorem 6 *If $0 \leq m < n \leq M$, for any k the kth natural frequency ω_{lmk} of lm-problem is no greater than kth natural frequency ω_{lnk} of ln-problem, i.e. $\omega^2_{lmk} \leq \omega^2_{lnk}$.*

Let us also investigate the natural frequencies under the changes of certain parameters of the problem. We will explicitly point these changes in the formulations of the theorems, and all the variables related to the modified problems will be marked with a star. As above, for the initial and modified problems the parameters that are not specified in the theorem formulations are assumed to be identical. Thus, the following theorems hold.

Theorem 7 *If the rigid clamped and contacting with punch pats of boundaries of two problems are such that $\Gamma_u \supset \Gamma_{*u}$, $\Gamma_{uj} \supset \Gamma_{*uj}$, $j = 0, 1, 2, ..., L$ we have $\omega^2_{lmk} \geq \omega^2_{*lmk}$ for all k.*

Theorem 8 *If the elastic moduli, piezomoduli, the constitutive constants characterizing the properties of porosity changes, the densities and the masses and the inertia moments of the punches of two problems are such that* $\tilde{c}_{lm}(\mathbf{v},\mathbf{v}) \geq \tilde{c}_{*lm}(\mathbf{v},\mathbf{v})$, $\rho(\mathbf{v},\mathbf{v}) \leq \tilde{\rho}_*(\mathbf{v},\mathbf{v})$ *for* $\forall\, \mathbf{v} \in \mathrm{H}^1_{ul}$, *then* $\omega^2_{lmk} \geq \omega^2_{*lmk}$ *for all k.*

Theorem 9 *If the electrode boundaries of two problems are such that* $\Gamma_\varphi \supset \Gamma_{*\varphi}$, $\Gamma_{\varphi j} \supset \Gamma_{*\varphi j}$, $j = 0, 1, 2, ..., M$ *we have* $\omega^2_{lmk} \leq \omega^2_{*lmk}$ *for all k.*

Theorem 10 *If the permittivities of two problems are such that* $\kappa(\chi,\chi) \geq \kappa_*(\chi,\chi)$ $\forall\, \chi \in \mathrm{H}^1_{\varphi m}$, *then* $\omega^2_{lmk} \leq \omega^2_{*lmk}$ *for all k.*

Theorem 11 *If the "porous" Dirichlet boundaries of two problems are such that* $\Gamma_{\psi 0} \supset \Gamma_{*\psi 0}$ *we have* $\omega^2_{lmk} \geq \omega^2_{*lmk}$ *for all k.*

Theorem 12 *If the diffusion coefficients for porosity change of two problems are such that* $a(\eta,\eta) \geq a_*(\eta,\eta)$ *for* $\forall\, \eta \in \mathrm{H}^1_{\psi}$, *then* $\omega^2_{lmk} \geq \omega^2_{*lmk}$ *for all k.*

We note that the results of Theorems 5–12 are valid for both problems with and without account for surface effects.

We will summarize the results of Theorems 5–12.

If on certain areas of Γ_{uk} we replace the boundary conditions of rigid fixing (16) by the contact boundary conditions (13)–(15), then, by Theorem 5, the natural frequencies can only decrease.

On the other hand, if on certain areas of $\Gamma_{\varphi k}$ we replace the boundary conditions for short-circuited electrodes (20) by the electric boundary conditions of contact type (18), (19) for open-circuited electrodes, then by Theorem 6 the natural frequencies can only increase.

Note that the natural frequencies in the problem with all operation short-circuited electrodes are usually named as *electric resonance frequencies*, while the natural frequencies in the problem with some open-circuited electrodes are *electric antiresonance frequencies*. Therefore Theorem 6 also asserts that *the electric antiresonance frequencies are not less than the electric resonance frequencies with the same order numbers*.

By Theorems 7 and 8 a reduction in the boundaries Γ_{uj} of rigid fixing or contacting with rigid punches or a specific reduction in the elastic moduli and an increase in the density or in the massive characteristics of punches can lead only to a decrease in the natural frequencies.

Conversely, by Theorems 9 and 10 a reduction in the electrode boundary $\Gamma_{\varphi j}$ or a specific reduction in the permittivity moduli can lead only to an increase in the natural frequencies.

Comparing the effects reflected in Theorems 5–10 we can conclude that a similar change in the mechanical and electric boundary conditions or in elastic and permittivity moduli leads to an opposite change in the natural frequencies.

Meanwhile, by Theorems 11 and 12 a reduction in the boundary $\Gamma_{\psi 0}$ or a specific reduction in the diffusion coefficients for porosity change can lead only to a decrease in the natural frequencies. Comparing the effects reflected in Theorems 7, 8, 11 and 12, we can conclude that similar changes in the elastic and porous boundary conditions or in the elastic moduli and in the constitutive constants characterizing the properties of porosity lead to an uniform changes in the natural frequencies.

It can be noted that the changes in mechanic conditions lead to the known changes in natural frequencies [18] which have clear physical explanations. Thus, if the body is more rigidly fixed, then its mechanic vibrations become more constrained, therefore his eigenvalues can only increase. If the changes in volumetric and surface moduli and physical properties are such that the total volumetric and surface potential energies increase and kinetic energy decreases then the natural frequencies also increase.

As it can be seen, the piezoelectric effect shows up so that the similar effects, caused by the change electric conditions, lead to be opposite in comparison to the change in mechanical conditions. On the contrary, the changes in the properties and boundary conditions related with porosity lead to the similar changes for eigenvalues as analogous changes of mechanical properties and boundary conditions.

5 Concluding Remarks

This chapter has considered the problems of natural oscillations of piezoelectric nanodimensional bodies with voids in the framework of the piezoelasticity theory with account for uncoupled surface effects induced by surface stresses and surface electric charges (dielectric films). We have investigated also the bodies with voids in the context of Cowin-Nunziato theory for porosity change function. Classical and generalized settings of the spectral problems were formulated in expanded and reduced forms. For generalized settings the corresponding functional spaces were introduced. It was proved that with the disregard of damping for bounded piezoelectric body with voids the spectrum was discrete and real and the eigenvectors were orthogonal.

The theorems that establish the dependencies of natural frequencies of piezoelectric nanodimensional bodies were formulated with account for uncoupled surface mechanical and electric effects, and the change of the rigidly fixed boundary, boundaries with punches and electrodes, and material parameters of piezoelectric nanosize bodies.

It was shown that under an increase of stiffness moduli and porosity change moduli bringing to an increase of total volumetric and surface potential energies the natural frequencies also increase. The natural frequencies also increase under an extension of the surfaces with the mechanical rigidly clamped or contacting with rigid punches boundaries and the boundaries with principal conditions for the porosity function. On the contrary, an increase of electric permittivity moduli and an increase of boundaries with electrodes can lead only to a decrease in the natural frequencies.

Appropriate results of changes of natural frequencies have been found also for the replacement of the boundary conditions of rigid fixing by contact boundary conditions and for the replacement of the boundary conditions for short-circuited electrodes by boundary conditions for open-circuited electrodes.

All these dependencies were established for the piezoelectric bodies with voids as without surface effects, as well as for the bodies with account for surface stresses and surface dielectric films.

The given results for natural frequencies generalize the known results for purely elastic and piezoelectric bodies [2, 26]. for elastic and piezoelectric bodies with voids without surface effects [14–16], and for the elastic and piezoelectric with surface uncoupled effects [1, 8, 19, 20] to the problems for bounded piezoelectric nanodimensional bodies with voids and with surface effects.

Acknowledgements This work for second author was supported by the Russian Science Foundation (grant number 15-19-10008).

References

1. Altenbach, H., Eremeyev, V.A., Lebedev, L.P.: On the spectrum and stiffness of an elastic body with surface stresses. ZAMM **91**(9), 699–710 (2011)
2. Belokon, A.V., Nasedkin, A.V.: Some properties of the natural frequencies of electroelastic bodies of bounded dimentions. J. Appl. Math. Mech. (PMM) **60**, 145–152 (1996)
3. Belokon, A.V., Vorovich, I.I.: Some mathematical problems of the theory of electroelastic solids. In: Current Problems in the Mechanics Of Deformable Media, Izv. Dnepropetr. Gos. Univ., Dnepropetrovsk, pp. 52–67 (1979) (in Russian)
4. Ciarletta, M., Iesan, D.: Some results in the dynamical theory of porous elastic bodies. J. Elasticity **50**, 3–14 (1998)
5. Ciarletta, M., Scarpetta, E.: Some results on thermoelasticity for porous piezoelectric materials. Mech. Res. Commun. **23**, 1–10 (1996)
6. Cowin, S.C., Nunziato, J.W.: Linear elastic materials with voids. J. Elasticity **13**, 125–147 (1983)
7. Dai, Sh., Gharbi, M., Sharma, P., Park, H.S.: Surface piezoelectricity: size effects in nanostructures and the emergence of piezoelectricity in non-piezoelectric materials. J. Appl. Phys. **110**, 104305-1–104305-7 (2011)
8. Eremeev, V.A., Nasedkin, A.V.: Natural vibrations of nanodimensional piezoelectric bodies with contact-type boundary conditions. Mech. Solids **50**(5), 495–507 (2015)
9. Gu, S.-T., Liu, J.-T.: He. Q.-C.: Piezoelectric composites: Imperfect interface models, weak formulations and benchmark problems. Comp. Mater. Sci. **94**, 182–190 (2014)
10. Gu, S.-T., Liu, J.-T.: The strong and weak forms of a general imperfect interface model for linear coupled multifield phenomena. Int. J. Eng. Sci. **85**, 31–46 (2014)
11. Gu, S.-T., Qin, L.: Variational principles and size-dependent bounds for piezoelectric inhomogeneous materials with piezoelectric coherent imperfect interfaces. Int. J. Eng. Sci. **78**, 89–102 (2014)
12. Gurtin, M.E., Murdoch, A.I.: A continuum theory of elastic material surfaces. Arch. Rat. Mech. Anal. **57**(4), 291–323 (1975)
13. Huang, G.Y., Yu, S.W.: Effect of surface piezoelectricity on the electromechanical behaviour of a piezoelectric ring. Phys. Status Solidi B **243**(4), R22–R24 (2006)
14. Iovane, G., Nasedkin, A.V.: Some theorems about spectrum and finite element approach for eigenvalue problems for elastic bodies with voids. Comput. Math. Appl. **53**(5), 789–802 (2007)
15. Iovane, G., Nasedkin, A.V.: Modal analysis of piezoelectric bodies with voids. I. Mathematical approaches. Appl. Math. Modell. **34**(1), 60–71 (2010)

16. Iovane, G., Nasedkin, A.V.: Modal analysis of piezoelectric bodies with voids. II. Finite element simulation. Appl. Math. Modell. **34**(1), 47–59 (2010)
17. Iovane, G., Nasedkin, A.V.: New model for piezoelectric medium with voids for application to analysis of ultrasonic piezoelectric transducers and porous piezocomposites. In: Parinov, I.A. (ed.) Advanced Nano- and Piezoelectric Materials and their Applications, pp. 145–170. Nova Science Publishers (2014)
18. Mikhlin, S.G.: Variational Methods in Mathematical Physics. Pergamon Press, Oxford (1964)
19. Nasedkin, A.V., Eremeyev, V.A.: Spectral properties of piezoelectric bodies with surface effects. In: Altenbach, H., Morozov, N.F. (eds.) Advanced Structured Materials, vol. 30, Surface Effects in Solid Mechanics—Models, Simulations and Applications, pp. 105–121. Springer (2013)
20. Nasedkin, A.V., Eremeyev, V.A.: Harmonic vibrations of nanosized piezoelectric bodies with surface effects. ZAMM **94**(10), 878–892 (2014)
21. Pan, X.H., Yu, S.W., Feng, X.Q.: A continuum theory of surface piezoelectricity for nanodielectrics. Sci. China: Phys. Mech. Astron. **54**(4), 564–573 (2011)
22. Riesz, F., Szokefalvi-Nagy, B.: Functional Analysis. Dover, New York (1990)
23. Wang, J., Huang, Z., Duan, H., Yu, S., Feng, X., Wang, G., Zhang, W., Wang, T.: Surface stress effect in mechanics of nanostructured materials. Acta Mech. Solida Sinica **24**(1), 52–82 (2011)
24. Wang, K.F., Wang, B.L., Kitamura, T.: A review on the application of modified continuum models in modeling and simulation of nanostructures. Acta Mech. Sinica **32**(1), 83–100 (2016)
25. Yan, Z., Jiang, L.Y.: Surface effects on the electromechanical coupling and bending behaviours of piezoelectric nanowires. J. Phys. D: Appl. Phys **44**, 075404 (2011)
26. Yang, J.S.: A few properties of the resonant frequencies of a piezoelectic body. Arch. Mech. **44**, 475–477 (1992)

A Review on Models for the 3D Statics and 2D Dynamics of Pantographic Fabrics

Emilio Barchiesi and Luca Placidi

Abstract A review on models for the statics of out-of-plane deformable pantographic fabrics is presented, along with a model describing the dynamics of in-plane-only deformable pantographic fabrics. We discuss those models able to describe the mechanical exotic properties conferred by the peculiar microstructure possessed by pantographic metamaterials, when three-dimensional deformations and in-plane dynamics are separately involved. For each approach, model formulation and modelling assumptions are discussed along with the presentation of numerical solutions in exemplary cases, and no attempt is made to model damage and failure phenomena.

Keywords Pantographic fabrics · Metamaterials · Higher gradient continua · Review · Microstructure · Dynamics · Out-of-plane deformations

1 Introduction

In the wide current research literature about mechanical and acoustic metamaterials, pantographic fabrics [1, 9, 16, 18, 20, 22, 23, 26, 28, 29, 37, 39, 40, 58, 67, 69, 71] (see Fig. 1) represent one of the very promising topics, being both of great practical and theoretical interest. Indeed, their potentially high strength-to-weight ratio yields pantographic fabrics particularly suitable for some class of applications (see [4, 36, 38]), while their great theoretical interest derives from the fact that, in order to describe their exotic phenomenology, one has to resort to higher gradient continuum theories [5, 45] or micromophic theories [2, 49] with the related problem of homogenization [59] and of different strategies for the numerical integration

E. Barchiesi
Department of Structural and Geotechnical Engineering, Universitá di Roma La Sapienza, Rome, Italy

L. Placidi (✉)
International Telematic University Uninettuno, C.so Vittorio Emanuele II, 39 00186 Rome, Italy
e-mail: lplacidi@gmail.com; luca.placidi@uninettunouniversity.net

M.A. Sumbatyan (ed.), *Wave Dynamics and Composite Mechanics for Microstructured Materials and Metamaterials*,
Advanced Structured Materials 59, DOI 10.1007/978-981-10-3797-9_14

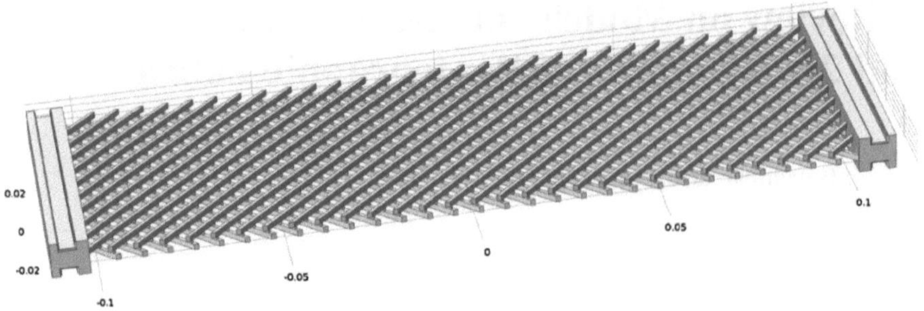

Fig. 1 A pantographic structure

[8, 10]. The field of fiber-reinforced materials [31, 44, 53, 54, 68] and of elastic sheets with bending stiffness [41–43, 55, 60] is certainly not new in the literature.

In the following sections we are going to present a concise survey of mathematical models which have been purposely developed for out-of-plane deformable pantographic structures, with regard to their statics, and for in-plane deformable ones, with regard to their dynamics. In [57] a review on models for the description of the mechanical behavior, at equilibrium, of pantographic sheets undergoing planar deformations only, i.e. models whose placement functions are $\chi : \mathbb{R}^2 \supseteq \mathscr{B} \rightarrow \mathbb{R}^2$, is presented. The present paper can hence be considered a sequel of [57] as, instead of considering only placement functions $\chi : \mathbb{R}^2 \supseteq \mathscr{B} \rightarrow \mathbb{R}^2$, we consider herein also placement functions $\chi : \mathbb{R}^2 \supseteq \mathscr{B} \rightarrow \mathbb{R}^3$ and $\chi : \mathbb{R}^3 \supseteq \mathscr{B} \rightarrow \mathbb{R}^3$, thus allowing out-of-plane deformations. Moreover, while [57] is only concerned with the quasi-static case, this paper also deals with models able to account for the dynamics and, in particular, suitable for the study of wave propagation. In Sect. 2, the standard formulation of a 3*D* Cauchy continuum, postulating isotropy and homogeneity of the material which the fibers are made of, is firstly sketched. The domain shape to be employed in order to accomplish a satisfactory predictivity with such a first gradient model and its numerical restrictions are successively discussed. We conclude by presenting some recently published results about the numerical identification of a 2*D* second gradient continuum model, i.e. a continuum model in which the strain-energy density function depends on the first and second gradients of the deformation, using a 3*D* standard Cauchy model. In Sect. 3, we show some results about the use of arrays of Euler beams in order to describe the dynamic behavior of in-plane deformable pantographic rectangular "long" wave-guides. Finally, in Sect. 4, we present the formulation of a second gradient continuum model in which pantographic sheets are regarded as elastic surfaces endowed with suitable kinematic descriptors. For each approach, numerical results are shown for some exemplary problems as those listed in [57].

2 Standard 3D Cauchy Continuum

In this section we recall the formulation of a standard Cauchy model. Let $\mathscr{B} \subset \mathbb{R}^3$ a regular domain which will be referred to as the reference shape. In the framework of a Lagrangian description of Kinematics, let $p \in \mathscr{B}$ be the position of a material particle in the reference configuration and let its image through the suitably regular map $\chi : \mathscr{B} \to \mathbb{R}^3$ be its position in the current configuration. Accordingly with classical notation, we introduce the deformation gradient tensor $\mathsf{F} = \nabla \chi$ and the Green–St-Venant strain tensor

$$\mathsf{E} = \frac{1}{2}\left(\mathsf{F}^{\mathsf{T}}\mathsf{F} - \mathtt{I}\right). \tag{1}$$

By introducing the displacement

$$\mathbf{u}(p) = \chi(p) - p \tag{2}$$

we can recast Eq. (1) as

$$\mathsf{E}(p) = \frac{1}{2}\left(\nabla \boldsymbol{u} + \nabla \mathbf{u}^{\mathsf{T}} + \nabla \boldsymbol{u}^{\mathsf{T}} \nabla \boldsymbol{u}\right).$$

Assuming the constitutive relation for isotropic and homogeneous materials, the Piola stress tensor can be evaluated as

$$\mathsf{T} = \lambda \mathrm{tr} \mathsf{E}\, \mathtt{I} + 2\mu\, \mathsf{E} \tag{3}$$

where λ and μ are the Lam parameters. Furthermore, the strain energy density is defined as

$$W_m(\mathsf{E}) = \frac{1}{2}\mathsf{T} : \mathsf{E} \tag{4}$$

where the symbol ":" stands for the double contraction of tensors. The governing equations are derived from Eq. (4) by means of a variational principle as follows

$$\delta \int_{\mathscr{B}} W_m(\mathsf{E})\, \mathrm{d}V = 0 \quad \forall\, \delta \boldsymbol{u}. \tag{5}$$

An interesting remark can be done about the domain (reference) shape. The domain we shall consider when dealing with the modelling of pantographic sheets is the one represented in Fig. 2, thus one resembling the microstructure of the specimen. Indeed, in the $3D$ Cauchy continuum case we can not consider the domain to be just a plain parallelepiped as we could do, e.g., when using higher gradient continuum models. This is because one basic assumption of Cauchy stress theory is that the mechanical contact interaction between two parts of the material is only due to

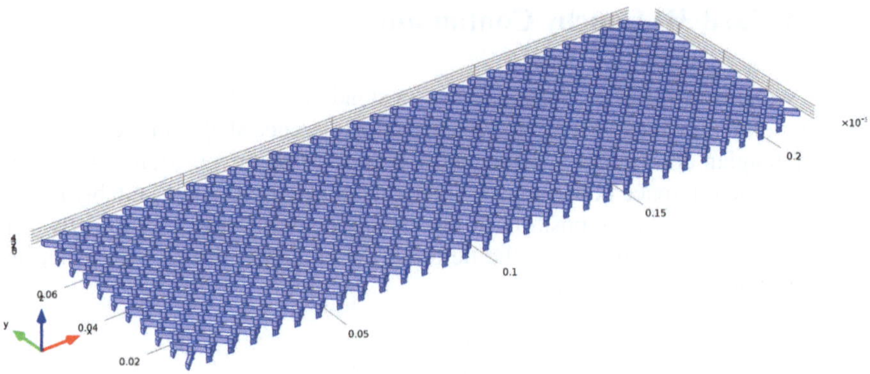

Fig. 2 3D domain employed in the standard Cauchy continuum model

a surface density of contact forces concentrated on the "cut", i.e. on the dividing surface. When dealing with materials exhibiting a pantographic microstructure, at the macroscopic level this basic assumption can not be considered any longer valid. The issues discussed above imply that, in order to correctly describe the mechanical behavior of pantographic structures with a standard 3*D* Cauchy model, a somewhat involved domain, resembling the microstructure of the specimen, must be employed. Moreover, the mesh size required to successfully solve numerically the field equations of the model with, e.g., the finite elements method is much smaller than the one required by higher-gradient continua. Hence, the domain complexity and the mesh size restrictions yield a heavy computational burden, resulting in the actual non feasibility of the use of a fully 3*D* Cauchy model to perform numerical simulations of pantographic structures, even for simple problems.

2.1 Numerical Identification

The above-discussed inappropriateness, mainly for numerical reasons, of a fully 3*D* Cauchy model for the description of pantographic structures can be overcome by employing, for example, higher-gradient reduced-order models. Recently, in [33] a fully 3*D* Cauchy continuum model is used as a reference model for the identification and validation of a higher-gradient model, accounting only for planar deformations, firstly proposed in [25] and successively reviewed also in [57]. Without going into much details, being the reader referred to the aforementioned papers, we recall that the higher-gradient model employed in [33] originates from a heuristic homogenization technique, analogue to the one first used by Piola to formulate a continuum fully non-linear beam model, applied to a discrete spring model. Moreover, by using the strain measures introduced in [33] and assuming the parameters $\mathbb{K}_a^\alpha$, $\mathbb{K}_b^\alpha$ and $\mathbb{K}_p$, appearing in the deformation energy, to be such that

1. $\mathbb{K}_a^1 = \mathbb{K}_a^2 :=K_I$
2. $\mathbb{K}_b^1 = \mathbb{K}_b^2 :=K_{II}$
3. $\mathbb{K}_p :=K_p$
4. K_I, K_{II} and K_p are independent of the position,

the strain energy density of the homogenized model conveniently recasts as

$$W_M(\varepsilon_\alpha, \kappa_\alpha, \gamma) = \sum_\alpha \left(\frac{1}{2} K_I \varepsilon_\alpha^2 + \frac{1}{2} K_{II} \kappa_\alpha^2 \right) + \frac{1}{2} K_p \gamma^2. \quad (6)$$

The parameter identification that we are going to show is a numerical one, in the sense that it requires fitting the constitutive parameters K_I, K_{II} and K_p of the homogenized model using several numerical simulations performed with the 3*D* Cauchy one. In more detail, a number of bias extension test simulations (see [57]) using both the standard Cauchy model and the higher-gradient model, for several displacements imposed on the shorter side of the specimen, are performed. Successively, for each simulation, the overall stored energy and two representative deformations at specific points are computed.

The two quantities are chosen to be the angles ψ_C and ψ_V, evaluated at the probe points shown in Fig. 3 i.e. at the center C of the specimen and at the corner V of the "quasi-rigid" triangle near a base of the specimen.

The material parameters of the macro model K_I, K_{II} and K_p are estimated solving a multi-objective optimization problem i.e. minimizing the squared errors for the overall stored energy and the two angles ψ_C, ψ_V, when computed both with the homogenized model and the standard Cauchy one.

The two angles ψ_C, ψ_V have been chosen among other possible control quantities because each of them is strongly related to one of the last two energy terms in (6), being in turn this two energy terms dependent only on one parameter each.

Indeed, the energy involved in the distortion angle at the center is mostly governed by the parameter K_p, while the distortion angle at the triangle vertex depends for the most part on the bending energy characterized by K_{II}, thus allowing to easily find the minimum of the squared error for the two angles by separately tuning K_p

Fig. 3 The two control angles employed in the identification procedure

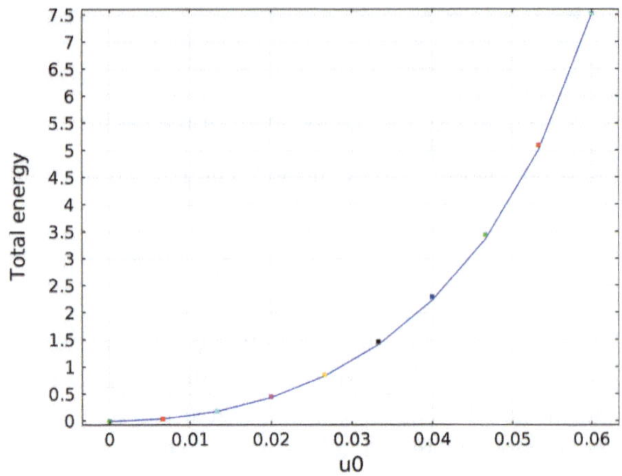

Fig. 4 Comparison of the total energy between the Cauchy model (points) and the second gradient model (*solid line*)

and K_{II} i.e. with separate regressions. Finally, the last parameter K_I is derived by considering the whole stored energy. Indeed a coherent estimate of the total deformation energy in the equilibrium configuration, which is clearly dependent on the expression of the postulated deformation energy density, as a function of the imposed relative displacement is of great importance in the applications, as the capability of the considered structure to resist to damage progress is clearly related to this physical quantity.

In Figs. 4 and 5 (left), the total energy and the angles ψ_C and ψ_V used for fitting the second gradient model are shown as the imposed displacement relative to the bias test is varying.

In Fig. 5 (right), in order to get a better insight into the regression results, a comparison between the total reaction of the micro model and the one evaluated with the macro model is plotted versus the imposed displacement.

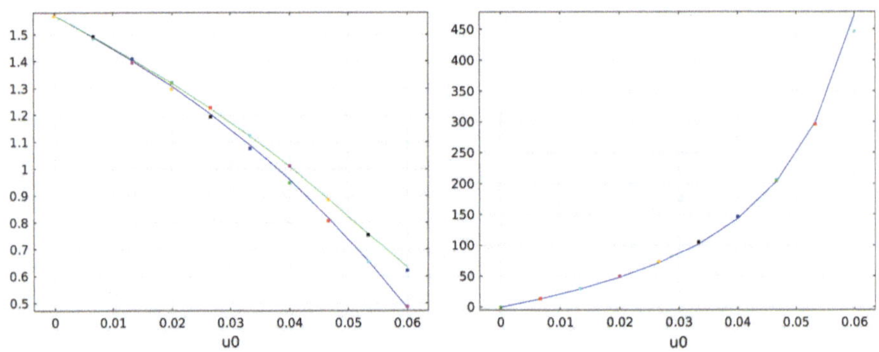

Fig. 5 Comparisons between the Cauchy model (points) and the regression with the second gradient model (*solid line*). Angle at the center ψ_C (*blue line*) and angle at the corner ψ_V (*green line*) on the *left*; total reaction on the *right*

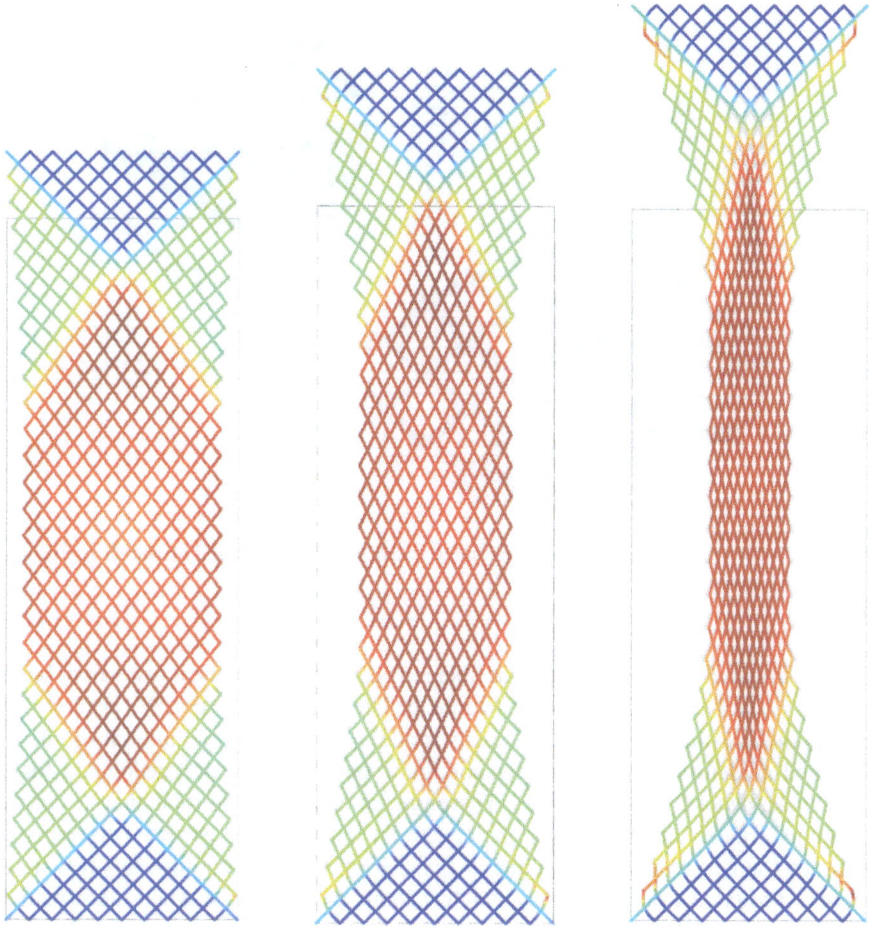

Fig. 6 Overlap between simulations of bias extension tests with the Cauchy model (*gray*) and with the second gradient model (*colors* indicate the shear strain relative to the initial fiber axes) for different imposed displacements

Indeed, in the context of the bias extension test, the total external resultant force which must be applied to get a given relative displacement, i.e. the constraint reaction, is a very important physical quantity, which was computed as a function of the imposed displacement by means of Castigliano's first theorem i.e. by computing the first derivative of the equilibrium total energy with respect to the imposed displacement.

As far as relatively small displacement are concerned, the accordance between the two models is quite satisfactory, while at $u_0 = 6$ cm a discrepancy is observed (see Fig. 6). This phenomenon can be explained by considering that the Cauchy model has a richer kinematics than the homogenized one. Indeed, the former model allows twist

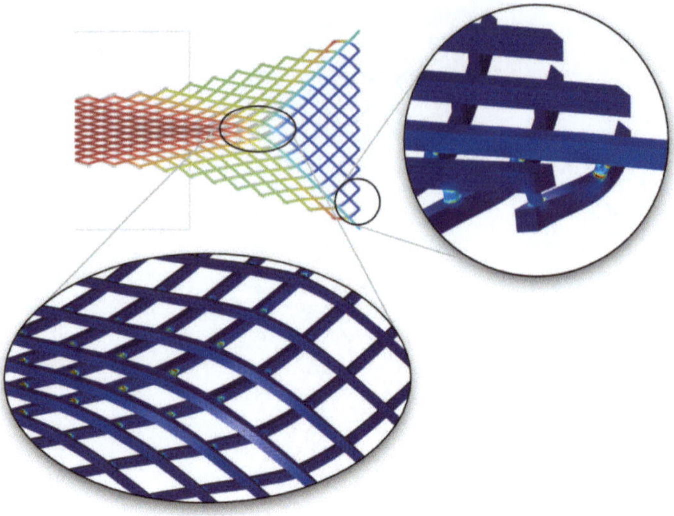

Fig. 7 3D deformation details; the colors in the zooms indicate the stored elastic energy density for the $3D$ Cauchy model

and possibly out-of-plane bending of fibers to occur as well as shear distortion, bending and possibly extension of pivots; while the latter model does not. We observe (see Fig. 7) that for the more refined Cauchy model a non-negligible amount of energy is stored as deformation energy for configurations which are not accounted for in the coarser second gradient model. Therefore, equating the two total energies of the examined models and evaluating the constraint reactions, we expect a discrepancy between the two reactions. Looking at Fig. 5 (right) we can confirm this last statement at the largest displacement, which is likely to be the one where deformation energies due to the richer kinematics of the refined model start to gain significance.

3 Wave Propagation

The model which is about to be introduced in this section was first presented in [24] and was thereafter employed in a number of papers, as [22, 48], in order to numerically study some dynamically interesting cases. In those papers pantographic rectangular "long" wave-guides are considered, in which imposed boundary displacements can induce the onset of traveling waves. In this model we consider two families of orthogonal $1D$ straight continua arranged in a rectangle in the reference configuration. Each continuum $\mathscr{C}_i$ has a standard linearised Euler elastic potential given by

$$U_i = \frac{1}{2}\int_{\mathscr{C}_i} k_M \left(u''(s)\right)^2 + k_N \left(w'(s)\right)^2 \mathrm{d}s \tag{7}$$

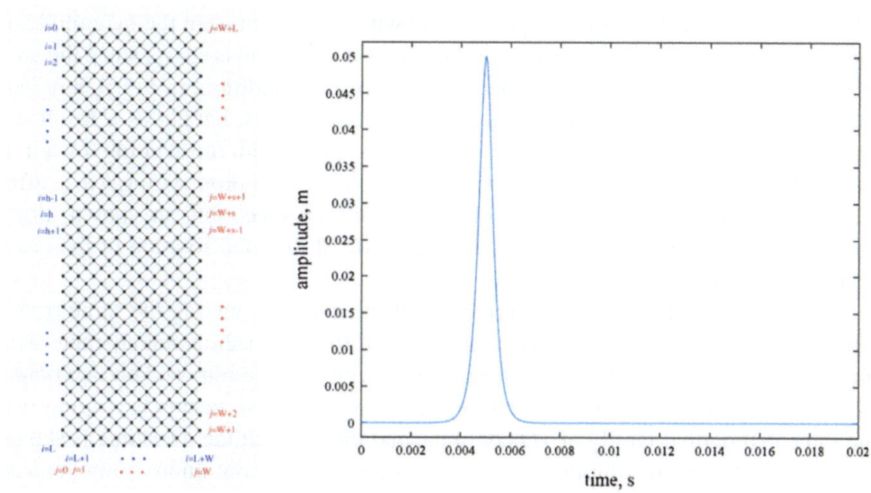

Fig. 8 Reference configuration (*left*) and time history of the impulse (*right*)

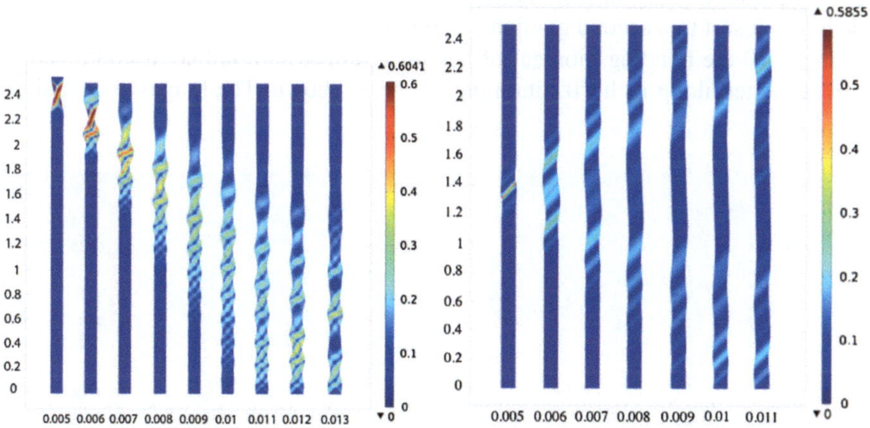

Fig. 9 Wave propagating after an imposed vertical displacement on the upper side (*left*). Wave propagating after double impulse (*right*)

with s an abscissa introduced on each $\mathscr{C}_i$, k_M a bending stiffness coefficient, k_N an axial stiffness coefficient, u and w, respectively, the magnitudes of the transverse and axial displacements. Dots in Fig. 8 (left) stand for the presence of hinges which do not oppose to rotations and do not interrupt the continuity of the beams.

In the examined literature, every kind of displacement imposed on the structure is analytically represented by an impulse function $\mathfrak{I} = u_0 * \text{sech}\left[\tau(t - t_0)\right]$ with τ being a parameter affecting the duration of the impulse. In Fig. 8 (right) the amplitude of the impulse is plotted as a function of time.

In Fig. 9 (left) some snapshots of the deformed shape of a propagating wave generated by a vertical impulse, uniformly applied at the upper side of the specimen while its lower side remains clamped, are shown. Colors represent the magnitude of the total rotation of the cross section of the beams. Dispersion is sharp, since the area of the perturbed region is clearly increasing in time. In addition, in Fig. 9, a reflection on the lower side is recognizable in the last snapshots, having as a result the summation of those waves departing from the lower side with those going toward it. In Fig. 9 (right) some snapshots of the deformed shape of a propagating wave, after a double impulse is applied at the middle height of the specimen, are shown, along with colors representing, as before, the magnitude of the total rotation of the cross section of the beams.

By double impulse we mean a couple of displacements, having the same orientation but opposite directions, oriented in one of the two orthogonal characteristic directions of the pantographic sheet. Such displacements are imposed on two points at the opposite ends of two adjacent beams, i.e. consecutive beams belonging two the same orthogonal family of $1D$ continua, and their amplitude over time is shown in Fig. 8. Such a double impulse should correspond, in the continuous homogenized limit case, to a double force, i.e. to a pair of forces with null resultant and moment. Since such a kind of forces can not be included in a first gradient continuum theory, in order to capture with a continuum model the dynamic features shown in Fig. 8, one has to resort to a second gradient continuum.

In Fig. 10 the bending moment of beams, arranged in a lattice discontinued by a set of hinges along its horizontal middle line, is shown. The hinges, which allow

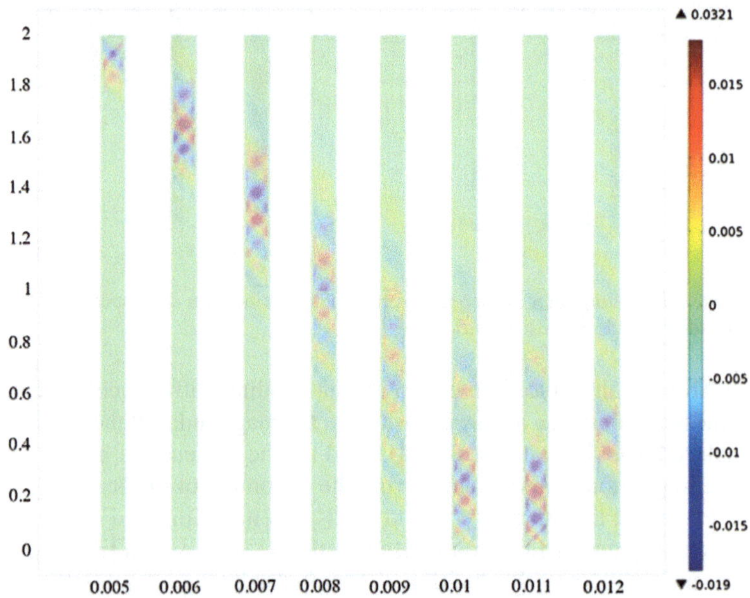

Fig. 10 Bending moment in a lattice discontinued by a set of hinges along its horizontal *middle line* (*left*)

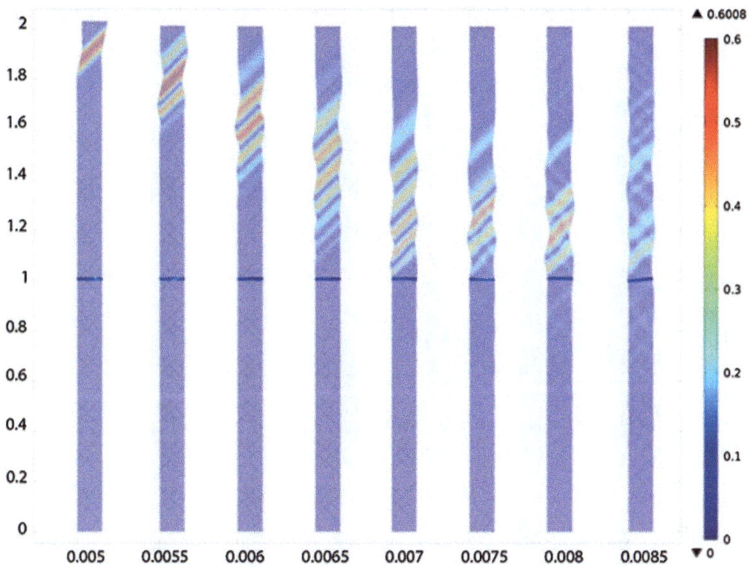

Fig. 11 Wave propagation in two identical lattices connected by an array of vertical beams

energy-free relative angular displacements between the upper and the lower part of each beam, in this case, do interrupt the continuity of the beams. Nevertheless, due to the kind of internal connections at nodes of the lattice, i.e. internal hinges not interrupting the continuity of beams, the bending moment at snapshots where the perturbation has reached the center of the specimen does not change its qualitative distribution. Also in this case reflection is observed.

In Fig. 11 some snapshots of the deformed shape of a propagating wave are shown, along with colors representing the magnitude of the total rotation of the cross section of the beams, which are arranged in a lattice discontinued by a set of vertical beams along its horizontal middle line. The amplitude of the displacement as a function of time is, as before, the same shown in Fig. 8 but, in this case, the displacement imposed at the upper side of the specimen is parallel to one of the two families of beams. We can clearly see that in Fig. 11 an interesting phenomenon arises. Indeed, the energy of the system remains substantially confined in the upper half of the waveguide and propagation of waves beyond the discontinuity is negligible. Therefore, such a kind of discontinued pantographic structures result in a simple, but still attractive, damping filter for this new type of metamaterial.

In Fig. 12, again, some snapshots of the deformed shape of a propagating wave are shown, along with colors representing the magnitude of the total rotation of the cross section of the beams. In this case, the displacement imposed at the upper and lower sides of the specimen is parallel to one of the two families of beams and results in two waves travelling toward each other which, after a while, end up summing together.

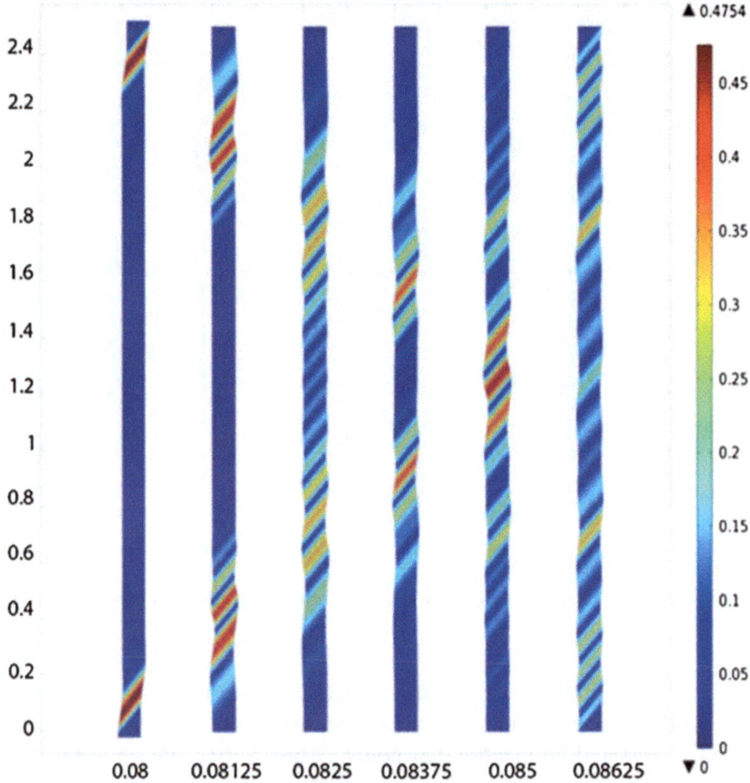

Fig. 12 Propagation of two waves traveling in opposite directions

4 Elastic Surfaces

In this section, following the approach employed in [27], we will present the formulation of a model for pantographic sheets, regarded as elastic surfaces, embedded in a three dimensional Euclidean space, endowed with suitable kinematic descriptors and an associated second-gradient areal strain-energy density which depends on the first and second gradients of the deformation (Figs. 13 and 14).

We will now proceed in a way very similar to what was done throughout [57]. In order to account for the fact that fibers are arranged in two material directions we introduce a Lagrangian Cartesian orthonormal coordinate system whose associated basis of unit vectors is $(\mathbf{D}_1, \mathbf{D}_2)$, which is constituted by two orthonormal vectors representing the directions of the families of fibers constituting the pantographic structure in the reference configuration. Let us now consider a $2D$ continuum whose reference shape is given by a rectangular domain $\mathscr{B} \subset \mathbb{R}^2$. The current shape of $\mathscr{B}$ is described by means of the suitably regular macro placement $\chi : \ \mathscr{B} \rightarrow \mathbb{R}^3$, whose target space is worth to be emphasized. Indeed, unlike the models presented

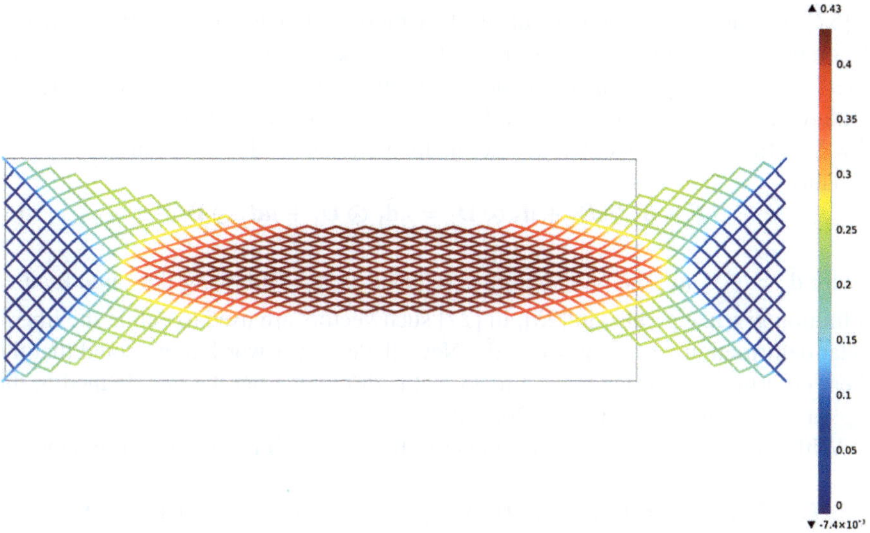

Fig. 13 Numerical simulation of the bias extension test using the elastic surface model presented above. Colors represent the shear strain

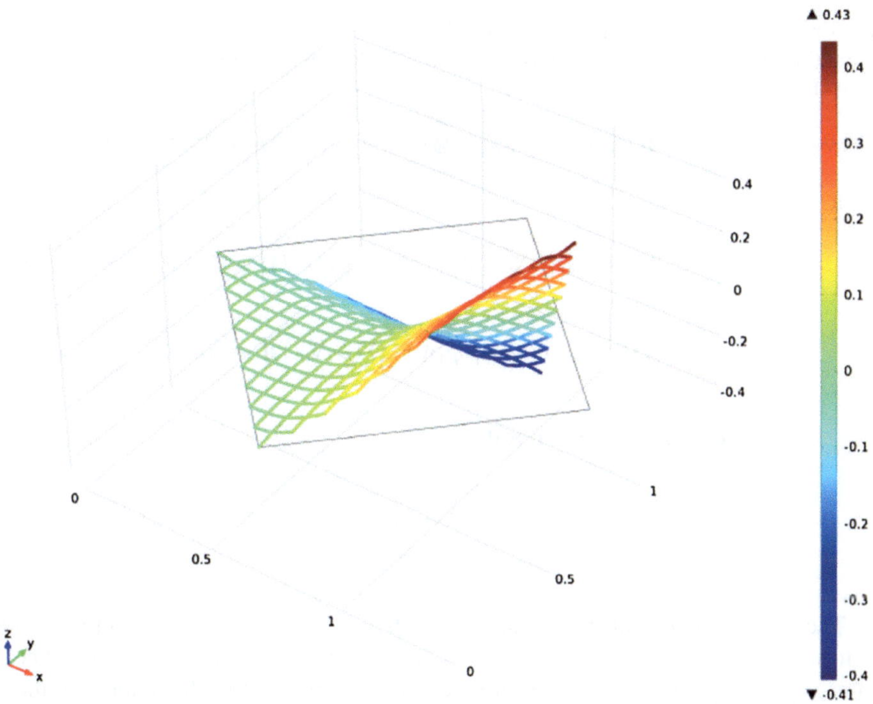

Fig. 14 Numerical simulation of torsion of a square sheet ($\theta = 60°$) using the elastic surface model presented above. Colors represent the out-of-plane component of the displacement, u_3

in [57], this model does include out-of-plane motions and therefore the Pipkin-Rivlin decomposition employed, e.g., in [21, 26, 58] does not hold even for inextensible fibers. As customary, $\mathbf{d}_1$ and $\mathbf{d}_2$ are defined as the push-forward vectors in the current configuration of the vectors $\mathbf{D}_1$ and $\mathbf{D}_2$ respectively, i.e. $\mathbf{d}_\alpha = \mathbf{F}\mathbf{D}_\alpha$, $\alpha = 1, 2$ where $\mathbf{F} = \nabla\chi$. In the sequel we will denote the fiber stretches $\|\mathbf{d}_\alpha\|$ as λ and μ.

Thus,

$$\mathbf{F} = \mathbf{d}_1 \otimes \mathbf{D}_1 + \mathbf{d}_2 \otimes \mathbf{D}_2 = \lambda\tilde{\mathbf{d}}_1 \otimes \mathbf{D}_1 + \mu\tilde{\mathbf{d}}_2 \otimes \mathbf{D}_2, \tag{8}$$

where $\tilde{\mathbf{d}}_\alpha = \frac{\mathbf{d}_\alpha}{\|\mathbf{d}_\alpha\|}$ are the unit vectors associated to $\mathbf{d}_\alpha$. Following the same kind of definition given, e.g., also in [25], in [27] such vectors are used to define the fibers' shear deformation γ as $\sin\gamma = \tilde{\mathbf{d}}_1 \cdot \tilde{\mathbf{d}}_2$. Nevertheless, it is worth to be noted that the shear deformation introduced in this model is different from the one defined in the Pipkin continuum model in [21, 26, 58].

With this notation, we have from (8) for the Cauchy-Green deformation tensor

$$\mathbf{C} = \mathbf{F}^{\mathsf{T}}\mathbf{F} = \lambda^2\mathbf{D}_1 \otimes \mathbf{D}_1 + \mu^2\mathbf{D}_2 \otimes \mathbf{D}_2 + \lambda\mu\sin\gamma\left(\mathbf{D}_1 \otimes \mathbf{D}_2 + \mathbf{D}_2 \otimes \mathbf{D}_1\right).$$

We also have

$$J\mathbf{n} = \mathbf{F}\mathbf{D}_1 \times \mathbf{F}\mathbf{D}_2 = \mathbf{d}_1 \times \mathbf{d}_2,$$

with $\mathbf{n}$ the unit normal of the deformed surface field and $J = \lambda\mu|\cos\gamma|$ the local areal dilation due to the deformation. In [27] the following representation formula is proven

$$\nabla\nabla\chi = \left(g_1 + K_1\mathbf{n}\right) \otimes \mathbf{D}_1 \otimes \mathbf{D}_1 + \left(g_2 + K_2\mathbf{n}\right) \otimes \mathbf{D}_2 \otimes \mathbf{D}_2 + (\Gamma + T\mathbf{n}) \otimes \left(\mathbf{D}_1 \otimes \mathbf{D}_2 + \mathbf{D}_2 \otimes \mathbf{D}_1\right) \tag{9}$$

with

$$g_1 = \lambda\eta_1\mathbf{p} + \left(\mathbf{D}_1 \cdot \nabla\lambda\right)\tilde{\mathbf{d}}_1;\; g_2 = \mu\eta_2\mathbf{q} + \left(\mathbf{D}_2 \cdot \nabla\mu\right)\tilde{\mathbf{d}}_2 \tag{10}$$

$$\Gamma = \left(\mathbf{D}_1 \cdot \nabla\mu\right)\tilde{\mathbf{d}}_2 + \lambda\mu\phi_1\mathbf{q} = \left(\mathbf{D}_2 \cdot \nabla\lambda\right)\tilde{\mathbf{d}}_1 + \lambda\mu\phi_2\mathbf{p} \tag{11}$$

$$\mathbf{q} = \mathbf{n} \times \tilde{\mathbf{d}}_2;\; \mathbf{p} = \mathbf{n} \times \tilde{\mathbf{d}}_1 \tag{12}$$

$$K_1 = \lambda^2\kappa_1;\; K_2 = \mu^2\kappa_2;\; T = \lambda\mu\tau. \tag{13}$$

The quantities η_1 and η_2 appearing in (10) are the geodesic curvatures of the deformed fibers, ϕ_1 and ϕ_2 appearing in (11) are the so-called Tchebychev curvatures, κ_1 and κ_2 appearing in (13) are the normal curvatures of the deformed fibers and τ measures the twist of the deformed surface. In [27] explicit expressions for geodesic and Tchebychev curvatures are provided and we have

$$
\begin{aligned}
J\eta_1 &= \mathbf{D}_1 \cdot \nabla(\mu \sin\gamma) - \mathbf{D}_2 \cdot \nabla\lambda \\
J\eta_2 &= \mathbf{D}_1 \cdot \nabla(\mu) - \mathbf{D}_2 \cdot \nabla(\lambda \sin\gamma) \\
J\phi_1 &= J\eta_2 + \lambda\mathbf{D}_2 \cdot \nabla(\sin\gamma) \\
J\phi_2 &= J\eta_1 + \mu\mathbf{D}_1 \cdot \nabla(\sin\gamma).
\end{aligned}
$$

Moreover, in [35] a strain-energy density function depending on the first and second gradients of the deformation and incorporating the orthotropic symmetry property conferred by the reference fibers arrangement is proposed and reads as

$$
W = w(\lambda, \mu, J) + \frac{1}{2}(A_1|g_1|^2 + A_2|g_2|^2 + A_\Gamma|\boldsymbol{\Gamma}|^2 + k_1K_1^2 + k_2K_2^2 + +k_TK_T^2)
$$

where $A_1,\ A_2,\ A_\Gamma,\ k_1,\ k_2,\ k_T$ are constitutive constants.

A linear elastic second gradient orthotropic two-dimensional model, invariant under rotation and for mirror transformation, is considered also in [56] where such anisotropy is exploited together with some original gedanken experiments to completely characterize the set of constitutive parameters in terms of the fibers' base material parameters (i.e. its Young modulus), of the fibers' cross section parameters (i.e. its area and its moment of inertia) and of the distance between the nearest pivots. Afterwards, a strain energy is derived in terms of the displacement fields and, remarkably, it closely resembles the strain energy of simple Euler beams.

5 Conclusions

The development of some advanced manufacturing techniques like, e.g., 3*D* printing, electrospinning and roll-to-roll processing could represent a turning point in the manufacturing industry, allowing the large-scale production of micro- and nano-enginereed materials exhibiting interesting macroscopic properties not found in natural materials. Successfully exploiting these new technological advancements calls for a highly multidisciplinary theoretical work in mathematical modelling, numerical analysis, computer-aided design and many other subjects.

The aim of this paper was to provide a review on some 3*D* models directed to the description of the statics of pantographic lattices and on a 2*D* model directed to the study of the dynamics of such structures, thus extending [57]. Our model selection was guided by some important features such as computational complexity and predictivity.

In Sect. 2, the formulation of a standard 3*D* Cauchy continuum, where isotropy and homogeneity of the material which the fibers are made of is assumed, was firstly presented. Subsequently we did a digression, discussing the domain shape to be employed in order to accomplish a satisfactory predictivity with such a first gradient model and the resulting computational unfeasible burdens. We concluded Sect. 2 by outlining some results obtained in [33], addressing the issue of numerically determining the parameters of the homogenized second-gradient model discussed in, e.g.,

[25, 57] using the standard Cauchy one. Comparing the total reaction of the Cauchy model and the one evaluated with the homogenized second-gradient model, as the imposed displacement on the shorter side of the specimen was varying, we finally remarked that, when the more detailed Cauchy model exhibits a significant three-dimensional deformation, the agreement between the two models is lost. In Sect. 3, we have shown some results about the use of arrays of Euler beams, i.e. two families of orthogonal $1D$ straight continua connected at their intersecting point by internal hinges, in order to describe the dynamic behavior of in-plane deformable pantographic rectangular "long" wave-guides, possibly interrupted by horizontal arrays of hinges or vertical beams. Some numerical simulations were presented, in cases where dispersion, reflection and damping filtering occurred. Finally, in Sect. 4, we sketched the formulation of a second gradient continuum model in which out-of-plane deformable pantographic sheets are regarded as elastic surfaces endowed with suitable kinematic descriptors. It is a commonly accepted fact that features exhibited by systems made up of simple interacting elements, like the system presented in Sect. 3, which involves Euler beams, can be very much richer than those shown by its constituting elements, when considered singularly.

Further developments in this research field include the investigation of damaged pantographic structures [70] with different methods [13–15] and a better understanding of their dynamical behaviour [7, 17, 30, 46, 47, 50], with a focus on eigenfrequency analysis [11, 12] accounting for all the necessary inertial properties. Buckling modes in pantographic lattices have already been analysed in [34]. However, the instability properties could be also investigated with methods presented in [3, 32, 51, 52, 61–66]. Finally, an interesting approach can be deduced also by analysing numerical results concerning particle systems, employing methods developed within the field of 2D swarm of robots interacting with closer neighbors [6, 19].

Hence, in conclusion, the study of the statics and of the dynamics in non-trivial structures like pantographic fabrics, especially in a $3D$ setting, still deserves further studies. Very likely, new exotic and interesting phenomenological behaviors could be unveiled, such as in the case of the above discussed damping filtering properties.

References

1. Alibert, J., Della Corte, A.: Second-gradient continua as homogenized limit of pantographic microstructured plates: a rigorous proof. Zeitschrift für angewandte Mathematik und Physik **66**(5), 2855–2870 (2015)
2. Altenbach, H., Eremeyev, V.: On the linear theory of micropolar plates. ZAMM J. Appl. Math. Mech./Zeitschrift für Angewandte Mathematik und Mechanik **89**(4), 242–256 (2009)
3. AminPour, H., Rizzi, N.: A one-dimensional continuum with microstructure for single-wall carbon nanotubes bifurcation analysis. Math. Mech. Solids **21**(2), 168–181 (2016)
4. Andreaus, U., Giorgio, I., Lekszycki, T.: A 2D continuum model of a mixture of bone tissue and bio-resorbable material for simulating mass density redistribution under load slowly variable in time. Zeitschrift für Angewandte Mathematik und Mechanik **13**, 7 (2013)

5. Auffray, N., dell'Isola, F., Eremeyev, V., Madeo, A., Rosi, G.: Analytical continuum mechanics à la Hamilton-Piola least action principle for second gradient continua and capillary fluids. Math. Mech. Solids **20**(4), 375–417 (2015)
6. Battista, A., Rosa, L., dell'Erba, R., Greco, L.: Numerical investigation of a particle system compared with first and second gradient continua: deformation and fracture phenomena. Math. Mech. Solids, 1081286516657889 (2016)
7. Bersani, A.M., Della Corte, A., Piccardo, G., Rizzi, N.L.: An explicit solution for the dynamics of a taut string of finite length carrying a traveling mass: the subsonic case. Zeitschrift für angewandte Mathematik und Physik **67**(4), 108 (2016)
8. Bilotta, A., Formica, G., Turco, E.: Performance of a high-continuity finite element in three-dimensional elasticity. Int. J. Numer. Methods Biomed. Eng. **26**(9), 1155–1175 (2010)
9. Boisse, P., Hamila, N., Guzman Maldonado, E., Madeo, A., Hivet, G., Dell'Isola, F.: The bias-extension test for the analysis of in-plane shear properties of textile composite reinforcements and prepregs: a review. Int. J. Mater. Form. 1–20 (2016)
10. Cazzani, A., Malagù, M., Turco, E.: Isogeometric analysis: a powerful numerical tool for the elastic analysis of historical masonry arches. Contin. Mech. Thermodyn. **28**(1–2), 139–156 (2016)
11. Cazzani, A., Stochino, F., Turco, E.: An analytical assessment of finite element and isogeometric analyses of the whole spectrum of Timoshenko beams. ZAMM J. Appl. Math. Mech./Zeitschrift für Angewandte Mathematik und Mechanik (2016)
12. Cazzani, A., Stochino, F., Turco, E.: On the whole spectrum of Timoshenko beams. Part I: a theoretical revisitation. Zeitschrift für angewandte Mathematik und Physik **67**(2), 1–30 (2016)
13. Contrafatto, L., Cuomo, M., Fazio, F.: An enriched finite element for crack opening and rebar slip in reinforced concrete members. Int. J. Fract. **178**(1–2), 33–50 (2012)
14. Contrafatto, L., Cuomo, M., Gazzo, S.: A concrete homogenisation technique at meso-scale level accounting for damaging behaviour of cement paste and aggregates. Comput. Struct. **173**, 1–18 (2016)
15. Cuomo, M., Contrafatto, L., Greco, L.: A variational model based on isogeometric interpolation for the analysis of cracked bodies. Int. J. Eng. Sci. **80**, 173–188 (2014)
16. Cuomo, M., dell'Isola, F., Greco, L., Rizzi, N.: First versus second gradient energies for planar sheets with two families of inextensible fibres: investigation on deformation boundary layers, discontinuities and geometrical instabilities. Compos, Part B Eng (2016)
17. D'Annibale, F., Rosi, G., Luongo, A.: Linear stability of piezoelectric-controlled discrete mechanical systems under nonconservative positional forces. Meccanica **50**(3), 825–839 (2015)
18. Del Vescovo, D., Giorgio, I.: Dynamic problems for metamaterials: review of existing models and ideas for further research. Int. J. Eng. Sci. **80**, 153–172 (2014)
19. Della Corte, A., Battista, A., dell'Isola, F.: Referential description of the evolution of a 2d swarm of robots interacting with the closer neighbors: perspectives of continuum modeling via higher gradient continua. Int. J. Non-Linear Mech. **80**, 209–220 (2016)
20. dell'Isola, F., Cuomo, M., Greco, L., Della Corte, A.: Bias extension test for pantographic sheets: numerical simulations based on second gradient shear energies. J. Eng. Math. 1–31 (2016)
21. dell'Isola, F., d'Agostino, M., Madeo, A., Boisse, P., Steigmann, D.: Minimization of shear energy in two dimensional continua with two orthogonal families of inextensible fibers: the case of standard bias extension test. J. Elast. **122**(2), 131–155 (2016)
22. dell'Isola, F., Della Corte, A., Giorgio, I., Scerrato, D.: Pantographic 2d sheets: discussion of some numerical investigations and potential applications. Int. J. Non-Linear Mech. **80**, 200–208 (2016)
23. dell'Isola, F., Della Corte, A., Greco, L., Luongo, A.: Plane bias extension test for a continuum with two inextensible families of fibers: A variational treatment with lagrange multipliers and a perturbation solution. Int. J, Solids Struct (2015)
24. dell'Isola, F., Giorgio, I., Andreaus, U.: Elastic pantographic 2d lattices: a numerical analysis on static response and wave propagation. Proceedings of the Estonian Academy of Sciences **64**, 219–225 (2015)

25. dell'Isola, F., Giorgio, I., Pawlikowski, M., Rizzi, N.: Large deformations of planar extensible beams and pantographic lattices: heuristic homogenization, experimental and numerical examples of equilibrium. In: Proceedings of the Royal Society A, vol. 472, p. 20150790. The Royal Society (2016)
26. dell'Isola, F., Lekszycki, T., Pawlikowski, M., Grygoruk, R., Greco, L.: Designing a light fabric metamaterial being highly macroscopically tough under directional extension: first experimental evidence. Zeitschrift für angewandte Mathematik und Physik **66**, 3473–3498 (2015)
27. dell'Isola, F., Steigmann, D.: A two-dimensional gradient-elasticity theory for woven fabrics. J. Elast. **18**, 113–125 (2015)
28. dell'Isola, F., Steigmann, D., Della Corte, A.: Synthesis of fibrous complex structures: designing microstructure to deliver targeted macroscale response. Appl. Mech. Rev. **67**(6), 060–804 (2015)
29. Di Carlo, A., Rizzi, N., Tatone, A.: Continuum modelling of a beam-like latticed truss: identification of the constitutive functions for the contact and inertial actions. Meccanica **25**(3), 168–174 (1990)
30. Di Egidio, A., Luongo, A., Paolone, A.: Linear and non-linear interactions between static and dynamic bifurcations of damped planar beams. Int. J. Non-Linear Mech. **42**(1), 88–98 (2007)
31. Everstine, G., Pipkin, A.: Boundary layers in fiber-reinforced materials. J. Appl. Mech. **40**, 518–522 (1973)
32. Gabriele, S., Rizzi, N., Varano, V.: On the postbuckling behaviour of thin walled beams with in-plane deformable cross-sections. In: Civil-comp proceedings (2013)
33. Giorgio, I.: Numerical identification procedure between a micro-cauchy model and a macro-second gradient model for planar pantographic structures. Zeitschrift für angewandte Mathematik und Physik **67(4)**(95) (2016)
34. Giorgio, I., Della Corte, A., dell'Isola, F., Steigmann, D.: Buckling modes in pantographic lattices. C. R, Mec (2016)
35. Giorgio, I., Grygoruk, R., dell'Isola, F., Steigmann, D.: Pattern formation in the three-dimensional deformations of fibered sheets. Mech. Res. Commun. **69**, 164–171 (2015)
36. Goda, I., Assidi, M., Ganghoffer, J.: A 3d elastic micropolar model of vertebral trabecular bone from lattice homogenization of the bone microstructure. Biomech. Model. Mechanobiol. **13**, 53–83 (2014)
37. Greco, L., Giorgio, I., Battista, A.: In plane shear and bending for first gradient inextesible pantographic sheets: numerical study of deformed shapes and global constraint reactions. Math. Mech. Solids, 1081286516651324 (2016)
38. Grillo, A., Wittum, G., Tomic, A., Federico, S.: Remodelling in statistically oriented fibre-reinforced composites and biological tissues. Math. Mech. Solids **20**, 1107–1129 (2015)
39. Harrison, P.: Modelling the forming mechanics of engineering fabrics using a mutually constrained pantographic beam and membrane mesh. Compos. Part A Appl. Sci. Manuf. **81**, 145–157 (2016)
40. Harrison, P., Anderson, D., Alvarez, M.F., Bali, E., Mateos, Y.: Measuring and modelling the in-plane bending stiffness and wrinkling behaviour of engineering fabrics (2016)
41. Hilgers, M., Pipkin, A.: Elastic sheets with bending stiffness. Q. J. Mech. Appl. Math. **45**, 57–75 (1992)
42. Hilgers, M., Pipkin, A.: Energy-minimizing deformations of elastic sheets with bending stiffness. J. Elast. **31**, 125–139 (1993)
43. Hilgers, M., Pipkin, A.: Bending energy of highly elastic membranes ii. Q. Appl. Math **54**, 307–316 (1996)
44. Hu, M., Kolsky, H., Pipkin, A.: Bending theory for fiber-reinforced beams. J. Compos. Mater, 235–249 (1985)
45. dell Isola, F., Seppecher, P., Della Corte, A.: The postulations á la d alembert and á la cauchy for higher gradient continuum theories are equivalent: a review of existing results. In: Proceedings of the Royal Society A, vol. 471, p. 20150415. The Royal Society (2015)
46. Luongo, A., D'annibale, F.: Bifurcation analysis of damped visco-elastic planar beams under simultaneous gravitational and follower forces. Int. J. Mod. Phys. B **26**(25), 1246,015 (2012)

47. Luongo, A., Zulli, D., Piccardo, G.: A linear curved-beam model for the analysis of galloping in suspended cables. J. Mech. Mater. Struct. **2**(4), 675–694 (2007)
48. Madeo, A., Della Corte, A., Greco, L., Neff, P.: Wave propagation in pantographic 2d lattices with internal discontinuities (2014). arXiv:1412.3926
49. Pietraszkiewicz, W., Eremeyev, V.: On natural strain measures of the non-linear micropolar continuum. Int. J. Solids Struct. **46**(3), 774–787 (2009)
50. Pignataro, M., Rizzi, N., Luongo, A.: Stability, Bifurcation and Postcritical Behaviour of Elastic Structures, vol. 39. Elsevier (2013)
51. Pignataro, M., Rizzi, N., Ruta, G., Varano, V.: The effects of warping constraints on the buckling of thin-walled structures. J. Mech. Mater. Struct. **4**(10), 1711–1727 (2010)
52. Pignataro, M., Ruta, G., Rizzi, N., Varano, V.: Effects of warping constraints and lateral restraint on the buckling of thin-walled frames. ASME Int. Mech. Eng. Congr. Expo. **10**(B), 803–810 (2010)
53. Pipkin, A.: Generalized plane deformations of ideal fiber-reinforced materials. Q. Appl. Math **32**, 253–263 (1974)
54. Pipkin, A.: Energy changes in ideal fiber-reinforced composites. Q. Appl. Math **35**, 455–463 (1978)
55. Pipkin, A.: Some developments in the theory of inextensible networks. Q. Appl. Math **38**, 343–355 (1980)
56. Placidi, L., Andreaus, U., Giorgio, I.: Identification of two-dimensional pantographic structure via a linear d4 orthotropic second gradient elastic model. J. Eng. Math. 1–21 (2016)
57. Placidi, L., Barchiesi, E., Turco, E., Rizzi, N.: A review on 2d models for the description of pantographic fabrics. Zeitschrift für angewandte Mathematik und Physik **67**(5)(121) (2016)
58. Placidi, L., Greco, L., Bucci, S., Turco, E., Rizzi, N.: A second gradient formulation for a 2d fabric sheet with inextensible fibres. Zeitschrift für angewandte Mathematik und Physik **67**(5)(114) (2016)
59. Rahali, Y., Giorgio, I., Ganghoffer, J., Dell'Isola, F.: Homogenization à la piola produces second gradient continuum models for linear pantographic lattices. Int. J. Eng. Sci. **97**, 148–172 (2015)
60. Rivlin, R.: Plane strain of a net formed by inextensible cords. In: Collected Papers of RS Rivlin, pp. 511–534. Springer (1997)
61. Rizzi, N., Varano, V.: The effects of warping on the postbuckling behaviour of thin-walled structures. Thin-Walled Struct. **49**(9), 1091–1097 (2011)
62. Rizzi, N., Varano, V.: On the postbuckling analysis of thin-walled frames. In: Thirteenth international conference on civil, structural and environmental engineering computing. Civil-Comp Press (2011)
63. Rizzi, N., Varano, V., Gabriele, S.: Initial postbuckling behavior of thin-walled frames under mode interaction. Thin-Walled Struct. **68**, 124–134 (2013)
64. Ruta, G., Pignataro, M., Rizzi, N.: A direct one-dimensional beam model for the flexural-torsional buckling of thin-walled beams. J. Mech. Mater. Struct. **1**(8), 1479–1496 (2006)
65. Ruta, G., Varano, V., Pignataro, M., Rizzi, N.: A beam model for the flexural-torsional buckling of thin-walled members with some applications. Thin-Walled Struct. **46**(7), 816–822 (2008)
66. Scerrato, D., Giorgio, I., Rizzi, N.: Three-dimensional instabilities of pantographic sheets with parabolic lattices: numerical investigations. Zeitschrift für angewandte Mathematik und Physik **67**(3), 1–19 (2016)
67. Scerrato, D., Zhurba Eremeeva, I., Lekszycki, T., Rizzi, N.: On the effect of shear stiffness on the plane deformation of linear second gradient pantographic sheets. ZAMM J. Appl. Math. Mech./Zeitschrift für Angewandte Mathematik und Mechanik (2016)
68. Soubestre, J., Boutin, C.: Non-local dynamic behavior of linear fiber reinforced materials. Mech. Mater. **55**, 16–32 (2012)
69. Turco, E., dell'Isola, F., Cazzani, A., Rizzi, N.: Hencky-type discrete model for pantographic structures: numerical comparison with second gradient continuum models. Zeitschrift für angewandte Mathematik und Physik **67** (2016)

70. Turco, E., dell'Isola, F., Rizzi, N.L., Grygoruk, R., Müller, W.H., Liebold, C.: Fiber rupture in sheared planar pantographic sheets: numerical and experimental evidence. Mech. Res. Commun. **76**, 86–90 (2016)
71. Zhu, B., Yu, T., Tao, X.: Large deformation and slippage mechanism of plain woven composite in bias extension. Compos. Part A Appl. Sci. Manuf. **38**(8), 1821–1828 (2007)

MIX
Papier aus verantwortungsvollen Quellen
Paper from responsible sources
FSC® C105338

If you have any concerns about our products,
you can contact us on
ProductSafety@springernature.com

In case Publisher is established outside the EU,
the EU authorized representative is:
Springer Nature Customer Service Center GmbH
Europaplatz 3, 69115 Heidelberg, Germany

Printed by Libri Plureos GmbH
in Hamburg, Germany